INTERNATIONAL UNION FOR CONSERVATION OF NATURE AND NATURAL RESOURCES
WORLD RESOURCES INSTITUTE
CONSERVATION INTERNATIONAL
WORLD WILDLIFE FUND-US
WORLD BANK

CONSERVING THE WORLD'S BIOLOGICAL DIVERSITY

By
Jeffrey A. McNeely
Kenton R. Miller
Walter V. Reid
Russell A. Mittermeier
Timothy B. Werner

Gland, Switzerland, and Washington, D.C.

Prepared and published by the International Union
for Conservation of Nature and Natural Resources,
World Resources Institute, Conservation International, World Wildlife Fund-US
and the World Bank

Citation: McNeely, Jeffrey A., Kenton R. Miller, Walter V. Reid, Russell A. Mittermeier and Timothy B. Werner
1990. CONSERVING THE WORLD'S BIOLOGICAL DIVERSITY.
IUCN, Gland, Switzerland; WRI, CI, WWF-US, and the World Bank, Washington, D.C.
ISBN: 0-915825-42-2
0-8213-1384-3 (The World Bank)
Library of Congress Catalog Card Number: 89-051386
Book design, layout and illustration: Stephen D. Nash
Printed by: Consolidated Business Forms, Lock Haven, PA
Available from:
IUCN Publications Services, 1196 Gland, Switzerland
WRI Publications, P.O. Box 4852 Hamden Station, Baltimore, MD 21211
World Bank Publications, P.O. Box 7247-8619, Philadelphia, PA 19170-8619

Cover photos (clockwise from upper-left): **Galapagos hawk** (by R. Mast); **coral reef** (by J. Post); **chameleon** (by R.A. Mittermeier); **muriqui** (by R.A. Mittermeier); **caterpillar** (by A. Young); **Papua New Guinean highlander** (by R.A. Mittermeier); **succulent plant from Campo Rupestre, Brazil** (by R.A. Mittermeier); (center) **agriculture in Thailand** (by Y. Hadar, World Bank).

Authors

Jeffrey A. McNeely is Chief Conservation Officer of IUCN, where he is responsible for designing the organization's conservation programs, providing the main liaison on technical matters with international institutions including various agencies of the United Nations, the World Bank and the World Wildlife Fund. Also with IUCN, Mr. McNeely served previously as Deputy Director General for Conservation, Director of the Policy and Programme Division, and Executive Officer of the Commission on National Parks and Protected Areas. Prior to joining IUCN in 1980, he was the World Wildlife Fund's representative in Indonesia. Mr. McNeely's more than 10 years field experience in Asia includes extensive scientific research, resource planning and nature management. He has published numerous books and technical articles on these subjects, particularly on the relationship between culture and natural resources. Mr. McNeely holds a Bachelors degree in Anthropology from the University of California, Los Angeles.

Kenton R. Miller is Director of WRI's Program in Forests and Biodiversity. Prior to joining WRI's staff in 1988, Dr. Miller served for five years as Director General of IUCN. Before that, he was Associate Professor of Natural Resources and Director, Center for Strategic Wildlife Management Studies, School of Natural Resources, University of Michigan. During this period, he served for seven years as Chairman of IUCN's Commission on National Parks and Protected Areas. Dr. Miller also acted as Secretary General for the Third World National Parks Congress in 1982 in Bali, Indonesia. Previously, he served for ten years with the Food and Agricultural Organization of the United Nations directing wildlands management activities in Latin America and the Caribbean. His work has carried him throughout the world and he has published extensively on national parks planning and wildland management. Dr. Miller holds degrees in Forest Management and a Ph.D. in Forest Economics from the State University of New York, Syracuse.

Walter V. Reid is Associate with the WRI Program in Forests and Biodiversity. Before joining WRI in June 1988, Dr. Reid was a Gilbert White Fellow with Resources for the Future where he studied the management of Southern Ocean fisheries and examined environmental issues related to sustainable development. He has taught courses in natural history and environmental sciences at the University of Washington and worked as a wildlife biologist with the California and Alaska departments of Fish and Game and the U.S. Forest Service in Alaska. Dr. Reid earned a Ph.D. in Zoology with specialization in Population and Community Ecology from the University of Washington.

Russell A. Mittermeier is President of Conservation International. Prior to this, he served as Vice-President for Science at World Wildlife Fund (1987-89) and as Director of that organization's programs for Brazil and the Guianas (1985-89), Madagascar (1985-89), Species Conservation (1986-89) and Primates (1979-89). He has served as Chairman of the Primate Specialist Group of the International Union for Conservation of Nature and Natural Resources' Species Survival Commission (IUCN/SSC) since 1977, as SSC's Vice-Chairman for International Programs since 1985, and as Chairman of the World Bank's Task Force on Biological Diversity in 1988 and 1989. He is also a member of the Board of Wildlife Preservation Trust International and the Belize Zoo, a Research Associate at the Museum of Comparative Zoology at Harvard University, and an Adjunct Associate Professor at the State University of New York at Stony Brook. Dr. Mittermeier's publications include five books and over 200 papers and popular articles on primates, reptiles, tropical forests and biodiversity, and he is responsible for publishing the journal *Primate Conservation.* He has conducted field work on three continents and more than 20 countries in the tropics, and his most recent field work has been on primates, protected areas and other conservation issues in the Atlantic forest region of eastern Brazil, in Suriname and on the island of Madagascar. Dr. Mittermeier received his B.A. (summa cum laude, Phi Beta Kappa) from Dartmouth in 1971, and his Ph.D. from Harvard in Biological Anthropology in 1977.

Timothy B. Werner is a Research Associate at Conservation International, where he specializes in biogeographical analyses and issues in conservation biology, and also coordinates new program initiatives in Oceania. Before joining CI, he worked at World Wildlife Fund in the Science Department as a Research Assistant, and prior to that as an intern on the Minimum Critical Size of Ecosystems Project, located in Brazil's Amazon region. He holds a degree in History (cum laude) from Boston University.

TABLE OF CONTENTS

FOREWORD, by M.S. Swaminathan 9
ACKNOWLEDGEMENTS 10
EXECUTIVE SUMMARY 11

I. BIOLOGICAL DIVERSITY: WHAT IT IS AND WHY IT IS IMPORTANT. 17
Biological Diversity and Development 19
Modern Approaches to Conserving Biological Diversity 20
Developing a Global Biodiversity Conservation Strategy 21

II. THE VALUES OF BIOLOGICAL DIVERSITY 25
Ethics, Economics, and Biological Diversity 25
Assessing the Value of Biological Resources 27
Direct Values of Biological Resources 28
Indirect Values of Biological Resources 31
Conclusions 34

III. HOW AND WHY BIOLOGICAL RESOURCES ARE THREATENED 37
The Dimensions of the Problem 39
Economic Factors Stimulating Overexploitation of Biological Resources 47
Social Factors that Threaten Biological Resources 49
Major Obstacles to Greater Progress in Conserving Biological Diversity 51

IV. APPROACHES TO CONSERVING BIOLOGICAL DIVERSITY 55
Policy Shifts, Integrated Land Use, and Biodiversity 55
Protecting Species and Habitats: The Need for an Integrated Approach 56
Contributions of Ex Situ Mechanisms to the Conservation of Biodiversity 62
Management Action in Response to Pollution and Climate Change 66
A New Global Convention on the Conservation of Biological Diversity 68

V. THE INFORMATION REQUIRED TO CONSERVE BIOLOGICAL DIVERSITY 71
Types of Information Needed 71
Information Needs at National and Local Levels 74
Information Management at the International Level 77
Information on Legislation 80
Conclusions 80

VI. ESTABLISHING PRIORITIES FOR CONSERVING BIOLOGICAL DIVERSITY 83
Establishing Priorities within a Nation 83
International Approaches to Determining Priorities 86
Conclusion: Guidelines for Determining Priorities for Conservation Action 104

VII. THE ROLE OF STRATEGIES AND ACTION PLANS IN PROMOTING CONSERVATION OF BIOLOGICAL DIVERSITY 109
Strategies and Action Plans for Conserving Species or Species Groups 109
Action Plans for Conserving Habitats 111
Cross-Sectoral Strategies and Action Plans 112
Conclusions 115

VIII. HOW TO PAY FOR CONSERVING BIOLOGICAL DIVERSITY ... 117
The Issue of Property Rights to Biological Resources ... 118
Mechanisms Useful Primarily at National and Local Levels ... 118
Mechanisms Useful Primarily at the International Level ... 123
Conclusions ... 125

IX. ENLISTING NEW PARTNERS FOR CONSERVATION OF BIOLOGICAL DIVERSITY ... 129
The Contributions of Biological Resources to "Non-Conservation" Sectors ... 129
The Special Case of the Military ... 131
New Approaches to Managing Areas for Sustainable Production of Biological Resources ... 132
Conclusions ... 132

ANNEX 1: CLASSIFICATION OF LIFE ON EARTH BY PHYLUM ... 133
ANNEX 2: THE WORLD CHARTER FOR NATURE ... 134
ANNEX 3: INTERNATIONAL LEGISLATION SUPPORTING CONSERVATION OF BIOLOGICAL DIVERSITY 137
ANNEX 4: THE BALI ACTION PLAN ... 140
ANNEX 5: THE WORLD BANK WILDLANDS POLICY ... 147
ANNEX 6: GLOSSARY ... 153
BIBLIOGRAPHY ... 157
INDEX ... 185
LIST OF ACRONYMS AND ABBREVIATIONS ... 192

FOREWORD

By
Dr. M.S. Swaminathan
President
International Union for Conservation of Nature and Natural Resources
President, World Wildlife Fund India
Trustee, World Resources Institute

When I began writing this Foreword on 7 February, 1989, the television screen in front of me was showing millions of pilgrims at Allahabad in India having their bath at the confluence of the rivers Ganges, Yamuna, and the invisible Saraswathi. Joseph Campbell in his book *Creative Mythology* wrote: "For those in whom a local mythology still works, there is an experience of both accord with the social order, and of harmony with the universe."

In the mythology of many civilizations around the world, living in harmony with nature is always a recurrent refrain. The mythology of the Ganges revolves around its glacier origin, the dual role of mountain forests as catchments and containments, and the estuarine mouth creating a swamp forest of rich genetic diversity in both flora and fauna. The pilgrims who have their holy bath at Allahabad believe that God manifesting as Gangadhara held the might of the torrential Ganges in the locks of his hair. In a way, this myth symbolizes the control of the gushing streams by the dense forests of Tehri Garhwal. Unfortunately, today the forests are disappearing and the gushing streams flow down unchecked, causing siltation of rivers and frequent floods. When forests disappear, the associated fauna and flora also disappear.

When scenes of severe floods appear on television screens, few people living comfortably in urban areas see the linkages between floods downstream and deforestation upstream. When markets are full with a wide range of food material, we tend to forget that we live on this earth as guests of the green plants that convert sunlight, nutrients, and water into food. If green plants cease to exist, animals cannot exist. In nature a delicate web of inter-dependence is spun among all living organisms as well as between the biosphere and the geosphere.

Biological diversity provides the foundation for further progress in enhancing the biological productivity of our planet, on a sustainable basis. The basic building blocks for this foundation are the genes contained in plants and animals, which by their diversity can enable the whole organisms to adapt to the changing environment.

Recent advances in molecular biology and genetic engineering have opened up new opportunities for moving genes across sexual barriers. Thus genetic engineering has enhanced the value of the rich genetic estate we have inherited. The need for conserving wild species of plants and animals has hence become even more urgent.

What are our blessings in terms of biological diversity? I would like to enumerate a few.

- All our food comes from wild species brought into domestication, and the cultivated varieties are fighting a constant evolutionary battle with the pests who find their fruits to be a tempting target. Continuous research, often drawing on wild species, is therefore essential to maintain the productivity of the plants that provide our main sources of sustenance (over half of human nutrition is provided by just three plants: rice, wheat, and maize).
- Our water is supplied by one of nature's most important processes, technically known as the hydrological cycle. Forested watersheds provide clear, high-quality water for domestic or industrial use, and healthy rivers provide water, transport, and fish.
- Species living and long-extinct support industrial processes. Oil and coal — from living creatures who captured the sun's energy before dying tens of millions of years ago — are major feedstocks for the chemical industry, keep us warm, and fuel our transportation systems. Cement comes from limestone, which is made up of the shells and skeletons of long-dead corals and other forms of marine life. Rubber, paper, wood, pesticides, and many other natural products support our industries, and forests and wetlands help clean up the pollutants afterward.
- Most of our medicines came originally from the wild, including our major painkillers, birth-control agents, and malaria drugs. While many are now produced synthetically, medicinal plants are still important in many parts of the world. In India, traditional doctors use 2,500 plants, and over 5,000 medicinal plants have been recorded in China. Quinine, digitalis, and morphine all still come from plants, and over 40 percent of all prescriptions in the USA still depend on natural sources.

These few examples demonstrate that abusing our limited stock of natural resources is self-destructive and irrational. But instead of nurturing these resources to provide benefits that can be sustained far into the future, too much of modern development is doing the opposite: abusing nature to provide excessive benefits for a generation or two of humans. The symptoms of this abuse are all around us, from local deforestation to global climate change.

Conserving the World's Biological Diversity is a guide to all who would like to turn the tide of destruction into a new, positive relationship between people and nature. A new form of civilization based on the sustainable use of renewable resources is not only possible, but essential. This book suggests the principles and tools that are available to promote the new civilization, based on community self-reliance, diversity in both nature and human cultures, economic systems that consider all costs and benefits of alternative actions, scientific research that is applied to the challenges of managing natural resources, and the use of modern information technology to ensure that decisions are based on full knowledge of the likely consequences.

Most of the major policy decisions that affect the use of natural resources are taken in the cities, far removed from the realities of the limitations imposed by nature's productivity. Policies on trade, international cooperation, land tenure, defense, agriculture, forestry, fisheries, education, health, and finance all affect the way biological resources are used or abused. This book can help ensure that urban decision-makers do not forget that the wellspring of human prosperity is in the countryside, and that new policies are required to ensure a continuing flow of benefits from biological resources to all of humanity.

ACKNOWLEDGEMENTS

This document has been prepared with the financial support of the Asian Development Bank, CI, HighGain, Inc., IUCN, John D. and Catherine T. MacArthur Foundation, the Pew Charitable Trusts, the Rockefeller Foundation, USAID, W. Alton Jones Foundation, Inc., Waste Management, Inc., the World Bank, WRI, WWF-US, and the Governments of Denmark, Finland, the Netherlands, Norway, Sweden, and Switzerland.

The ideas contained herein have benefitted from discussions held in Nairobi, Kenya, on 29-31 August 1988 with Martin Holdgate, Tom Lovejoy, David Munro, Reuben Olembo, Perez Olindo, Peter Raven, and Michael Soulé, at the invitation of UNEP Executive Director Mostafa Tolba. The document has drawn heavily from previous work by IUCN, particularly in regards to economics and biological diversity (McNeely, 1988), and the draft white paper on biodiversity prepared by WRI under a grant from the Rockefeller Foundation. Many of the issues discussed were originally called to the world's attention by Norman Myers, who has often been years ahead of his time; the paper also owes intellectual debts to Ray Dasmann, Jared Diamond, David Pearce, Duncan Poore, Robert and Christine Prescott-Allen, and Jerry Warford. A number of IUCN staff, including Steve Davis, Pat Dugan, Jerry Harrison, Vernon Heywood, Peter Hislaire, Martin Holdgate, Robin Pellew, Jeff Sayer, Hugh Synge, Simon Stuart, and Jim Thorsell, have also contributed. Joanna Erfani and Jeanne Colombo have been tireless in their secretarial support.

The first draft was discussed at a meeting of the World Bank's Biological Diversity Task Force, chaired by Ken Piddington and John Spears and organized by Mary Dyson and Russ Mittermeier. A large number of comments were received during that session and are reflected in this document. Gerardo Budowski, Bill Conway, Promila Kapoor, Tom Lovejoy, Larry Mason, Norman Myers, George Rabb, John Seidensticker, Michael Soulé, Tom Stoel, Bill Weber, and many others made important contributions.

The first draft was also sent to a large number of experts in all parts of the world. Helpful comments were received from: Warren Brockelman (Thailand); David Chivers (Cambridge, UK); J.C. Daniel (India); Doug Fuller (Washington, D.C.); Jorg Ganzhorn (Tubingen, Federal Republic of Germany); Colin Groves (Canberra, Australia); Barbara Harrisson (Netherlands); K. Elaine Hoagland (Washington, D.C.); Cynthia Jensen (Washington, D.C.); Molly Kux (Washington, D.C.); David Langdon (Australia); Jeremy Mallinson (Jersey, UK); Clive Marsh (Sabah, Malaysia); Sharon Matola (Belize); Norman Myers (Cambridge, UK); J.C. Ogden (St. Petersburg, Florida); Junaidi Payne (Sabah, Malaysia); Jordi Sabater Pi (Barcelona, Spain); Ajay Rastogi (India); Pramote Saiwichian (Bangkok, Thailand); Shirley Strum (Kenya); and Judith Weis (Washington, D.C.).

EXECUTIVE SUMMARY

Our species entered the industrial age with a population of one billion and with biological diversity — the total of genes, species, and ecosystems on earth — possibly at an all-time high. Biological resources — the portion of diversity of actual or potential use to people — were freely available for exploitation to support development.

In the late 20th century, we are coming to realize that *biological resources have limits,* and that *we are exceeding those limits* and thereby reducing biological diversity. This is therefore a time of extraordinary change in the relationship between people and the biological resources upon which their welfare depends. Each year, more people are added to the human population than ever before, species are becoming extinct at the fastest rate known in geological history, and climate appears to be changing more rapidly than ever.

Human activities are progressively eroding the earth's capacity to support life at the same time that growing numbers of people and increasing levels of consumption are making ever greater demands on the planet's resources. The combined destructive impacts of a poor majority struggling to stay alive and an affluent resource-consuming minority are inexorably and rapidly destroying the buffer that has always existed, at least on a global scale, between human resource consumption and the planet's productive capacity.

The erosion of the planet's life-support systems is likely to continue until human aspirations come more into line with the realities of the earth's resource capacities and processes, so that activities become sustainable over the long term. The problems of conserving biological diversity therefore cannot be separated from the larger issues of social and economic development.

Maintaining maximum biological diversity assumes far greater urgency as rates of environmental change increase. Diversity in genes, species, and ecosystems provides the raw materials with which different human communities will adapt to change, and the loss of each additional species reduces the options for nature — and people — to respond to changing conditions.

The tropics harbor a major proportion of the planet's biological diversity. The industrialized countries also depend on tropical resources, as industrial materials, sources of breeding material, pharmaceuticals, tourism sites, and a wide range of other tangible and intangible benefits. So far, however, the exploitation of the tropics by the industrialized societies has yielded great benefits without making commensurate investments in conservation and without paying the environmental costs of over-exploitation. Cheap labor, raw materials with low prices that do not reflect their true value, inappropriate development aid, and the control of commodity prices and interest rates, among other factors, have encouraged much more rapid levels of resource depletion and destruction than would otherwise be the case. The situation is continually worsening through the ramifications of the developing world's debt crisis and related high interest rates.

Governments, industry, development agencies, and the general public are therefore becoming increasingly concerned about the depletion of biological resources, with the growing awareness that development depends on their maintenance.

How can the scientific knowledge be mobilized that will best enable the planet's biological diversity to be conserved? How can the process of change be managed so that biological resources can make their best contribution to sustainable development? What information is required to address the problems of conserving biological diversity? Which problems need to be addressed first? How can the many initiatives in conservation of biological diversity be coordinated most effectively? Where can the financial resources be found to respond to these issues at a scale that will be commensurate with the problems?

This document seeks answers to these questions.

The Values of Biological Diversity

Biological resources provide the basis for life on earth, including that of humans. The fundamental social, ethical, cultural, and economic values of these resources have been recognized in religion, art, and literature from the earliest days of recorded history. The great interest that children have in nature, the numerous wildlife clubs, the generous donations made to non-governmental conservation organizations, the political support for "Green Parties," and the popularity of zoos and wildlife films are economic expressions of preference and show that the general public does not think of biological resources merely in terms of a cash value.

But in order to compete for the attention of government and commercial decision-makers in today's world, policies regarding biological diversity first need to *demonstrate in economic terms the contribution biological resources make to the country's social and economic development.* Even partial valuation in monetary terms of the benefits of conserving biological resources can provide at least a lower limit to the full range of benefits and demonstrate that conservation can yield a profit in terms that are meaningful to national accounts.

Three main approaches have been used for determining the value of biological resources:

- assessing the value of nature's products — such as firewood, fodder, and game meat — that are consumed directly, without passing through a market ("consumptive use value");
- assessing the value of products that are commercially

harvested, such as timber, fish, game meat sold in a market, ivory, and medicinal plants ("productive use value"); and,
- assessing indirect values of ecosystem functions, such as watershed protection, photosynthesis, regulation of climate, and production of soil ("non-consumptive use value"), along with the intangible values of keeping options open for the future ("option value") and simply knowing that certain species exist ("existence value").

How and Why Biological Resources are Threatened

The proximate causes of the loss of biological resources are clear. Biological resources are degraded and lost through such activities as the large-scale clearing and burning of forests, overharvesting of plants and animals, indiscriminate use of pesticides, draining and filling of wetlands, destructive fishing practices, air pollution, and the conversion of wildlands to agricultural and urban uses.

When the problem of biodiversity loss is defined in terms of its immediate causes, the response is to take defensive and often confrontational actions, such as enacting laws, closing access to resources, and declaring additional protected areas. Such responses are necessary in times of rampant over-exploitation. But they are *seldom really sufficient to change the social and economic causes of the threats to biological diversity.*

The foundations of over-exploitation include demands for commodities such as tropical hardwoods, wildlife, fiber, and agricultural products. The growing human population, even without accompanying economic growth and development, places increasing demands on natural resources and ecosystem processes that are already impoverished and stressed. Settlement policies promote the movement of the growing unemployed labor forces to frontier zones. The debt burden forces governments to encourage the production of commodities that can earn foreign exchange. Energy policies encourage inefficiency in many nations, and in so doing add to the burden of air pollutants and the risk of substantial global climate change. Inappropriate land tenure arrangements discourage rural people from making the investments that would enable sustainable use of the available biological resources.

When the problem is defined in terms of its root causes, a more constructive response can be stimulated that seeks *cooperative efforts to address the social and economic foundations of resource depletion.*

Six main obstacles to greater progress in conserving biological diversity need to be addressed:
- National development objectives give insufficient value to biological resources.
- Exploiting biological resources yields the greatest profit for traders and manufacturers (who can externalize environmental costs), not for the local people who have few other sources of livelihood, and who must pay the environmental costs of over-exploitation.
- The species and ecosystems upon which human survival depends are still poorly known.
- The available science is insufficiently applied to solving management problems.
- Conservation activities by most organizations have focused too narrowly.
- Institutions assigned responsibility for conserving biodiversity have lacked sufficient financial and organizational resources to do the job.

Approaches to Conserving Biological Diversity

Conserving biological diversity needs to address both proximate and ultimate causes. The complex threats to biological diversity call for a wide range of responses across a large number of private and public sectors. All are necessary, with the mix of responses adjusted to the local conditions. Since government policies are often responsible for depleting biological resources, it stands to reason that policy changes are often a necessary first step toward conservation. National policies dealing directly with wildlands management or forestry, or influencing biodiversity indirectly through land tenure, rural development, family planning, and subsidies for food, pesticides, or energy can have significant impacts on the conservation of biodiversity. National and sub-national conservation strategies can often provide the mechanism for carrying out such reviews.

Protecting species can best be done through *protecting habitats.* Most national governments have established legal means for protecting habitats that are important for conserving biological resources. These can include: national parks and other categories of reserves (some 4,500 major reserves exist, covering nearly 500 million hectares); local laws protecting particular forests, reefs, or wetlands; regulations incorporated within concession agreements; planning restrictions on certain types of land; and customary laws protecting sacred groves or other special sites. The responsibility for such management is often spread widely among public and private institutions. While accomplishments to date are impressive, the amount of habitat protected needs to be increased by a factor of three if these areas are to make the necessary contribution to conserving biological diversity; these new areas may need more flexible approaches to management than is usual in national parks.

In addition, the protected areas will succeed in realizing their conservation objectives only to the extent that the areas themselves are effectively managed, and to the extent that the management of the land surrounding them is compatible with the objectives of the protected areas. This will typically involve protected areas becoming parts of larger regional schemes to ensure biological and social sustainability, and to deliver appropriate benefits to the rural population.

Ex situ conservation programs — zoos, aquaria, seed banks, botanic gardens, and so forth — supplement *in situ* conservation by providing for the long-term storage, analysis, testing, and propagation of threatened and rare species of plants and animals and their propagules. They are particularly important for wild species whose populations are highly reduced in numbers, serving as a backup to *in situ* conservation, as a source of material for reintroductions, and as a major repository of genetic material for future breeding programs of domestic species. Some *ex situ* facilities — notably zoos and botanic gardens — provide important opportunities for public education, and many make important contributions to taxonomy and field research.

Measures to *curb the pollution of the biosphere*, perhaps the most widespread conservation measures, are the most expensive, and have attracted the greatest attention from both the public and government. Biological diversity is threatened by various forms of chemical pollution, but the gravest threat may be climate change brought about by air pollution and the increase in atmospheric carbon dioxide due to deforestation and the burning of fossil fuels. Mean world temperatures could increase by about 2 °C and mean sea levels rise by around 30-50 centimeters in the next 40 years. While the species and ecosystems contained within protected areas will certainly be affected by climate change, it is unrealistic to expect the boundaries of existing protected areas to change very much, because they are usually surrounded by more intensive human land uses. Instead, new forms of management intervention will be required to maintain systems deemed desirable.

Many of the responses just discussed have been supported by *international legislation* that has fostered useful cooperation in conserving biological diversity. However, species and ecosystems are still being exploited at rates that far exceed their sustainable yield. Recognizing the growing severity of threats to biological diversity and the increasingly international nature of the actions required to address the threats, IUCN and UNEP have embarked on the preparation of an International Convention on the Conservation of Biological Diversity. This effort has gained the broad support of governments, including a joint resolution from the US Congress.

People form the foundation for the sustainable use of biological resources. Local communities need to be more involved in the management of biological resources, and to benefit from their sustainable use. Because groups of *indigenous people* in many parts of the world regard natural resources, particularly wildlife, as essential to their cultural continuity and economic well-being, they should be given particular attention in all conservation programs. Local people should be closely associated with the authorities responsible for the management of biological resources and for the establishment and management of protected areas. However, the tension between local interests and national interests in conservation requires great sensitivity and site-specific solutions.

The Information Required to Conserve Biological Diversity

Effective action must be based on accurate information, and the more widely shared the information, the more likely it is that individuals and institutions will agree on the definition of problems and solutions. Developing and using information is therefore an essential part of conservation at all levels, from the local to the global community.

The current state of knowledge about species and ecosystems is woefully inadequate; detailed knowledge is still lacking on the distribution and population sizes of even such large and well-studied animals as African primates. It seems self-evident that increasing knowledge about the kind and variety of organisms that inhabit the earth — and the ways that these organisms relate to each other and to humans — must be a foundation of conservation action. Therefore, a major effort is required to:

- document the wealth of the world's species of plants and animals, involving museums, zoos, aquaria, botanic gardens, universities, and research stations;
- carry out ecological fieldwork to show how the various pieces fit together, discover the population dynamics of species of particular concern, assess the effects of fragmentation of natural habitats, and determine what management steps are required to enable ecosystems to flourish with their full complements of species;
- develop new mechanisms for *ex situ* conservation, including both captive propagation and eventual release into ''natural'' ecosystems;
- monitor the changes in ecosystem diversity and function as the influences of humans become more pervasive, including climate change, deforestation, and various forms of pollution;
- assess the ecological differences between relatively large but minimally disturbed ecosystems and ecosystems that have been heavily affected by humans, as a basis for enhancing productivity and restoring degraded ecosystems to a more productive state; and to
- carry out research in the social sciences to determine how local people manage their resources, how changes in resource availability and land use affect human behavior, and how people decide how to use their biological resources.

Such basic inventory and fundamental research work should be carried out simultaneously with field action, with the two forms of activity reinforcing each other.

Government agencies, local communities, and conservation organizations all need information to enable them to manage their biological resources more effectively. Information tools that can help meet this need include basic descriptions of fauna and flora, practical handbooks for field identification, rapid inventory techniques, and basic computer programs for use with micro-computers.

The *information needs in the tropics are particularly important,* because these areas hold the majority of the world's biological diversity and they are losing species at rates that far exceed the world's capacity to record them. Highest priority for basic inventory work should be given to the sites of greatest diversity and local endemism coupled with the greatest threat, for the information contained by the species in these areas could disappear before humanity even knows what it is losing.

Development agencies should support national efforts to establish local, sectoral, and national information management systems, through demonstrating methodologies, providing training opportunities for taxonomists and biologists, and subsidizing the publication of status reports. Universities, research institutions, and non-governmental organizations (NGOs) need to be strengthened so that they can help governments assess their biological resources. Closer working relationships should be established between museums and other taxonomic-oriented institutions and those concerned with conservation of biological diversity.

Establishing Priorities for Conserving Biological Diversity

When governments approved the World Charter for Nature in the United Nations in 1982, they agreed that *all species and habitats should be safeguarded to the extent that it is technically, economically, and politically feasible.* But resources for conservation are always limited, so efforts spent in deciding what to do first are usually well repaid in savings of time, finances, and personnel.

Determining priorities is a complex task. The genetic landscape is constantly changing through evolutionary processes, and the world contains more variability than can be expected to be protected by explicit conservation programs; further, the capacity of governments or private organizations to deal with environmental problems is limited and many urgent demands compete for their attention. So governments, international organizations, and conservation agencies seeking to conserve biological diversity must be selective, and ask *which species and habitats most merit a public involvement in protective measures.*

No generally accepted scheme exists for establishing priorities for the conservation of biological diversity, nor is it either possible or advisable for such a scheme to be devised. *Different organizations and institutions can be expected to have different ways of establishing priorities because of their differing goals.* For example, from a global perspective on biodiversity, regions with high species diversity may be most "valuable"; from the perspective of a pastoral community in the Sahel, however, the diversity of life available in the local ecosystem will be of highest value even though it exhibits relatively little diversity on a global scale.

The various methods of establishing priorities suggest different types of conservation action and will result in the conservation of different subsets of the world's biological resources. Each system has its own strengths and weaknesses, with the major point of difference being the *objective for which the system was devised.*

The Role of Strategies and Action Plans in Promoting Conservation of Biological Diversity

One of the best ways to ensure that the various institutions involved in conservation are in general agreement on priorities is to prepare a *strategy that defines the basic problems and agrees to appropriate objectives.* Strategies are turned into action through a more tactical process of planning specific activities to address the broad strategies; this often involves the preparation of an action plan.

A global strategy is required to provide the framework for local and regional efforts, and to give concise guidance on the options and opportunities for action capable of achieving global goals while addressing local priorities. The strategy needs to be supported by regional, national, local, and sectoral strategies, and by action designed to meet specific needs.

Such a strategy, dealing with all aspects of biodiversity, including both marine and terrestrial ecosystems at all latitudes, is currently being prepared by a coalition of the World Resources Institute, IUCN, and UNEP, in close collaboration with WWF, CI, the World Bank, the Asian Development Bank, and other key governmental and non-governmental institutions in both tropical and temperate nations. It is expected that FAO and Unesco will also participate in the process. It aims to:

- establish a common perspective, foster international cooperation, and agree to priorities for action at the international level;
- examine the major obstacles to progress and analyze the needs for national and international policy reform;
- specify how conservation of biological resources can be integrated with development more effectively and identify the linkages with other related issues facing humanity; and to
- promote the further development of regional, national, and thematic action plans for the conservation of biological diversity, and promote their implementation.

How to Pay for Conserving Biological Diversity

Innovative funding mechanisms will be required to support conservation efforts. These mechanisms should be based on the principle that *those who benefit from biological resources should pay more of the costs of ensuring that such resources are used sustainably.* Efforts are required at the community level to provide economic incentives for conservation, at the national level to ensure that government policies are compatible with such incentives, and at the international level to ensure that the wealthy nations benefitting from the biological resources of the tropics are able to invest in conserving the productive capacity of those resources.

Approaches useful primarily at the national level include charging entry and other fees to national parks, levying charges for ecological services, collecting special taxes, building funding linkages with large development projects, returning profits from exploitation of biological resources, building conditionality into concession agreements, seeking support from the private sector, and establishing foundations for conservation. Approaches useful at the international level include international conventions that provide financial support, direct assistance from international conservation organizations, debt-for-nature swaps, restricted currency holdings, and conservation concessions.

In general, conservation should be supported to the maximum extent possible through the marketplace, but the marketplace needs to be established through appropriate policies from the central government. One problem faced by all the funding mechanisms described in this book is opportunity costs; any funds earned might be used by the government in other ways that the government considers of higher priority. The attraction of the methods suggested is that the *income is being earned by the biological resources*, and some of the funding is being provided by the public in expression of their support for non-consumptive uses of biological resources.

In many countries, funding is not the major constraint to conservation achievement. While conservation agencies never have sufficient funding, and additional funding is certainly called for, even generous budgets will not lead to conservation if government policies in other sectors are incompatible with conservation. Therefore, any new funding mechanisms need to be part of a package that includes necessary *policy changes in national security, land tenure, energy, frontier settlement, foreign trade, transportation, and so on.*

The major requirement from government policymakers is that they recognize the many values of biological resources, and take advantage of *opportunities to invest in the continued productivity* that such resources require. They also need to be persuaded to create conditions whereby the local community or the private or NGO sector can assume total management control of certain important biological resources or areas, and can seek their own funding in an attractive tax and investment climate. Through the use of innovative funding mechanisms backed by compatible government policies, one of the major obstacles to progress in conservation can be overcome.

Conclusion

The elements now exist that will reverse the trend toward the biotic impoverishment of the world. Novel approaches, new financial mechanisms, and new policies need to be applied at the appropriate level of responsibility to translate the new approaches into a *reality of improved human well-being and a secure biotic heritage.* New partners in conservation need to be found, involving all ministries, departments and private institutions that are directly dependent on biological resources. National parks departments, for example, should be joined in habitat management by a wide range of other institutions to represent all interests. Furthermore, other line agencies need to develop the capacity to manage biodiversity of particular relevance to their respective missions.

The 1990s may be the last decade during which constructive and creative decisions, activities, and investments — rather than emergency rescue efforts — can be made to ensure that many of the world's species and ecosystems are maintained, examined for their material and ecological value, and promoted for sustainable use to support new and innovative approaches to development. The combination of maintaining the *maximum possible biological diversity,* the *maximum possible cultural diversity,* and the *greatest possible scientific endeavor* would seem the most sensible approach toward dealing with the dynamic future facing humanity.

We are at a crossroads in the history of human civilization. Our actions in the next few years will determine whether we take a road toward a chaotic future characterized by overexploitation and abuse of our biological resources, or take the opposite road — toward maintaining great biological diversity and using biological resources sustainably. *The future well-being of human civilization hangs in the balance.*

Following page: A southern elephant seal *(Mirounga leonina)* pup (photo by W.V. Reid).

CHAPTER I

BIOLOGICAL DIVERSITY: WHAT IT IS AND WHY IT IS IMPORTANT

As the fundamental building blocks for development, biological resources provide the basis for local self-sufficiency. At the same time, biological diversity is a global asset, bringing benefits to people in all parts of the world. Efforts to maintain the diversity of biological resources are urgently required at local, national, and international levels.

Our generation has a great opportunity, and a great responsibility. We have inherited the most diverse community of living creatures that has ever occupied our planet (Wilson, 1988b), and we have the most sophisticated technology that has ever existed. Using our modern technology to exploit the resources our planet provides, people living in the 20th century have also witnessed the greatest social and demographic changes our species has ever experienced. We live in momentous times indeed.

But the prosperity of our future is far from assured, and if present trends continue our generation will be responsible for destroying much of the natural wealth we have inherited. The decisions we make in the next few years about how natural resources will be used will determine the future evolution of both human civilization and life on earth.

The combination of energy from the sun and natural resources on the earth provides the basis for human prosperity. Some of these resources, such as oil, coal, gold, and

Box 1: What is Biological Diversity?

"Biological diversity" encompasses all species of plants, animals, and microorganisms and the ecosystems and ecological processes of which they are parts. It is an umbrella term for the degree of nature's variety, including both the number and frequency of ecosystems, species, or genes in a given assemblage. It is usually considered at three different levels: genetic diversity, species diversity, and ecosystem diversity. Genetic diversity is the sum total of genetic information, contained in the genes of individuals of plants, animals, and microorganisms that inhabit the earth. Species diversity refers to the variety of living organisms on earth and has been variously estimated to be between 5 and 50 million or more, though only about 1.4 million have actually been described. Ecosystem diversity relates to the variety of habitats, biotic communities, and ecological processes in the biosphere, as well as the tremendous diversity within ecosystems in terms of habitat differences and the variety of ecological processes. Ecosystems cycle nutrients (from production to consumption to decomposition), water, oxygen, methane, and carbon dioxide (thereby affecting the climate), and other chemicals such as sulphur, nitrogen, and carbon.

Biologists classify life on earth into a widely accepted hierarchical system that reflects evolutionary relationships among organisms. In ascending order, the main categories, or taxa, of living things are: Species, Genus, Family, Order, Class, Phylum, Kingdom. Humans, for example, are classified as follows: Animalia (Kingdom), Chordata (Phylum), Mammalia (Class), Primates (Order), Hominidae (Family), *Homo* (Genus), *sapiens* (Species). These last two designations, together referred to as the Latin binomial, are used to identify an organism, and distinguish it from any other. Species differ from one another in at least one characteristic and generally do not interbreed (Raven and Johnson, 1989). In general, the higher the category ranking of an organism, the more ancient the evolutionary divergence. Thus, with *Homo sapiens,* it was more recently that the species became established than the genus, and more recently that the genus evolved than did the family (Hominidae), and so on up to the Kingdom level. Most biologists recognize five kingdoms of organisms: Prokaryotae (bacteria), Protoctista (includes algae and protozoans), Fungi (mushrooms, molds, and lichens), Animalia (animals), and Plantae (plants) (Margulis and Schwartz, 1982). Currently, approximately 100 phyla are recognized (see Annex 1 for listing).

Box 2: The Dimensions of the Issue: How Many Species Exist?

The foundation for assessing the importance of biological diversity is an inventory of how many species exist, and which species exist where. At the global level, the plants and vertebrates are relatively well known, though major discoveries are reported regularly among fish and some groups of plants. But scientists can only guess at the numbers of many groups of insects (especially the beetles of the tropical forest). Erwin (1982), for example, suggests as many as 30 million species in total, with most undescribed species living in tropical forests. Mites and nematodes could also number in the hundreds of thousands, if not millions, of species. Since most estimates of extinctions are based on extrapolations, the lack of precise estimates of total numbers has led to considerable imprecision regarding extinction rates. May (1988) suggests that research on food webs, relative abundance, and the relationship between numbers and physical size of organisms could reveal patterns that would enable the total diversity of plants, animals, and microorganisms to be deduced from appropriate rules.

But the fact remains that basic knowledge of the organisms that make up most ecosystems, especially in the tropics, is woefully inadequate. The Committee on Research Priorities in Tropical Biology (NAS, 1980) concludes that at least a five-fold increase in the number of systematists (above the current estimated 1,500 trained professional systematists competent to deal with any of the tropical organisms) is necessary to deal with a significant proportion of the estimated diversity while it is still available for study. For convenience, many assume that about 10 million species exist, though the final figure is likely to be 30-50 million.

Given these limitations, the following represents a summary of the current state of knowledge (from Wilson, 1988a, except where otherwise noted):

Group	**No. of described species**
Bacteria and blue-green algae:	4,760
Fungi:	46,983
Algae:	26,900
Bryophytes (mosses and liverworts:)	17,000 (WCMC, 1988)
Gymnosperms (conifers):	750 (Raven *et al.*, 1986)
Angiosperms (flowering plants):	250,000 (Raven *et al.*, 1986)
Protozoans:	30,800
Sponges:	5,000
Corals & Jellyfish:	9,000
Roundworms & earthworms:	24,000
Crustaceans:	38,000
Insects:	751,000
Other arthropods and minor invertebrates:	132,461
Mollusks:	50,000
Starfish:	6,100
Fishes (Teleosts):	19,056
Amphibians:	4,184
Reptiles:	6,300
Birds:	9,198 (Clements, 1981)
Mammals:	4,170 (Honacki *et al.*, 1982)
Total	**1,435,662 species**

iron, are non-renewable; once they have been consumed, they cannot be replaced in time frames meaningful to us. Other resources are renewable; water can be recycled repeatedly, and wildlife, forests, and crops reproduce themselves and even increase when managed appropriately.

Considerable care needs to be given to decisions on how non-renewable resources shall be consumed, and considerable efforts are being devoted to finding substitutes for those which are being depleted (Borman, 1976), to seeking more effective means of recycling, and ensuring the most efficient practical forms of use (including reduction of waste). But far more attention needs to be given to the management of renewable resources, because they provide the basis for long-term sustainable production of goods and services essential for human welfare.

Biological resources — genes, species, and ecosystems that have actual or potential value to people — are the physical manifestation of the globe's **biological diversity** (sometimes shortened to "biodiversity"), which simply stated is **the variety and variability among living organisms and the ecological complexes in which they occur** (Boxes 1 and 2) (OTA, 1987). Species are the building blocks of ecosystems, and ecosystems provide the life-support systems for humans. Modern technologies, capital investments, infrastructure, social organization, and so forth can enhance or deplete these life-support systems, but such recent phenomena as carbon dioxide increase, global warming, and the depletion of the globe's ozone shield demonstrate that nature has limits to her capacity to absorb environmental abuse.

Biological diversity is an umbrella term covering the totality of species, genes, and ecosystems, but biological resources can actually be managed; they can be consumed or replenished, and they can be the subject of directed conservation action. The way biological resources are managed can enhance or reduce biological diversity. Effective systems of management can ensure that biological resources not only survive, but increase while they are being used, thus providing the foundation for sustainable development. Practical applications will need to involve actions to address both the abstract biological diversity and the tangible biological resources.

Many development plans fail to recognize that the retention of natural systems often constitutes the optimal use of the land in question, in economic as well as ecological terms. Transforming natural areas often brings greater risks than benefits, because in their natural state these systems equilibriate water runoff, are reservoirs of valuable plants and animals, can yield timber on a sustainable basis, build soils and prevent erosion, and attract tourist revenues. Improving management of species and habitats could preserve for societies the resources available in their ecosystems, while producing sufficient surplus to support better standards of living.

But instead of conserving the rich resources of forest, wetland, and sea, current processes of development are depleting many biological resources at such a rate that they are rendered essentially non-renewable. Once a tropical forest is cleared of its trees, for example, the nutrients are removed from the system and it may take millennia for the system to recover (Gomez-Pompa *et al.*, 1972; Whitmore, 1984). The benefits to society are therefore substantially less than could be realized if the resources were managed on a sustainable basis. Experience has shown that market forces alone will often lead to such overexploitation, largely because many of the costs are external to those doing the exploiting; they gain the benefits without paying the costs. Since insufficient biological resources will be conserved by current market mechanisms alone, the conservation needs of society must to be met by a combination of international cooperation, effective government intervention and greatly increased participation by business, industry, local communities, universities, and other institutions.

Biological Diversity and Development

The contribution of conservation to development was acknowledged by the World Commission on Environment and Development, drawing on a decade of work in this field (Box 3). "The challenge facing nations today," said the WCED (1987), "is no longer deciding whether conservation is a good idea, but rather how it can be implemented in the national interest and within the means available to each country."

Conservation in the modern sense is part of development. As defined by the *World Conservation Strategy,* it means: "The management of human use of the biosphere so that it may yield the greatest sustainable benefit to present generations while maintaining its potential to meet the needs and aspirations of future generations. Thus, conservation is positive, embracing preservation, maintenance, sustainable utilization, restoration, and enhancement of the natural environment" (IUCN, 1980).

The *World Conservation Strategy* has provided a useful rationale for people involved in both conservation and development, providing policy guidance on how conservation can support sustainable development. It concentrates on the main problems directly affecting the achievement of conservation objectives, and identifies the action needed both to improve conservation efficiency and to integrate conservation and development. It specifically identifies the preservation of biological diversity as one of the three main foundations of conservation; the second — to maintain essential ecological processes and life-support systems — provides support to biological diversity; the third objective — to ensure that any utilization of species and ecosystems is sustainable — deals with the uses to which biological resources are put.

Box 3: Recent Advances in Concepts of Conserving Biological Diversity.

Considerable scientific work has been done to address the needs for conserving biodiversity, and to describe the technology available for doing so. Notable examples of the 1980s include:

- *Planning National Parks for Ecodevelopment* (Miller, 1980).
- *Conservation Biology: An Evolutionary-Ecological Approach* (Soulé and Wilcox, 1980)
- *Conservation and Evolution* (Frankel and Soulé, 1981)
- *Genetics and Conservation: A Reference for Managing Wild Animal and Plant Populations* (Schonewald-Cox *et al.*, 1983)
- *National Parks, Conservation, and Development* (McNeely and Miller, 1984)
- *Marine and Coastal Protected Areas: A Guide for Planners and Managers* (Salm and Clark, 1984)
- *The Value of Conserving Genetic Resources* (Oldfield, 1984)
- *Plant Genetic Resources: A Conservation Imperative* (Yeatman *et al.*, 1984)
- *The Gaia Atlas of Planet Management* (Myers, 1985)
- *Managing Protected Areas in the Tropics* (MacKinnon *et al.*, 1986)
- *Conservation Biology: The Science of Scarcity and Diversity* (Soulé, 1986)
- *Technologies to Maintain Biological Diversity* (OTA, 1987)
- *Gene Banks and the World's Food* (Plucknett *et al.*, 1987)
- *Biodiversity* (Wilson and Peter, 1988)
- *Economics and Biological Diversity* (McNeely, 1988)
- *Wildlands: Their Protection and Management in Economic Development* (World Bank, 1988)
- *Keeping Options Alive: The Scientific Basis for Conserving Biodiversity* (Reid and Miller, 1989)

While government institutions responsible for wildlife and protected areas need strengthening, even the most successful species conservation programs and protected area systems are only part of a larger package of appropriate conservation policies and programs in other sectors. Progress in sustainable approaches to forestry, agriculture, rural development, international trade, disaster prevention, energy, climate

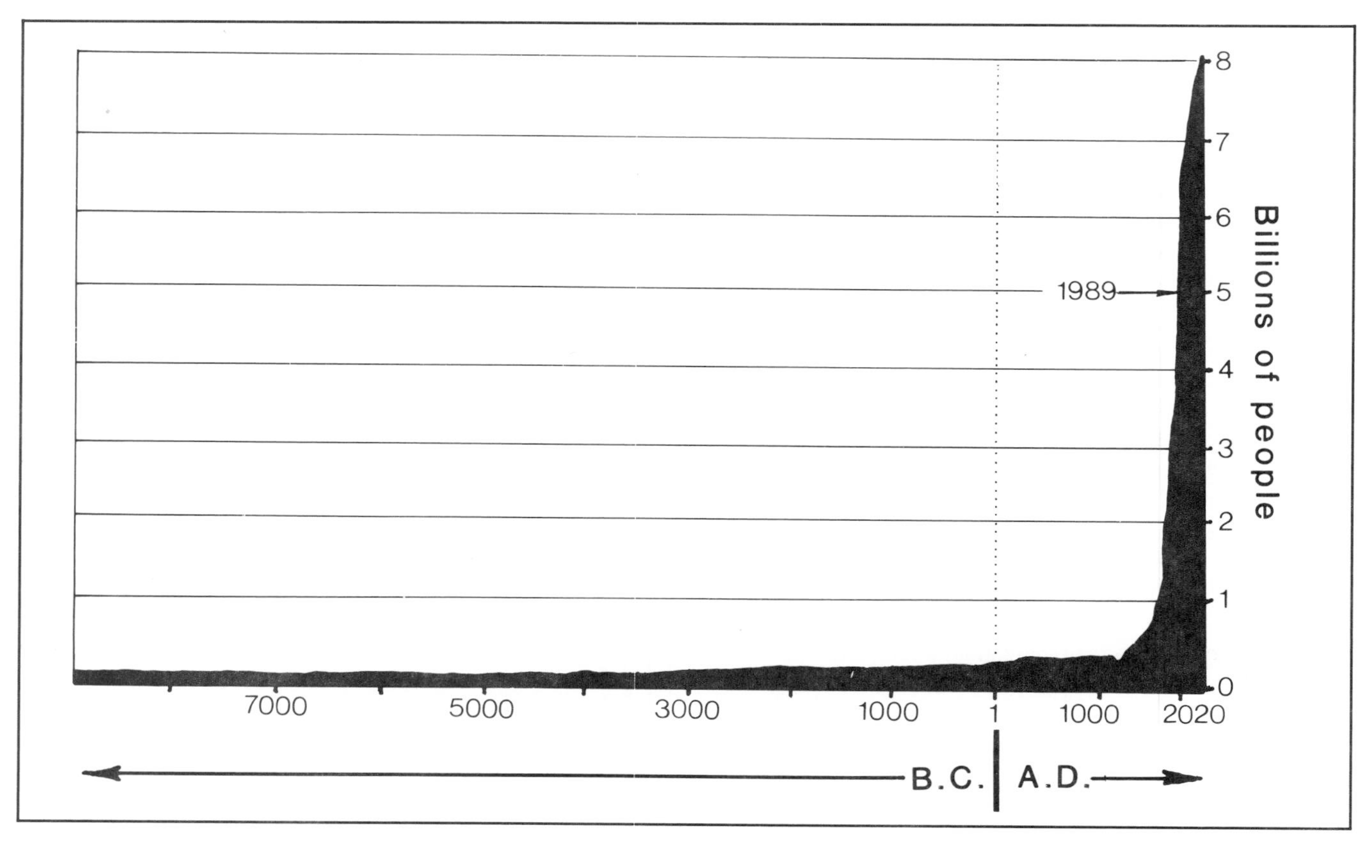

Fig. 1. Human population growth.

change, population, national security, and other areas are essential to the success of efforts to conserve biological diversity. This will involve policy shifts and improved management in a wide variety of sectors that have impacts on biological resources, often calling for line agencies — ministries of forestry, agriculture, fisheries, tourism, communications, health, and defense — to assume new responsibilities for conservation.

Modern Approaches to Conserving Biological Diversity

For most of human history, the natural world has been protected from the most disruptive human influences by relatively humble technology, cultural/ecological factors such as taboos preventing over-exploitation, tribal warfare that kept wide areas as wilderness "buffer zones" between groups, land ownership by ancestors or lineages rather than individuals, relatively sparse human populations, and many other factors. But the "Garden of Eden" vision contains its weeds. Where human hunters moved into new habitats filled with game animals that had no prior experience with humans, major extinctions often occurred; the Americas, Australia, Madagascar, and New Zealand are well-known examples (Martin, 1984). Humans are also implicated in the extinction of some 90 percent of the endemic mammalian genera of the Mediterranean after the development of agriculture (Sondaar, 1977). Despite these illustrations of the power of hunting and agricultural societies to drive species to extinction, the natural world at the dawn of the industrial age was characterized by highly diverse ecosystems and human cultures.

But in our era, the past few generations or so, economic growth based on the conversion of fossil fuels to energy, greatly expanded international trade, and improved public health measures has spurred such a rapid expansion of human numbers (Figure 1) that new approaches to resource management have been required. Within the past 100 years, governments have established explicit policies aimed at conserving wild living resources. Today, all but a small handful of countries have national parks and national legislation promoting conservation. Most governments have joined international conservation conventions, and built environmental considerations into the national education system. Non-governmental organizations are active in promoting public awareness of conservation issues, including those dealing with biological diversity.

But still the devastation continues, and even accelerates. Why?

Part of the problem may be that conservation has not yet involved the right institutions. The conservation movement has been led by naturalists, including both interested amateurs and trained biologists. While their contributions have been fundamental, they are unable to address fully the basic problems of conservation because the problems are not biological, but rather political, economic, social, and even ethical. The decisions affecting the natural environment are influenced by pressures and incentives that go far beyond the relatively straightforward technical considerations of what might in theory be best for biological resources.

Conservation action therefore needs to be based on the best available scientific information and implemented by development practitioners, engineers, politicians, rural sociologists, agronomists, and economists. Local resource users are often the ones who make local-level decisions, and their decisions are affected above all by enlightened self-interest. Those seeking to conserve biological diversity need to be able to identify the legitimate self-interests of rural people, and design ways of ensuring that the interests of conservation and community self-interest coincide.

No simple recipe exists for determining how biological resources in each locality can best be conserved, and how land should be used to best achieve the objectives of conservation. Ecological, social, political, economic, and technological factors all enter into the decisions made, and each of these factors can change over time; because these factors are interrelated, a change in one can have effects — sometimes unpredictable — on all the others. In the final analysis, decisions need to be taken by people exercising their best judgement at the current state of knowledge. The dynamic state of development throughout the world is likely to continue, and building the capacity to adapt to constant change will require concerted action.

Developing a Global Biodiversity Conservation Strategy

The increasing international interest in biodiversity stems from the growing dangers of species extinctions, depletion of genetic diversity, and disruptions to the atmosphere, water supplies, fisheries, and forests. As climatic, political, and economic conditions change over the coming decades, the various populations of *Homo sapiens* are going to be challenged to live up to their name. Biological diversity provides the building blocks with which each human group can use its intelligence and acquired wisdom to adapt to change, and having more blocks available will provide more options for adapting to new conditions.

The growing awareness about the importance of biodiversity on the part of both governments and the general public has resulted in a desire to ensure that no part of the world's natural heritage is lost through inadvertence or ignorance. Biodiversity brings together a variety of constituencies: forestry, agronomy, biotechnology, pharmaceuticals, and international trade, to name but a few. All these different constituencies look at biodiversity in their own ways, but all approaches are founded on a common perception of the variety of life as a raw material, a resource, and a priceless heritage in its own right.

In order to implement new action for conserving biological diversity in a time when many tropical governments are feeling the squeeze of external debt, the activities of the various interested agencies — both national and international, governmental and non-governmental — need to reinforce each other rather than work in opposition born of ignorance. International agencies need to support government action, and NGO activities need to stimulate new approaches at both national and local levels.

A global strategy for conserving the greatest possible biological diversity is required to provide the framework for local and regional efforts, and to give concise guidance on the options and opportunities for action capable of achieving global goals while addressing local priorities. Such a strategy needs to be supported by regional, national, local, and sectoral strategies and action designed to meet specific needs.

The global and regional strategies need to seek solutions to the problems facing biological diversity in several ways:

- seeking appropriate policy reform and management action in areas outside the "conservation sector," as traditionally perceived, that have major impacts on biological diversity (e.g., agriculture, forestry, tourism, transport and communications, education, defense, etc.);
- ensuring that "traditional" development activities are carried out in such a way that they contribute to conserving biological diversity (i.e., implementing sustainable development in the sense used by the World Commission on Environment and Development);
- enhancing the role of development agencies in contributing directly to the conservation of biological diversity;
- providing a strong legal basis for international cooperation in conserving biological diversity, and for support of national initiatives;
- strengthening the institutions in the "conservation sector," through enhanced training, new financial mechanisms, and stronger mandates, and building greater public support for conserving biological diversity.

Such a global strategy, dealing with all aspects of biodiversity, including both marine and terrestrial ecosystems at all latitudes, is currently being prepared by a coalition of the World Resources Institute, IUCN, and UNEP, in close collaboration with WWF, CI, the World Bank, the Asian Development Bank, and other key governmental and non-governmental institutions in both tropical and temperate nations. It is expected that FAO and Unesco will also participate in the process. As an important part of the new *World Conservation Strategy,* it aims to:

- establish a common perspective, foster international cooperation, and agree to priorities for action at the international level;
- examine the major obstacles to progress at the international level and analyze the needs for national and international policy reform;
- specify how conservation of biological resources can be

integrated with development more effectively and identify the linkages with other related issues facing humanity; and

- promote the further development of regional, national, and thematic action plans for the maintenance, study, and sustainable use of biological diversity, and promote their implementation.

The 1990s must be a time of intensive action, involving major national and international investments in conserving biological diversity. As eminent Harvard biologist Edward O. Wilson has said, "How the human species will treat life on Earth, so as to shape this greatest of legacies, good or bad, for all time to come, will be settled during the next 10 years" (Wilson, 1988b). This document suggests the kinds of approaches that will enable this generation of humans to enrich rather than impoverish the earth.

Following page, overleaf: Boy and lemurs, Madagascar (photo by R.A. Mittermeier).

الامارات العربية المتحدة

" الشارقة " - " العاقبة "

١٩٩٣

CHAPTER II
THE VALUES OF BIOLOGICAL DIVERSITY

While the values of biological resources are not always represented in the marketplace, they are nonetheless significant. New approaches are required for ensuring that these values are incorporated in national development planning so that costs and benefits come into closer balance.

Biological resources provide the basis for life on earth. The fundamental social, ethical, cultural, and economic values of these resources have been recognized in religion, art, and literature from the earliest days of recorded history. Given these multiple values, it is not surprising that most cultures (and governments) have embraced the principles of conservation. The great interest that children have in nature, the numerous wildlife clubs, the generous donations made to non-governmental conservation organizations, the political support for "Green Parties," the popularity of zoos and wildlife films, and many other intangible indicators are strong evidence that the general public does not think of biological resources merely in terms of a cash value.

But in order to compete for the attention of government decision-makers in today's world, policies regarding biological diversity first need to demonstrate in economic terms the value of biological resources to a country's social and economic development. Some have argued that biological resources are in one sense beyond value because they provide the biotic raw materials that underpin every major type of economic endeavour at its most fundamental level (Oldfield, 1984). But ample economic justification can be marshalled by those seeking to exploit biological resources, so the same kinds of reasoning need to be used to support alternative uses of the resources.

However, some serious problems in economic analysis remain: The standard models do not give sufficient weight to long-term benefits; approaches to assessing the economic values of natural processes such as watershed protection or amelioration of climate remain rudimentary at best; and the aesthetic, ethical, cultural, and scientific considerations that must be part of the economic equation are usually ignored.

New approaches to economic assessment would ensure that economic values incorporate both monetary and non-monetary expressions of preference, and not be limited to simply attempting to put a price tag on nature. Assigning these qualitative and quantitative values would provide a justification for more effective government action, often through the use of economic incentives for conserving biological resources (McNeely, 1988).

Ethics, Economics, and Biological Diversity

Before examining the economics of conserving biological resources, it is worth noting that the governments of the world have already made an important, but little noticed, ethical commitment to nature. The World Charter for Nature, "adopted and solemnly proclaimed" by the General Assembly of the United Nations on 28 October, 1982 (Annex 2), expresses absolute support by governments of the principles of conserving biodiversity. It recognizes that humankind is part of nature, that every form of life is unique and warrants respect regardless of its worth to human beings, and that lasting benefits from nature depend upon the maintenance of essential ecological processes and life-support systems and upon the diversity of life forms. It calls for strategies for conserving nature, scientific research, monitoring of species and ecosystems, and international cooperation in conservation action. But the World Charter for Nature has been all but forgotten by both governments and conservationists, and needs to be given far greater exposure in the future.

Drawing on the principles of the World Charter for Nature and the World Conservation Strategy (IUCN, 1980), IUCN's Working Group on Ethics and Conservation has produced an ethical foundation for conservation (Box 4). It concluded that the ethical basis for conserving biological diversity needs to be consistent with ecological principles and that it is important to promote activities that are sustainable in the long run. People need to recognize that the reasons for the existence of species and ecosystems may be more subtle and inscrutable than simply supporting the economic desires of the current generation of consumers. When a gene pool is driven to extinction by the current generation of humans who are maximizing their personal benefit, all future generations pay the cost (Rolston, 1985b; Norton, 1986; Ehrenfeld, 1972, 1988).

Nature also has considerable abstract importance, such as symbolizing the wild world of nature, the opposite of the urban life that many people find so stressful. This symbolism is communicated to the public through films, television,

Box 4: An Ethical Basis for Conserving Biological Diversity.

- The world is an interdependent whole made up of natural and human communities. The well-being and health of any one part depends upon the well-being and health of the other parts.
- Humanity is part of nature, and humans are subject to the same immutable ecological laws as all other species on the planet. All life depends on the uninterrupted functioning of natural systems that ensure the supply of energy and nutrients, so ecological responsibility among all people is necessary for the survival, security, equity, and dignity of the world's communities. Human culture must be built upon a profound respect for nature, a sense of being at one with nature and a recognition that human affairs must proceed in harmony and balance with nature.
- The ecological limits within which we must work are not limits to human endeavour; instead, they give direction and guidance as to how human affairs can sustain environmental stability and diversity.
- All species have an inherent right to exist. The ecological processes that support the integrity of the biosphere and its diverse species, landscapes, and habitats are to be maintained. Similarly, the full range of human culture adaptations to local environments is to be enabled to prosper.
- Sustainability is the basic principle of all social and economic development. Personal and social values should be chosen to accentuate the richness of flora, fauna, and human experience. This moral foundation will enable the many utilitarian values of nature — for food, health, science, technology, industry, and recreation — to be equitably distributed and sustained for future generations.
- The well-being of future generations is a social responsibility of the present generation. Therefore, the present generation should limit its consumption of non-renewable resources to the level that is necessary to meet the basic needs of society, and ensure that renewable resources are nurtured for their sustainable productivity.
- All persons must be empowered to exercise responsibility for their own lives and for the life of the earth. They must therefore have full access to educational opportunities, political enfranchisement, and sustaining livelihoods.
- Diversity in ethical and cultural outlooks toward nature and human life is to be encouraged by promoting relationships that respect and enhance the diversity of life, irrespective of the political, economic, or religious ideology dominant in a society.

books, commercials, photographs, calendars, and many other media. Judging from the popularity of these symbolic representations of wild nature, they must be helping to keep the stresses of urban dwelling within bearable bounds.

Ehrenfeld (1988) cautions that arguments for conservation should not be based simply upon economic considerations: "It is certain that if we persist in this crusade to determine value where value ought to be evident, we will be left with nothing but our greed when the dust finally settles. I should make it clear that I am referring not just to the effort to put an actual price on biological diversity but also to the attempt to rephrase the price in terms of a nebulous survival value . . . As shown by the example of the faltering search for new drugs in the tropics, economic criteria of value are shifting, fluid, and utterly opportunistic in their practical application. This is the opposite of the value system needed to conserve biological diversity over the course of decades and centuries."

Further, many scientists will argue, nobody knows enough about any gene, species, or ecosystem to be able to calculate its ecological and economic worth in the larger scheme of things. And, Ehrenfeld (1988) adds, "the species whose members are the fewest in number, the rarest, the most narrowly distributed — in short, the ones most likely to become extinct — are obviously the ones least likely to be missed by the biosphere." On the other hand, many of these may be greatly missed by people; one dramatic example is the population of the wild rice *(Oryza nivara)* which is the only source of resistance to grassy stunt virus. (Many other examples are contained in Myers, 1983a, and Prescott-Allen and Prescott-Allen, 1982a.)

Various strains of rice that have been bred at the International Rice Research Institute in Los Banos, Philippines (World Bank photo by Edwin G. Huffman).

Such perspectives are well worth bearing in mind, but the fact remains that major decisions affecting the status and trends of biological resources are based on economic factors, including the establishment of their value. Even partial valuation in monetary terms of the benefits of conserving biological resources can provide at least a lower limit to the full range of benefits and demonstrate to governments that conservation can yield a profit in terms that are meaningful to national accounts. Effective management of biological resources cannot avoid addressing issues of economic value, even realizing the ethical limitations of these issues. Food for the stomach comes before nourishment for the spirit, and the rural people who must worry most about where their next meal is coming from often live in the midst of the greatest diversity.

However, economic methods have their limitations, and should not always be assumed to be highly accurate. As World Bank economist Jeremy Warford (1987b) states, "If economic methods are to be successful, it is crucial that their limitations be understood and continually kept in mind. In particular, it should be recognized that value judgments about distributional and irreversible effects are unavoidable, but quantification in monetary terms of as many variables as possible is important in crystallizing those issues involving implicit value judgments which may otherwise be ignored."

The mainstream economic approach today, as exemplified by USAID (1987), is to compile a utilitarian calculation expressed in money values and including (in raw or modified form) the commercial values that are expressed in markets. However, it expands the account to include considerations that enter human preference structures but are not exchanged in organized markets. This extension and completion of a utilitarian account, where conservation of biological resources is at issue, is useful because it demonstrates that commercial interests do not always prevail over arguments based on broader economic considerations (Randall, 1988).

Completing such a utilitarian account does not depend on any prior claim that the utilitarian framework is itself the preferred ethical system. Ethical goals may also be served by completing a utilitarian account that demonstrates the value implications of human preferences that extend beyond commercial goods to include biodiversity. While some people might argue that a complete discussion of the value of biodiversity should extend beyond utilitarian concerns, "even these people would, presumably, prefer a reasonably complete and balanced utilitarian analysis to the truncated and distorted utilitarian analysis that emerges from commercial accounts" (Randall, 1988).

The ethical commitment contained in the World Charter for Nature provides a powerful justification for conserving biological diversity, but it is only "soft law" that does not bind governments. It has therefore seldom been invoked or quoted; indeed, it has been honored more in the breach than the observance. Nor can the ethical principles in Box 4 be expected to lead by themselves to major changes in human behavior. Additional justification is required to change the way governments take decisions, and this will usually require economic arguments. Biological diversity has fundamental values in material, aesthetic, and ethical terms and while the general public often recognizes the more intangible values, the processes of development tend to stress material benefits.

Assessing the Value of Biological Resources

In order for governments to assess the priority they will give to investments in conservation of biological resources, they need to have a firm indication of what contribution these resources make to their national economy. Economists have devised a variety of methods for assigning values to natural biological resources (see Barrett, 1988; Brown and Goldstein, 1984; Cooper, 1981; Fisher, 1981b; Hufschmidt *et al.*, 1983; Johansson, 1987; Krutilla and Fisher, 1975; Pearce, 1976; Peterson and Randall, 1984; and Sinden and Worrell, 1979 for details). This multiplicity of approaches is to be expected, because the benefits derived from a biological resource may be measured for one purpose by methods that may not be appropriate for other objectives, and the ways to measure one resource may not be the same for others. The value of a forest in terms of logs, for example, would be measured in quite a different way from the value of the forest for recreation or for watershed protection.

Three main approaches have been used for determining the value of biological resources:

- assessing the value of nature's products — such as firewood, fodder, and game meat — that are consumed directly, without passing through a market ("consumptive use value");
- assessing the value of products that are commercially harvested, such as game meat sold in a market, timber, fish, ivory, and medicinal plants ("productive use value"); and
- assessing indirect values of ecosystem functions, such as watershed protection, photosynthesis, regulation of climate, and production of soil ("non-consumptive use value"), along with the intangible values of keeping options open for the future and simply knowing that certain species exist ("option value" and "existence value," respectively).

Some biological resources can be easily transformed into revenue through harvesting, while others provide flows of services that do not carry an obvious price tag. Therefore, in order for governments to base decisions on allocating scarce resources on the best available information, a number of different methods are required to quantify the magnitude and value of the positive and negative impacts. Governments should be seeking means of determining total valuation, which requires a wide range of assessment methods. The major approaches are summarized in Box 5, and discussed below (drawn from McNeely, 1988).

Assessing benefits and costs of protecting biological resources provides a basis for determining the total value of any protected area or other system of biological resources. Since the value of conserving biological resources can be considerable, conservation should be seen as a form of economic development. And since biological resources have economic values, investments in conservation should be judged in economic terms, requiring reliable and credible means of measuring the benefits of conservation.

Box 5: Classification of Values of Biological Resources.

Direct Values
- Consumptive Use Value (non-market value of firewood, game, etc.)
- Productive Use Value (commercial value of timber, fish, etc.

Indirect Values
- Non-consumptive Use Value (scientific research, bird-watching, etc.)
- Option Value (value of maintaining options available for the future)
- Existence Value (value of ethical feelings of existence of wildlife)

Direct Values of Biological Resources

Direct values are concerned with the enjoyment or satisfaction received directly by consumers of biological resources. They can be relatively easily observed and measured, often by assigning prices to them.

Consumptive Use Value

This is the value placed on nature's products that are consumed directly, without passing through a market. These values can be considerable; for example, some 84 percent of the Canadian population participates in wildlife-related recreational activities in a given year, providing Canadians with benefits that they declare to be worth $800 million annually (Fillon *et al.*, 1985).

While relatively few detailed studies have been carried out on the consumptive use value of species in developing countries, the available information has been well summarized by Myers (1983b), Oldfield (1984), Krutilla and Fisher (1975), and Fitter (1986). Of particular interest is the study by Prance *et al.* (1987), which presented quantitative data on the use of trees by four indigenous Amazonian Indian groups. ''Use'' was defined rather narrowly, including uses as food, construction material, raw material for other technology, medicinals, and trade goods; uses as firewood or as food for harvested animals were not included. The percentage of tree species used by the four groups varied from 48.6 to 78.7, indicating that the rain forests of Amazonia contain an exceptionally large number of species that are useful to local people.

Consumptive use values seldom appear in national income accounts, but no serious obstacles appear to prevent the inclusion of at least some consumptive use values in such measures as Gross Domestic Product (GDP) (Repetto *et al.*, 1989). For example, firewood and dung provide over 90 percent of the total primary energy needs in Nepal, Tanzania, and Malawi and more than 80 percent in many other countries (Pearce, 1987a); this contribution to the economy could be assigned a financial value.

In Africa, harvested species make a considerable contribution to human welfare in the form of food for rural people, and especially to the poorest villagers living in the most remote areas. Much of this is consumed directly rather than being sold in the marketplace, but the value is nonetheless significant and economic values can be assigned. In Botswana, over 50 species of wild animals provide animal protein exceeding 90 kg per person per annum in some areas (some 40 percent of their diet); over 3 million kg of meat is obtained yearly from springhare alone. In Ghana, about 75 percent of the population depends largely on traditional sources of protein, mainly wildlife, including fish, insects, caterpillars, maggots, and snails. In Nigeria, game constitutes about 20 percent of the mean annual consumption of animal protein by people in rural areas (including 100,000 tons of the giant rats known as ''grasscutters,'' per Myers, 1988b), while 75 percent of the animal protein consumed in Zaire comes from wild sources. Senegal's population of 5 million consumes at least 373,631 metric tons of wild mammals and birds per year (Sale, 1981).

Indigenous use of forest resources in Siberut, Indonesia, where monkeys are hunted for food (photo by R. Tenaza).

Consumptive use value can be assigned a price through such mechanisms as estimating market value if the product were sold on the market instead of being consumed. In Sarawak, Malaysia, for example, a detailed field study found that wild pigs harvested by hunters had a market value of some $100 million per year (Caldecott, 1988).

A cautionary note: In many tropical countries, the consumptive use value of wildlife has stimulated over-exploitation (see, for example, Davies, 1987, for details on Sierra Leone). While wildlife has been consumed by humans for hundreds of thousands of years, today's increasing populations and changing social and political structures have removed most traditional controls on hunting. If wildlife is to continue to make its important contribution to economies, more effective controls are often required to ensure that wildlife populations are maintained at productive levels. The means of doing this will vary from place to place, but the foundations must be based on sound economic and ecological principles.

In terms of economic development, perhaps the most important point of consumptive use is that some rural communities closest to the forests or other natural areas can prosper through the sustainable harvesting of wild species, and almost all rural communities can gain at least some development benefit through proper management of biological resources that are consumed directly. Relatively small investments aimed at ensuring that such management systems continue to prosper can help avoid the much larger investments often required when biological resources have been so severely degraded that expensive environmental rehabilitation projects are needed.

Economically efficient and productive systems exist outside the market economy, drawing on biological resources to support basic human needs. These systems should be well understood before attempts are made to replace them by modern approaches that may not be as sustainable or productive in the long run.

Wood-gathering in Burkina Faso (World Bank photo by Y. Hadar).

Productive Use Value

This value is assigned to products that are commercially harvested for exchange in formal markets, and is therefore often the only value of biological resources that is reflected in national income accounts. Productive use of such biological resource products as fuelwood, timber, fish, animal skins, musk, ivory, medicinal plants, honey, beeswax, fibers, gums, resins, rattans, construction materials, ornamentals, animals harvested for game meat, fodder, mushrooms, fruits, dyes, and so forth can have a major impact on national economies. Estimates of such values are usually made at the production end (landed value, harvest value, farmgate value, etc.) rather than at the retail end, where values are much higher because of the costs and value added through transport, processing, and packaging; for example, the estimated production value of cascara (a laxative derived from tree bark) in the United States is $1 million per year, but the retail value is $75 million per year (Prescott-Allen and Prescott-Allen, 1986).

Wild biological resources also contribute to the production of domesticated resources in several ways:

- wild genetic resources are used to improve established domesticates (a contribution valued as billions of dollars per year);
- rangeland and wild forage species contribute to livestock production;
- wild species — especially of plants — serve as sources of new domesticates (Plotkin, 1988);
- wild pollinators are essential to many crops, and wild enemies of pests help control their depredations on crops.

In this regard, a clear distinction needs to be drawn between products that are continuously taken from nature, such as ivory or medicinal plants, and ones that are harvested once or infrequently to provide a small "founder stock" that is then propagated or used as a genetic blueprint. With plants, the latter is often much the most important, as virtually any plant can be propagated and cultivated. The value of these plants as genetic resources may be compared to an intellectual property right (Williams, 1984; de Klemm, 1985).

Prescott-Allen (1986) concluded that the productive use value of wild genetic resources demonstrates that genetic resources are indispensable to modern agriculture, that most of them come from a country other than where they are utilized, that the turnover of domestic genetic resources is rapid, and that use of new genetic resources is increasing (therefore requiring the lines of supply from other countries to be kept open and a great diversity of genetic resources to be maintained). The wild relatives of domestic plants will be an essential component of ensuring food security for the next century (Hoyt, 1988). IUCN, WWF, and IBPGR have called for a "search and rescue operation" to locate and conserve wild crop relatives, both *in situ* and *ex situ,* to complement the existing and equally vital work on conserving the cultivated diversity, mainly the land-races, of the major agricultural crops.

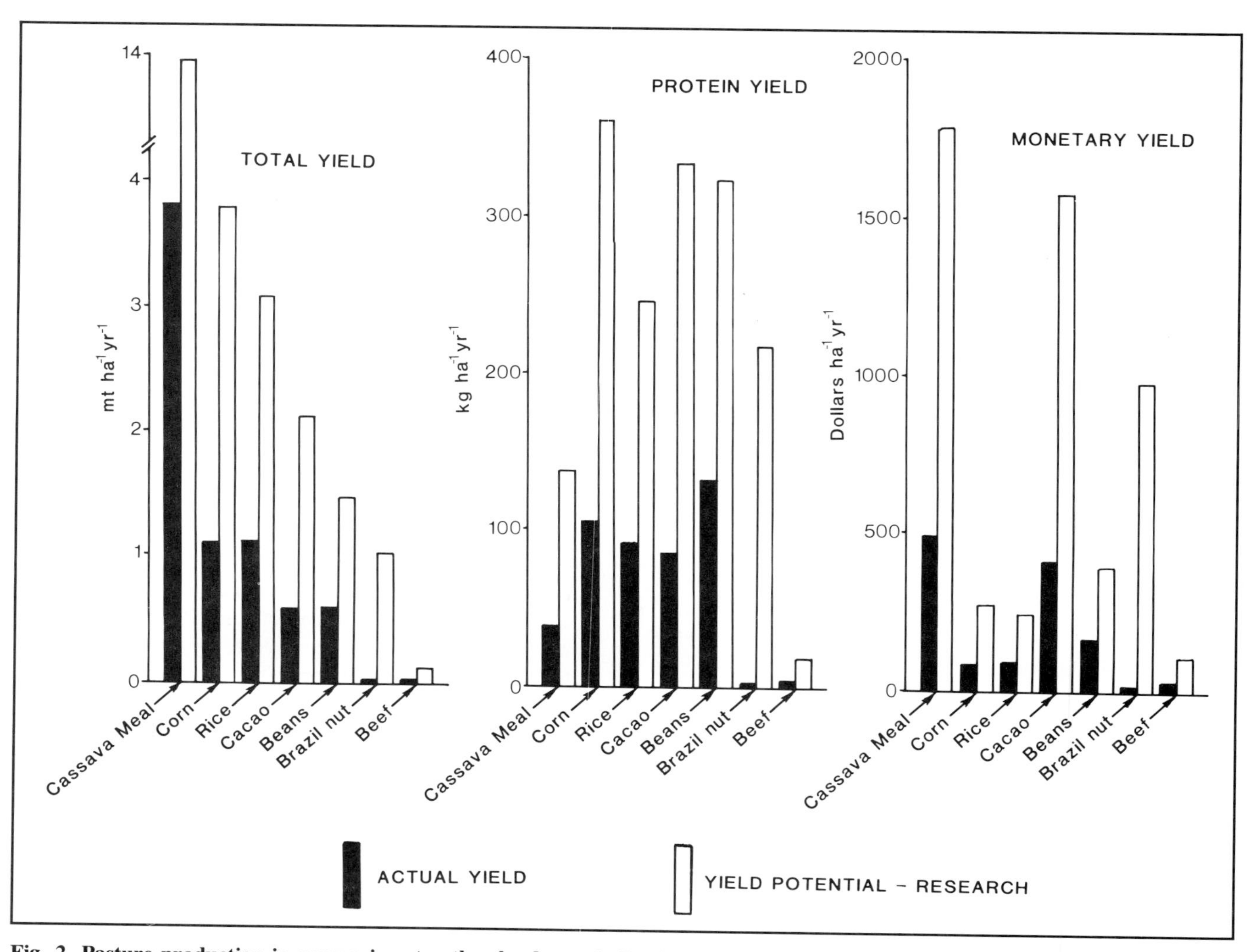

Fig. 2. Pasture production in comparison to other land uses in Pará State, Brazil (Source: Chris Uhl, unpublished data, 1989).

Exchanges of genetic material among developing countries are particularly important in view of the economic importance of such perennial crops as rubber, oil palm, and cocoa, and annuals such as cassava and sugarcane. Such crops are often far more productive outside their native habitats, but they are also subject to attacks from various pests and diseases that can be counterattacked by fresh infusions of genetic material (Frankel and Bennett, 1970; Frankel and Hawkes, 1974; Plucknett *et al.*, 1987).

Productive use value can be derived directly from the market demand curve for the resources consumed, which measures consumers' willingness to pay for various quantities of the resource. Where close substitutes are available, the demand curve will be fairly flat and the productive use value can be approximated by market price. Where close substitutes are not available, a "consumers' surplus" exists over and above the market price. In this case, use of price data may severely underestimate productive use value.

Prescott-Allen and Prescott-Allen (1986), in a groundbreaking study that demonstrated how the dollar value of biological resources can be estimated, carried out a detailed analysis of the contribution wild species of plants and animals made to the American economy, concluding that some 4.5 percent of GDP is attributable to wild species. The combined contribution to GDP of wild harvested resources averaged some $87 billion per year over the period 1976 to 1980.

The percentage contribution of wild species and ecosystems to the economies of developing countries is usually far greater than it is for an industrialized country like the USA, especially if consumptive use value is included. Timber from wild forests, for example, is the second leading foreign exchange earner for Indonesia (after petroleum), and throughout the humid tropics governments have based their economies on the harvest of wild trees; total exports of wood products from Asia, Africa, and South America averaged $8.1 billion per year between 1981 and 1983 (WRI/IIED, 1986).

Non-wood forest products can also be of considerable value. Indonesia, for example, earned some $200 million in foreign exchange from non-wood forest products in 1982 (Gillis, 1986), while non-wood forest products in a recent year provided 40 percent of the total net revenues accruing to the Indian government from the forestry sector, and 63 percent of the forestry exports (Gupta and Guleria, 1982). In comparing wood and non-wood forest resources, Myers (1988b) concludes that a tropical forest tract of 50,000 hectares could, with effective management, "produce a self-renewing crop of wildlife with a potential value of at least $10 million per year, or slightly more than $200 per hectare. These revenues contrast with the return from commercial logging in the area of only a little over $150 per hectare. Moveover, with present timber-harvesting practices, commercial logging tends to be an ecologically disruptive procedure, whereas wildlife harvesting can leave forest ecosystems virtually undisturbed."

Similar results came from a study that attempted to assess market values from all economic trees in a one-hectare tract of species-rich forest in Peruvian Amazonia (Peters *et al.*, 1989). The hectare of forest, located on the Rio Nanay about 30 km from the city of Iquitos, contained 275 species of trees, with a total count of 842 trees greater than 10 cm in diameter. Over 41 percent of the trees yielded fruit, timber, or latex with a local market value. Fruit and latex yield about $700 per year ($422 net of labor and transport), but given that both these resources are renewable and can be harvested annually, the present net value of fruit and latex is estimated at $8,400. The merchantable timber in the tract amounts to about 94 cubic meters, but the maximum sustainable harvest would amount to about 30 cubic meters every 20 years, yielding a present net value of $490 for timber. Fruit and latex therefore represent over 90 percent of the total market value of the forest; sustainable yields of wildlife and medicinal plants would add considerably to the non-timber yield of the forest.

In comparison to other uses of Amazonian forest, a one-hectare tract in Brazil harvested for pulpwood was similarly valued at $3,184, and valued even less as a cattle pasture at $2,960. These values are considerably lower than the one calculated for the sustainable harvest of fruit and latex, and also assume — contrary to the knowledge of most tropical systems — that such uses are sustainable.

The returns from wildlife usually will be far less in drier habitats, though often exceeding alternative uses. In Zimbabwe's Zambezi Valley, for example, Cumming (1985) estimates that potential gross returns from wildlife utilization amount to $12 per hectare. "These returns," he states, "are as good if not better than returns from the best-run commercial beef ranches in the country and the profit margins are probably higher."

While sport hunting by foreigners certainly has some problems of image, tending to appear imperialistic to some local people who are themselves prohibited from hunting, these problems can be overcome when local people are also able to benefit through consumptive uses of "surplus" meat that can be harvested on a sustainable-yield basis.

In conclusion, market prices represented by productive use value can be an important indicator of value. However, as will be demonstrated by the discussion of indirect values below, the market price is not always an accurate representation of the true economic value of the resource, and does not deal effectively with questions of distribution and equity. It is also apparent that consumers may value resources in ways different from producers; tropical forests are valued by consumers of scenic beauty differently than by consumers of lumber products, but no market is available to mediate these claims.

Indirect Values of Biological Resources

Indirect values, which deal primarily with the functions of ecosystems ("environmental services"), do not normally appear in national accounting systems, but they may far outweigh direct values when they are computed. They tend to reflect the value of biological diversity to society locally or at large rather than to individuals or corporate entities.

Direct values often derive from indirect values because harvested species of plants and animals are supported by the goods and services provided by their environments. Species without consumptive or productive use values may play important roles in the ecosystem, supporting species that are valued for their productive or consumptive use. In Sabah, for example, recent studies suggest that high densities of wild birds in commercial *Albizia* (silk tree) plantations limit the abundance of caterpillars that would otherwise defoliate the trees; the birds require natural forest for nesting (Fitter, 1986).

As another example, the U.S. National Marine Fisheries Service estimates that the destruction of U.S. coastal estuaries between 1954 and 1978 cost the nation over $200 million annually in revenues lost from commercial and sport fisheries. The commercial fisheries provide productive use value and the sport fisheries provide consumptive use value, to which the estuaries contribute without being consumed.

Non-consumptive Use Value

Environmental resources — generally speaking, nature's services rather than her goods — often provide value without being consumed, traded in the marketplace, or reflected in national income accounts. Still, efforts are being developed to evaluate economically the benefits provided by these resources (Oldfield, 1984, Peterson and Randall, 1984; Sinden and Worrell, 1979; de Groot, 1986). It is apparent that the benefits of environmental services are much easier to measure at the local level than at the global level. Quantifying the hydrological benefits of a watershed, for example, is relatively straightforward, while measuring the value of the global carbon cycle would be a daunting exercise and would in any case be of little practical value. Box 6 and the following paragraphs summarize some of the indirect non-consumptive values of biological resources, particularly of ecological services provided by nature.

Box 6: Non-Consumptive Benefits of Conserving Environmental Resources.

The benefits accruing to society in return for investments in conserving environmental resources will vary considerably from area to area and from resource to resource. Most such benefits will fall into one or another of the following categories:

- Photosynthetic fixation of solar energy, transferring this energy through green plants into natural food chains, and thereby providing the support system for species that are harvested;
- Ecosystem functions involving reproduction, including pollination, gene flow, cross-fertilization; maintenance of environmental forces and species that influence the acquisition of useful genetic traits in economic species; and maintenance of evolutionary processes, leading to constant dynamic tension among competitors in ecosystems;
- Maintaining water cycles, including recharging groundwater, protecting watersheds, and buffering extreme water conditions (such as flood and drought);
- Regulation of climate, at both macro- and microclimatic levels (including influences on temperature, precipitation, and air turbulence);
- Soil production and protection of soil from erosion, including protecting coastlines from erosion by the sea;
- Storage and cycling of essential nutrients, e.g., carbon, nitrogen, and oxygen, and maintenance of the oxygen-carbon dioxide balance;
- Absorption and breakdown of pollutants, including the decomposition of organic wastes, pesticides, and air and water pollutants; and
- Provision of recreational-aesthetic, sociocultural, scientific, educational, spiritual, and historical values of natural environments.

Stabilizing hydrological functions. Natural vegetation cover on water catchments regulates and stabilizes water runoff. Deep penetration by tree roots or other vegetation makes the soil more permeable to rainwater so that runoff is slower and more uniform than on cleared land. As a consequence, streams in forested regions continue to flow in dry weather and floods are minimized in rainy weather. Daniel and Kulasingham (1974) showed that the peak runoff per unit area of forested catchments in Malaysia is about half that of rubber and oil palm plantations, while the low flows are roughly double. Watershed protection has helped justify many valuable reserves that otherwise might not have been established, so irrigation and energy agencies can make powerful potential allies for protected areas that safeguard watersheds (McNeely, 1987).

In Honduras, for example, La Tigra National Park, a 7,500-ha area consisting mainly of cloud forest, produces a high quality, well-regulated water flow throughout the year, producing over 40 percent of the water supply to Tegucigalpa (the capital city). Some 25 small collection facilities scattered throughout the park require only limited maintenance because the water is so pure and free of sediments. Because of its value for watershed protection, La Tigra is the focus of a major investment program involving a series of economic incentives for villagers living in the surrounding regions.

Another estimate placed the economic value of a hectare of Atlantic *Spartina* marsh at over $72,000 a year. According to the U.S. Army Corps of Engineers, retaining a wetlands complex outside of Boston, Massachusetts realized an annual cost savings of $17 million in flood protection alone (a figure that did not include the many other benefits — such as sediment reduction, fish and wildlife production, and aesthetic values — that the wetlands afforded area residents) (Hair, 1988).

In many cases, the total costs of establishing and managing reserves that protect catchment areas can be met and justified as part of the hydrological investment. In Thailand, Hufschmidt and Srivardhana (1986) have shown that an annual expenditure for watershed protection related to the Nam Pong Reservoir of about $1.5 million per year would be justified in terms of benefits to the reservoir. And in Indonesia, the Dumoga-Bone National Park was established by a loan of $1.2 million from the World Bank, justified on the basis of the protection the park provided to a major irrigation project in the lowlands below.

Protecting soils. Good soil protection by natural vegetation cover and litter can preserve the productive capacity of land, prevent dangerous landslides, safeguard coastlines and riverbanks, and prevent the destruction of coral reefs and freshwater and coastal fisheries by siltation. In Malaysia, the suspended sediment load following logging increased 70 to 97 percent in comparison with a non-logged area (Kasran, 1988). Thus, management of watershed as a protected area can greatly reduce sediment loads (and can therefore contribute significantly to the longevity of reservoirs and irrigation systems downstream). A startling example of soil conservation is provided by Nepal's Royal Chitwan National Park, where villagers have cleared and grazed the north bank of the Rapti River (which forms the park boundary) so intensively that it has been subject to rapid erosion. On the south bank, within the park, the protected vegetation binds the soil so that when monsoon rains swell the Rapti it is the north bank that is washed away. As a result, the course of the river has shifted and in less than a decade roughly one hundred hectares has been taken from villagers and added to the park by natural forces (Roberts and Johnson, 1985).

Maintaining the natural balance of the environment. The existence of a protected area may help maintain a more natural balance of the ecosystem over a much wider area. Natural habitats afford sanctuary to breeding populations of birds that control insect and mammal pests in agricultural areas. Bats, birds, and bees that nest, roost, and breed in reserves may range far outside their boundaries and pollinate fruit trees in the surrounding areas. Ledec and Goodland (1986) have shown how the production of Brazil nuts depends on a variety of poorly known forest plants and animals. Male euglossine bees that pollinate the flowers of the Brazil nut tree gather certain organic compounds from epiphytic orchids to attract females for mating. The hard shell covering the nut is opened naturally only by the forest-dwelling agouti (a large rodent), thereby enabling the tree to disperse seeds. Thus maintaining Brazil nut production appears to require conserving enough natural forest to protect bee nesting habitat, other bee food plants, certain orchids and the trees upon which they grow, the insects or hummingbirds that pollinate the orchids (and all their necessities in turn), and agoutis.

Another good example comes from Tanzania, where the poaching and uncontrolled hunting of elephants and rhinos to the south-east of Tarangire National Park led to bush encroachment because the main browsers no longer had a major influence on the vegetation. This in turn caused an increase in tsetse flies, which reduced the population of domestic livestock in the area. Far from being just an ethical action, the conservation of elephants and rhinos would have enhanced the productivity of the livestock industry (Peterson, 1976).

Species can also have non-consumptive use value, as in bird watching and some scientific research (especially ecological field studies). And people derive indirect non-consumptive use value from species through media such as film, video, and literature.

Option Value

The future is uncertain, and extinction is forever. Prescott-Allen and Prescott-Allen (1986) suggest that society "should prepare for unpredictable events, both biological and socio-economic. The best preparation in the context of wildlife use is to have a safety net of diversity — maintaining as many gene pools as possible, particularly within those wild species that are economically significant or are likely to be." Option value is a means of assigning a value to risk aversion in the face of uncertainty.

Natural habitats preserve a reservoir of continually evolving genetic material — irrespective of whether the values of that material have yet been recognized — that enables the various species to adapt to changing conditions. The plants and animals conserved may spread into surrounding areas where they may be able to be cropped at some future date, or may eventually contribute genetic material to domestic crops or livestock. Protecting natural habitats can therefore

Protected areas are not only havens for wild species, but also maintain ecological balances. The resulting change in water flow from a hillside that has been deforested could ruin a lowland irrigation system (World Bank photo by E.G. Hoffman).

The African elephant acts as a keystone species within its habitat. Its browsing on vegetation modifies the habitat and creates conditions favorable for other species (photo by N. Myers).

be seen as a means for nations, especially those in the species-rich tropics, to keep at least part of their biological resources intact for the future benefit of their populace.

As a result, society as a whole may be willing to pay to retain the option of having future access to a given species or level of diversity. As the demand for biological resources grows while the supply continues to dwindle (if current trends continue), their value is likely to increase. Therefore, some economists suggest that conventional cost-benefit relationships need to incorporate mechanisms to deal with the probability of higher future values and the irretrievability of lost opportunities to preserve natural environments and genetic material.

Existence Value

Many people, especially in the industrial nations, also attach value to the existence of a species or habitat that they have no intention of ever visiting or using; they might hope that their descendants (or future generations in general) may derive some benefit from the existence of these species, or may just find satisfaction in knowing that the oceans hold whales, the Himalayas have snow leopards, and the Serengeti has antelope. The ethical dimension is therefore important in determining "existence value," which reflects the sympathy, responsibility, and concern that some people may feel toward species and ecosystems. An accurate cost-benefit analysis of such values is clearly impossible, but the magnitude of these values is suggested by the sizeable voluntary contributions to private conservation agencies in the industrial world by people who do not expect to visit or use the resource they are helping to conserve. (WWF alone receives nearly $100 million per year in such donations on a worldwide basis.)

Conclusions

Wild species and the genetic variation within them make contributions to agriculture, medicine, and industry worth many billions of dollars per year. Perhaps even more important are the essential life processes that are carried out by nature, including stabilization of climate, protection of watersheds, protection of soil, protection of nurseries and breeding grounds, and so on. Conserving these processes cannot be divorced from conserving the individual species that constitute natural ecosystems.

The developing countries are particularly vulnerable to abuses of biological resources because they tend to be agrarian societies with the bulk of their populations living on the land rather than in cities. Biological resources make a far greater contribution to these local economies (at least in percentage terms) than they do to the national and international industrial economies. Species that are important to human welfare in both industrial and developing countries are not limited to wild plants that are relatives of agricultural crops, or to animals or plants that are harvested for food, fuel, or medicine. They also include species such as earthworms, bees, and termites that may make even more important contributions to society in terms of the role they play in maintaining healthy and productive ecosystems.

Biological resources have multiple values in all societies, but different approaches to valuation are relevant at different levels. At the local level, consumptive use value is often the most relevant, while national governments tend to be most interested in productive use value, often in terms of the foreign exchange earned. Although many products from biological resources are traded internationally, the world community is also likely to be interested in existence value and non-consumptive use value (ways of converting this interest into financial support are presented in Chapter VIII). Wealthy individuals or nations may be more concerned about option value than nations that are carrying a heavy debt burden and that may be forced into unsustainable productive uses.

But whatever methodology is used, valuation is only a fundamental first step. It informs planners, resource managers, and local people about how important biological diversity may be to national development objectives, it demonstrates how important an area is for the biological resources it contains, it reveals common interests in conservation among various sectors, and it facilitates comparison of costs and benefits of different development proposals.

The second step is to determine how these species and areas can be conserved. It is here that economic incentives and disincentives can play their important role in ensuring that the benefits suggested above are in fact delivered to the community, and that the community in turn is enabled to protect the resources upon which its continued prosperity depends (McNeely, 1988).

Following page: Forest destruction in Brazilian Amazonia (photo by R.O. Bierregaard).

Erosion in Ethiopia (photo by IIED).

CHAPTER III

HOW AND WHY BIOLOGICAL RESOURCES ARE THREATENED

Today's threats to species and ecosystems are the greatest in recorded history. Virtually all of them are caused by human mismanagement of biological resources, often stimulated by misguided economic policies and faulty institutions that enable the exploiters to avoid paying the full costs of their exploitation.

In seeking ways to conserve biological resources, it is necessary to have a clear understanding of the major threats to biological resources on the ground and in the water. Solutions depend above all on how the problem is defined, and it appears that the *problems facing the conservation of biological diversity have tended to be defined in ways that do not lead to acceptable solutions.*

When the problems are defined in terms of insufficient protected areas, excess poaching, poor law enforcement, land encroachment, and illegal trade, possible responses include establishing more protected areas, improving standards of managing species and protected areas, and enacting international legislation controlling trade in endangered species. All of these measures are necessary. But they respond to only part of the problem. Biological diversity will be conserved only partially by protected areas, wildlife management, and international conservation legislation. Fundamental problems lie beyond protected areas in sectors such as agriculture, mining, pollution, settlement patterns, capital flows, and other factors relating to the larger international economy.

This chapter attempts to define the problems of conserving biological diversity in a more comprehensive way that will lead to more effective solutions being developed.

Two scenes of the advance of human settlement on wild lands: The Pantanal region of Brazil, where a dike project to produce year round agricultural land has created significant changes in the seasonally flooded ecosystem, and the Impenetrable Forest in Uganda, where agricultural activity extends right up to the park boundary (photos by R.A. Mittermeier).

Dead trees on Mt. Mitchell, North Carolina, 1988. Acid rain is believed to have contributed to the decline of Appalachian forests in the United States (photo by Jim MacKenzie, WRI).

Both problems and solutions are built on economic foundations. Major threats to biodiversity include:

- Habitat alteration, usually from highly diverse natural ecosystems to far less diverse (often monoculture) agroecosystems. This is clearly the most important threat, often related to land-use changes on a regional scale that involve great reduction in the area of natural vegetation. Such reductions in area — often involving fragmentation of species habitats — inevitably mean reductions in populations of species, with a resulting loss in genetic diversity and an increase in vulnerability of species and populations to disease, hunting, and random population changes (Soulé and Wilcox, 1980).
- Over-harvesting, the taking of individuals at a higher rate than can be sustained by the natural reproductive capacity of the population being harvested. When species are protected by law, harvesting is called "poaching."
- Chemical pollution, which has been implicated in the dying forests of Europe, deformities in birds (Anderson, 1987), and premature births in seals (DeLong *et al.*, 1973), has become a major threat in virtually all parts of the world. Chemical pollution is complex and all-pervasive. It is expressed in such different forms as: atmospheric pollution with sulphur and nitrogen oxides and with oxidants, directly damaging vegetation and harming fresh waters through the deposition of "acid rain"; excessive use of agricultural chemicals, contaminating watercourses and causing ecological imbalance in wetlands and shallow seas through the runoff of nitrate and phosphate and harming wildlife through the accumulation of persistent pesticides; and the release of many compounds of heavy metals and other toxic substances from industrial sources, with an impact on the life of land, fresh waters, and inshore seas.
- Climatic change, often related to changing regional vegetation patterns; this problem involves such factors as global carbon dioxide build-up, regional effects such as El Niño (Graham and White, 1988) and monsoon systems, and local effects, often involving fire management. Climate change, which appears to be taking place at the fastest rate in history, could have drastic effects on boreal forests, coral reefs, mangroves, and wetlands, as well as change the boundaries of the world's biomes.
- Introduced species, which on many oceanic islands have virtually replaced the native species of plants (Fosberg, 1988). Even reasonably well protected islands such as the Galapagos have as many introduced species of plants as native ones (Adsersen, 1989). Continental areas are also affected, and the problem of introduced species of plants has been identified as the most serious threat facing the U.S. national park system. Animals are not immune to such threats; for example, in some of the African Rift Valley lakes, which have remarkably high levels of endemism, introduced species of fish have threatened most native species with extinction (Miller, 1989). Mongooses, snakes, and other introduced animals can rather quickly lead to the extinction of the native fauna, while introduced herbivores such as goats and even reindeer can extinguish the native flora (Savidge, 1987; Pimm, 1987; Mooney, 1985).

A herd of goats introduced on Santa Catalina Island, California, roam the denuded landscape they helped create (photo by B. Coblentz).

- Increase in population, accompanying the industrial revolution, global trade, harnessing of fossil fuels, and more effective public health measures. Our species reached a population of 1 billion at the beginning of the 19th century, reached 2 billion in the 1920s, and totals over 5 billion today. Optimists predict that a combination of development, education, the provision of reproductive health services, and intelligent self-control will cause the population to level off at around 8 to 10 billion in the latter part of the next century. A dispassionate external observer must question whether such a population is sustainable, given the degradation in the resource base that has accompanied the recent increase. The danger that the raw forces of nature — drought, flood, famine, strife, and disease — will dominate in at least some regions will certainly continue to place very heavy demands on biological diversity. It is apparent that the longer it takes people to limit their fertility, the more certain it is that misery will prevail (Holdgate, 1989).

The above list of major threats is primarily a list of the symptoms rather than a description of the fundamental problems that lead to these threats. While the specifics of the problems will vary from place to place, the main source of all these symptoms can be found in the distribution of costs and benefits of both exploitation and conservation. Those who have reaped the benefits from exploitation have not paid the full costs, and those who have paid most of the costs of conservation (especially opportunity costs) have gained few of the benefits.

Ultimately, the solution is to redress this imbalance through *ensuring that exploiters pay the full costs of their exploitation, and that conservers earn more of the benefits of their actions.* This requires a more comprehensive perspective on conservation and development, and a more integrated approach to decision-making.

The Dimensions of the Problem

While the various threats to biodiversity tend to be cumulative in their effects, it is informative to look more closely at the manifestations of these threats on species and habitats (realizing how closely intertwined species are with their habitats). It is important to bear in mind that from tropical habitats — the most species-rich formations on earth — only 10 percent of the total number of species has even been described; without understanding the parts of the system, it is difficult to understand the systems themselves. Our ignorance of tropical organisms and ecosystems is vast.

Species

Extinction has been a fact of life since life first emerged from the primordial ooze (Figure 3). The present few million

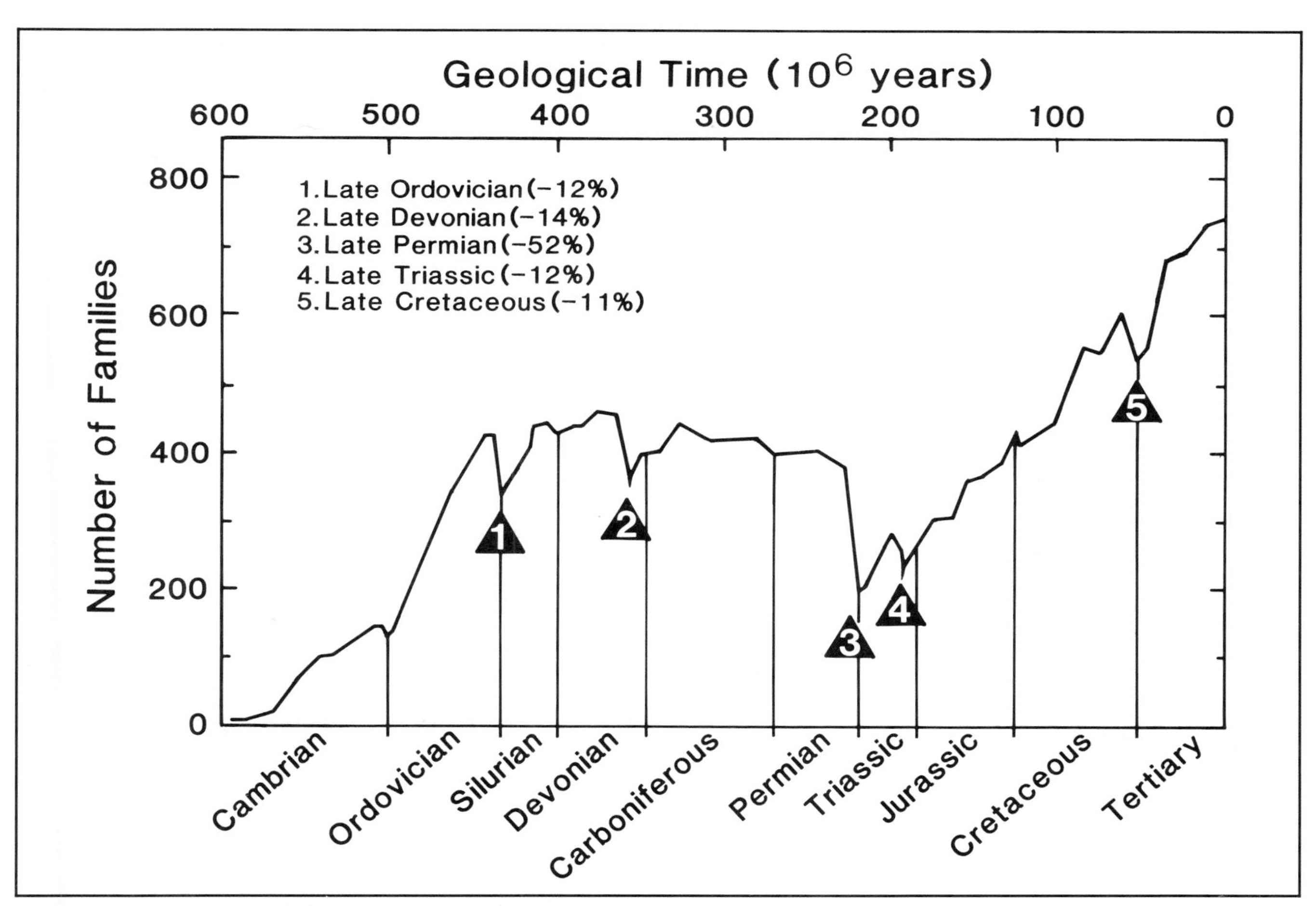

Fig. 3. The five major extinction episodes of life on earth as exemplified by the changes through geological time in family diversity of marine vertebrates and invertebrates (Source: E. Wilson, 1988b).

species are the modern-day survivors of the estimated several billion species that have ever existed. All past extinctions have occurred by natural processes, but today humans are overwhelmingly the main cause of extinctions.

The average duration of a vertebrate species is some 5 million years. The best current estimates are that on average 900,000 vertebrate species have become extinct every 1 million years during the last 200 million years, so the average "background rate" of extinction has been very roughly 90 species of vertebrates each century (Raup, 1986). Myers (1988c) quotes a crude estimate for higher plants of about 1 species becoming extinct every 27 years over the past 400 million years, with the rate increasing in more recent times as the number of species of higher plants has increased. Within the last few hundred years, major waves of human-caused extinctions have washed over oceanic islands, in large part due to the destruction of lowland forests and to the introduction of predators, mammalian herbivores, diseases, and aggressive, weedy plants. About 75 percent of the mammals and birds that have become extinct in recent history were island-dwelling species (Frankel and Soulé, 1981), and even more island extinctions are likely. Over 10 percent of the

High rises and *favelas* (shanty towns) encroach on remaining habitat in Rio de Janeiro, Brazil (photo by P. Almasy).

world's species of birds are confined to individual islands. Similarly, the island floras tend to be far more endangered than the continental ones, and on several islands (Ascension, Lord Howe, Norfolk, Rodrigues, and St. Helena) more than 90 percent of the endemic vascular plant species are rare, threatened, or extinct (Table 1).

Table 1: Status of Endemic Vascular Plant Taxa on Selected Oceanic Islands.

Island	Total	Not Threatened	Insufficiently Known	Rare, Threatened or Extinct
Ascension Island	11	0	1	10 (91%)
Azores	56	14	10	32 (57%)
Canary Islands	612	169	36	407 (67%)
Galapagos	222	89	3	130 (59%)
Juan Fernandez	119	6	17	95 (81%)
Lord Howe Island	78	2	1	75 (96%)
Madeira	129	23	19	87 (67%)
Mauritius	280	31	18	194 (69%)
Norfolk Island	48	1	2	45 (94%)
Rodrigues	55	3	2	50 (91%)
St Helena	49	0	2	47 (96%)
Seychelles*	90	0	1	72 (81%)
Socotra	215	81	2	132 (61%)

* Refers to granitic islands only.

Source: Davis *et al.*, 1986.

The rapid destruction of the world's most diverse ecosystems, especially in the tropics, has led most experts to conclude that perhaps a quarter of the earth's total biological diversity is at serious risk of extinction during the next 20-30 years (Raven, 1988). By many indications, the world is already experiencing extinction rates of greater scale and impact than at any previous time in the earth's history (Wilson, 1988a). More species than ever before are threatened with extinction, with thousands — mostly insects — disappearing each year, many before they are even described. A recent comprehensive review of the world's avifauna concluded that of the globe's 9,000 birds, over 1,000 (11 percent) were at some risk of extinction, up from just 290 bird species threatened in 1978 (an increase at least partially due to more complete information becoming available in the past several years)(Collar and Andrew, 1988).

The World Conservation Monitoring Centre (WCMC) is the major repository of data on threatened species (see Chapter VI). Using the Red Data Book categories established by IUCN (Box 7), it has recorded the degree of threat to some 60,000 plants and 2,000 animals. These categories have received some criticism (Fitter and Fitter, 1987), particularly on the grounds that they can only be used where full data are available on the decline of a species, and on the threats to its survival throughout its entire range.

Box 7: IUCN Categories of Threat.

Extinct (Ex): Species not definitely located in the wild during the past 50 years (criterion as used by CITES).
Endangered (E): Taxa (species and sub-species) in danger of extinction and whose survival is unlikely if the causal factors continue operating. Included are taxa whose numbers have been reduced to a critical level or whose habitats have been so drastically reduced that they are deemed to be in immediate danger of extinction. Also included are taxa that are possibly already extinct but have definitely been seen in the wild in the past 50 years.
Vulnerable (V): Taxa believed likely to move into the "Endangered" category in the near future if the causal factors continue operating. Included are taxa of which most or all the populations are decreasing because of over-exploitation, extensive destruction of habitat, or other environmental disturbance; taxa with populations that have been seriously depleted and whose ultimate security has not yet been assured; and taxa with populations that are still abundant but are under threat from severe adverse factors throughout their range.
Rare (R): Taxa with small world populations that are not at present "Endangered" or "Vulnerable" but are at risk. N.B. in practice, "Endangered" and "Vulnerable" categories may include, temporarily, taxa whose populations are beginning to recover as a result of remedial action, but whose recovery is insufficient to justify their transfer to another category. These taxa are usually localized within restricted geographical areas or habitats or are thinly scattered over a more extensive range.
Indeterminate (I): Taxa known to be "Endangered," "Vulnerable," or "Rare" but where there is not enough information to say which of the three categories is appropriate.

Such knowledge is available for relatively few taxa. Whitten *et al.* (1987), for example, discovered in the course of their work in compiling information on the natural history of Sulawesi (Indonesia) that the Caerulean paradise-flycatcher (*Eutrichomyias rowleyi*) had not been seen in several decades, nor were recent records to be found for many of the endemic species of the fish family Adrianichtyidae; at least seven other species of endemic Sulawesi birds had apparently not been observed in over a decade, but had not found their way into Red Data Books. Further, the Red Data Books cannot be expected to deal with tropical forest invertebrates, of which millions of species are undescribed but are certainly under threat as their habitats are cleared out from under them. Diamond (1987) has pointed out that even the lists that do exist include primarily species known to be threatened and suggests instead that species must be presumed extinct or endangered unless shown to be extant and secure. Such a

"Green List" might be much shorter than Red Data Book lists.

To cope with tropical plants, the IUCN Plant Information Plan (TPU, 1988) proposed the listing of extinction-prone species, defined as species confined to endangered vegetation types, as well as of threatened species falling into Red Data Book categories. It also proposed the identification of plant-rich sites for conservation, as a further way of identifying threatened plant diversity (see Chapter VI). The information in Red Data Books and threatened species lists should, therefore, be taken as only indicating part of the problem. The full picture is far worse.

Recent work has indicated that the concept of rarity is far more complex than is represented in the Red Data Books. Rabinowitz *et al.* (1986) suggest no less than seven forms of rarity for plants, based on three factors:

- Geographic range: Does a species occur over a broad area or is it endemic to a particular small area?
- Habitat specificity: Does a species occur in a variety of habitats or is it restricted to one or a few specialized sites?
- Local population size: Is a species found in large populations somewhere within its range or does it have small populations wherever it is found?

While these factors are really continuous variables, Rabinowitz *et al.* (1986) for convenience constructed the dichotomous table in Box 8. In this model, the only set that can be considered common in the ordinary sense are those with wide ranges, many habitats, and large population sizes; all others are rare. Species with narrow distribution, specialized habitat, and small numbers (type G in Box 8) are the ones that are "rare" in the public mind, but species sharing six other combinations of attributes should also be considered rare and deserving of special management attention.

Box 8: Forms of Rarity.

Geographic distribution:	Wide		Narrow	
Habitat specificity:	Broad	Restricted	Broad	Restricted
Local population size: Somewhere large	COMMON	RARE (A)	RARE (B)	RARE (C)
Everywhere small	RARE (D)	RARE (E)	RARE (F)	RARE (G)

Source: Rabinowitz *et al.*, 1986.

The different forms of rarity have considerable practical relevance for conservation biology, helping to determine the management strategy employed and the priority allocated to certain species. For example, management of "endemic rarities" of type C in Box 8 might focus on protecting the specific habitat where the species occurs, that of endemic rarities of type G might call for attempting to reintroduce the species to appropriate habitats elsewhere, and the strategy for "patchily distributed rarities" of type E might focus on legal restrictions on trade and direct consumption. Patchily distributed rarities of type D, which occur in small populations over a wide geographic range in a variety of habitats, are likely to become endangered only in the face of widespread habitat destruction and therefore deserve relatively low priority for management attention. Rabinowitz *et al.* (1986) conclude that "the preponderant attention which conservationists pay to endemic species is well justified," as these narrowly distributed species are easily threatened by habitat destruction or over-exploitation. They found that conserving habitats remains the most effective way to conserve species, and that conservationists concerned with rare species need to consider geographic range, habitat specificity, and local abundance in their assessments.

The very real limitations in the level of current understanding about the concept of rarity and its causes can be increasingly overcome by advances in knowledge and field techniques. In the meantime, the concept of threatened species has been a very effective instrument in promoting conservation of biological diversity. Keeping the limitations in mind, Table 2 presents the current state of knowledge of threatened species.

Table 2: Current Status of Threatened Species.

	Ex	E	V	R	I	Total Globally Threatened Taxa
Plants	384	3325	3022	6749	5598	19078
Fish	23	81	135	83	21	343
Amphibians	2	9	9	20	10	50
Reptiles	21	37	39	41	32	170
Invertebrates	98	221	234	188	614	1355
Birds	113	111	67	122	624	1037
Mammals	83	172	141	37	64	497

Key: Ex = Extinct (post-1600), E = Endangered, V = Vulnerable, R = Rare, I = Indeterminate.

Source: Reid and Miller, 1989; WCMC, unpublished data, Jan. 1989.

Even many animal species not in immediate danger of extinction are suffering from declining populations and declining genetic variability. While some wild species — sparrows, starlings, opossums, rats, raccoons, coyotes, white-tailed deer, and other opportunists — are expanding their ranges and populations, far more are suffering catastrophic population crashes. Low populations make species far more vulnerable to disease, climate change, habitat alteration, inbreeding, and many other factors that can threaten their survival. Declining populations also have important implications for development, as reduced populations have less potential for utilization. Where heavy hunting pressures, for example, have reduced populations of game animals to levels far below the carrying capacity of the habitat, the economic benefits of harvesting are much less than they would be with harvesting at a sustainable yield level that maintains the harvested population at the carrying capacity of the habitat.

The Madagascar flat-shelled spider tortoise (*Acinixys planicauda*) is a very rare endemic species found only in a highly restricted area in the dry forest region of southwestern Madagascar (photo by R.A. Mittermeier).

The planet is also being impoverished by the loss of races and varieties within domesticated species. The variety of genetic riches inherent in one single species can be seen in the variability manifested in the many races of dogs, cats, cattle, or horses, or the many specialized types of potatoes, apples, or maize developed by breeders. But whole races or cultivars are being lost at a rate that quickly reduces their genetic variability and thus their ability to adapt to climatic change, disease (O'Brien and Evermann, 1988), or other forms of environmental adversity. The remaining cultivated gene pools in the major crop plants such as maize and rice amount to only a fraction of the genetic diversity they harbored only a few decades ago, even though the species themselves are anything but threatened and the various seed banks still retain many of the previously cultivated forms. But little evolution and adaptation can take place in a seed bank. Thus for biological resources, both loss of species and loss of gene reservoirs are significant, and many agriculturalists argue that the loss of genetic diversity among domestic plants and animals looms as an even greater threat to human welfare than does the loss of wild species (Plucknett *et al.*, 1987; Frankel and Hawkes, 1974).

The hidden danger of ever-growing lists of threatened species is that individual recovery efforts are diluted each time a new plant or animal is added to the list (Scott *et al.*, 1987). Some have called for greater attention to be given to a more broad-based ecosystem approach aimed at preventing species from becoming endangered, because it is easier and more cost-effective to protect intact, functioning ecosystems with all their species than to initiate emergency conservation measures for one endangered species after another, or to wait until common species become endangered before acting to save them (Scott *et al.*, 1987).

On the other hand, the ecosystem approach can sometimes ignore the role of individual species in favor of processes and community organization; therefore, a species-specific approach is required to address the needs of taxa that might otherwise be neglected. The Red Data Books have been very important in drawing public attention to the conservation needs of a number of such species.

Habitats

According to one estimate, almost 40 percent of the net primary terrestrial productivity (associated with plants, algae, and photosynthetic bacteria) is directly consumed, diverted, or wasted as a result of human activities (Vitousek *et al.*, 1986). This estimate provides an excellent indication of how powerful the ecological influence of humans is on our planet. For many centuries, landscapes have been altered and simplified by humans through deforestation, fire, and pastoralism.

Tropical moist forests cover only 7 percent of the earth's land surface but contain at least half of the earth's species. If estimates of the millions of undescribed forest beetles are accurate, they could contain 90 percent or even more of all species. Some sites are extraordinarily rich; Whitmore *et al.* (1985) counted 233 species of vascular plants in just 100 square meters of a lowland tropical rain forest in Costa Rica, equivalent to about one-sixth the total flora of the British Isles on half the area of a singles tennis court.

Tropical rainforest in eastern coastal Brazil, a highly endangered biome that now covers only an estimated 1.5% of its original extent (photo by R.A. Mittermeier).

Surprisingly, no generally agreed estimate on the amount of tropical forest remaining has been produced, with figures ranging from 800 million to 1,200 million hectares. However, it is apparent that deforestation is continuing at a rapid pace, with very conservative estimates suggesting rates as high as 6.5 percent per year in Côte d'Ivoire and averaging about 0.6 percent per year (about 7.3 million ha) for all tropical countries (Table 3). At this rate, which is a net figure incorporating reforestation and natural regrowth, all closed tropical forests would be cleared within 177 years (FAO, 1981). Including both closed and open tropical forests (woodlands), FAO/UNEP (1982), estimate that 11.1 million ha are eliminated outright each year, and at least a further 10 million ha are grossly disrupted annually. But even this may be far too conservative. The Brazilian Space Research Institute has reported that forest fires in 1987 destroyed 20 million ha of Brazilian forest, including 8 million ha of primary rain forest; these figures exceed the FAO figures for the entire world.

In short, estimates of world forest cover and deforestation rates suffer from a surprising lack of firm statistics. Since so much conservation action depends on sound data, and because remote sensing technology is available for providing fairly precise estimates, a global study would seem a very high priority. A systematic assessment of current forests and deforestation rates for the entire tropics could be carried out for about $5 million per year (Booth, 1989).

Since the information base is so poor, figures on how long it will take for all tropical forest to disappear can only be estimates. Raven (1988), for example, suggests that about 48 percent of the world's plant species occur in or around forest areas that are going to be destroyed over more than 90 percent of their area during the next 20 years, leading to about a quarter of those species being lost. Further, as deforestation becomes a more severe problem and the accessible forests are exploited, harvesting rates (and income from forestry exports) tend to slow down. Many major tropical timber exporters of the 1960s and 1970s have stopped exporting, and some — such as Thailand — are now even net importers.

But given the projected growth in both human population and economic activity, the rate of deforestation is far more likely to increase than stabilize. The World Commission on Environment and Development (1987) concluded that by the end of the century, or shortly thereafter, little virgin tropical moist forest outside of protected areas may remain outside of the Zaire Basin, the extreme northeast Brazilian Amazonia adjacent to the southern Guianas, western Amazonia, the Guianan tract of forest in northern South America, and parts of the island of New Guinea (see Chapter VI). The accessible forests in these zones are unlikely to survive beyond a few further decades, as world demand for their produce continues to expand. Forests on steep slopes, on the other hand, are quite likely to endure even very dense human populations because of their inaccessibility and their important economic functions in protecting watersheds.

The dimensions of these habitat changes have been assessed for sub-Saharan Africa (IUCN/UNEP, 1986b) and tropical Asia (IUCN/UNEP, 1986c) (Tables 4 and 5). The implications of these habitat changes for primates in tropical Asia are summarized in Table 6 (IUCN/UNEP, 1986c). In these tables, "original habitat" was determined on the basis of vegetation maps prepared by Unesco for Africa (White, 1983) and tropical Asia (van Steenis, 1958). These maps depict the ideal climax vegetation based on climatic, elevation, and edaphic factors, without significant human interven-

Deforestation in Peruvian Amazonia (photo by R.A. Mittermeier).

tion, and usually correspond to the area of the country; they are therefore an ideal that needs to be tempered by reality. The estimates of natural habitat remaining were derived from a wide variety of sources of variable accuracy, so the figures should be taken as indicative rather than definitive.

Despite these disclaimers, it is apparent from the figures presented that original wildlife habitat has been greatly reduced in virtually all nations in the Old World Tropics. Only Angola, Congo, Djibouti, Gabon, Kenya, Namibia, Somalia, Tanzania, and Zambia in Africa and Bhutan, Brunei, and Malaysia in Asia have lost less than 50 percent of their wildlife habitat. From the species point of view, habitat losses for Southeast Asian primates (which can be taken as reasonable indicators of the other fauna and flora in the region) have been significant. But the impacts on species varies considerably; compare the primates from densely

Table 3: Estimates of Forest Areas and Deforestation Rates in the Tropics.

Country	Closed Forest area (1,000 ha)	Percent deforested per year
Tropical Africa:		
Côte d'Ivoire	4,458	6.5
Nigeria	5,950	5.0
Rwanda	120	2.7
Burundi	26	2.7
Benin	47	2.6
Guinea-Bissau	660	2.6
Liberia	2,000	2.3
Guinea	2,050	1.8
Kenya	1,105	1.7
Madagascar	10,300	1.5
Angola	2,900	1.5
Uganda	765	1.3
Zambia	3,010	1.3
Ghana	1,718	1.3
Mozambique	935	1.1
Sierra Leone	740	0.8
Tanzania	1,440	0.7
Togo	304	0.7
Sudan	650	0.6
Chad	500	0.4
Cameroon	17,920	0.4
Ethiopia	4,350	0.2
Somalia	1,540	0.2
Equatorial Guinea	1,295	0.2
Zaire	105,750	0.2
Central African Republic	3,590	0.1
Gabon	20,500	0.1
Congo	21,340	0.1
Zimbabwe	200	(a)
Namibia	(a)	(a)
Botswana	(a)	(a)
Mali	(a)	(a)
Burkina Faso	(a)	(a)
Niger	(a)	(a)
Senegal	220	(a)
Malawi	186	(a)
Gambia	65	(a)
TOTALS	**216,634**	**0.61**
Tropical America:		
Paraguay	4,070	4.7
Costa Rica	1,638	4.0
Haiti	48	3.8
El Salvador	141	3.2
Jamaica	67	3.0
Nicaragua	4,496	2.7
Ecuador	14,250	2.4
Honduras	3,797	2.4
Guatemala	4,442	2.0
Colombia	46,400	1.8
Mexico	46,250	1.3
Panama	4,165	0.9
Belize	1,354	0.7
Dominican Republic	629	0.6
Trinidad & Tobago	208	0.4
Peru	69,680	0.4
Brazil	357,480	0.4
Venezuela	31,870	0.4
Bolivia	44,010	0.2
Cuba	1,455	0.1
French Guiana	8,900	(a)
Suriname	14,830	(a)
Guyana	18,475	(a)
TOTALS	**678,655**	**0.6**
Tropical Asia:		
Nepal	1,941	4.3
Sri Lanka	1,659	3.5
Thailand	9,235	2.7
Brunei	323	1.5
Malaysia	20,995	1.2
Laos	8,410	1.2
Philippines	9,510	1.0
Bangladesh	927	0.9
Viet Nam	8,770	0.7
Indonesia	113,895	0.5
Pakistan	2,185	0.3
Burma	31,941	0.3
Kampuchea	7,548	0.3
India	51,841	0.3
Bhutan	2,100	0.1
Papua New Guinea	34,230	0.1
TOTALS	**305,510**	**0.6**

(a) No data; in most cases this is where the areas are very small.
Source: FAO, 1981; most other sources consider these figures to be the best available, but far below actual rates of deforestation.

populated Java (Javan Gibbon and Javan Lutong) and Indochina (Francois' Leaf Monkey) with those from the sparsely populated Mentawai islands (Mentawai Gibbon).

If adequate information on the status and value of forest land is available, the governments of tropical countries — out of a sense of enlightened self-interest — will wish to stabilize the area of forest at an amount that enables them to meet national development goals of watershed protection, tourism, firewood, construction, and species conservation. Responsible governments today are constantly seeking ways to ensure that forestry can contribute to the development goals of the nation. The Tropical Forestry Action Plan, prepared by FAO in collaboration with World Bank, UNDP, and World Resources Institute in cooperation with other institutions, specifies the kinds of actions that are required (FAO *et al.*, 1987; see also Chapters VI and VII, this volume).

But tropical forests are far from being the only highly diverse ecosystems. Mediterranean-climate regions (that is, with a cool, wet winter and a hot, dry summer) also have very rich floras with high levels of endemism. For example, the Cape Region of South Africa has about 8,600 species of plants, of which 68 percent are endemic; California has 5,000 plants (30 percent endemic); and southwest Australia has 3,600 plants (with about 68 percent endemic to the region) (Reid and Miller, 1989). In temperate woodlands soils, species diversity may approach one thousand species of animals per square meter, with populations exceeding 2 million individuals. When microfloral communities are added, the numbers are even more impressive (Stanton and Lattin, 1989).

While wetlands are not noted for high species diversity or local endemism (in part because they tend to be somewhat more ephemeral than most other ecosystems), they do comprise very complex ecosystems and some old lakes display very high diversity indeed. Each of the great lakes of the African Rift Valley contains more species than any other lake in the world, with very high levels of endemism. Lake Tanganyika has more than 140 endemic species, Lake Victoria has over 200 endemics, and Lake Malawi has at least 500 endemic species (with estimates indicating that nearly as many more still need to be described) (Miller, 1989; Ribbink *et al.*, 1983).

Biological diversity in marine ecosystems is also remarkable, and indeed coral reefs are sometimes compared with tropical forests in terms of diversity (Connell, 1978). Marine ecosystems are far more diverse than terrestrial ones at the higher taxonomic levels. For example, of the 33 animal phyla, only 11 occur on land (one endemic) while 28 (13 endemic) are found in the seas (May, 1988). Further, Ray (1988) points out that filter feeders, especially zooplankton, create extra levels in aquatic food chains that do not exist on land, and the oceans contain far greater diversity in body size — from whales to picoplankton — than is found on land. Consequently, aquatic food webs tend to be more complex than terrestrial ones and aquatic food chains contain more trophic levels. In addition, marine organisms are highly diverse at the genetic level, with individuals in many taxa being heterozygous at 5 to 15 percent of their genetic loci

Table 4: Wildlife Habitat Loss in Africa South of the Sahara.

Country	Original Wildlife Habitat (1000 hectares)	Amount Remaining (1000 hectares)	Habitat Loss (percent)
Angola	124,670	76,085	39
Benin	11,580	4,632	60
Botswana	58,540	25,758	56
Burkina Faso	27,380	5,476	80
Burundi	2,570	359	86
Cameroon	46,940	19,245	59
Central African Republic	62,300	27,412	56
Chad	72,080	17,299	76
Congo	34,200	17,442	49
Côte d'Ivoire	31,800	6,678	79
Djibouti	2,180	1,112	49
Equatorial Guinea	2,500	920	63
Ethiopia	110,100	3,030	70
Gabon	26,700	17,355	35
Gambia	1,130	124	89
Ghana	23,000	4,600	80
Guinea	24,590	7,377	70
Guinea Bissau	3,610	794	78
Kenya	56,950	29,614	48
Lesotho	3,040	973	68
Liberia	11,140	1,448	87
Madagascar	59,521	14,880	75
Malawi	9,410	4,046	57
Mali	75,410	15,836	79
Mauritania	38,860	7,383	81
Mozambique	78,320	3,678	57
Namibia	82,320	44,453	46
Niger	56,600	12,788	77
Nigeria	91,980	22,995	75
Rwanda	2,510	326	87
Senegal	19,620	3,532	82
Sierra Leone	7,170	1,076	85
Somalia	63,770	37,624	41
South Africa	123,650	53,170	57
Sudan	170,300	51,090	70
Swaziland	1,740	766	56
Tanzania	88,620	50,513	43
Togo	5,600	1,904	66
Uganda	19,370	4,261	78
Zaire	233,590	105,116	55
Zambia	75,260	53,435	29
Zimbabwe	39,020	17,169	56
TOTAL	**2,079,641**	**773,774**	**65**

Note: Data for Mauritania, Mali, Niger, Chad, and Sudan cover only the sub-Saharan portion of those countries. Islands other than Madagascar are not included.

Source: IUCN/UNEP, 1986b.

(as compared with the average of 3.6 percent for mammals and 4.3 percent for birds) (Polunin, 1983). All these factors give coastal and marine ecosystems a form of diversity that differs from terrestrial systems, often requiring different approaches to conservation.

A view of diversity on a coral reef (photo by J. Post).

In conclusion, highly diverse ecosystems are found in many parts of the world, and all ecosystems make important contributions to human welfare. Effective conservation of these ecosystems is unlikely to come only from direct protection of small samples of them; instead, governments seeking to carry out their conservation programs more effectively also require improved policies that deal with other resource management issues that have major impacts on management of species and ecosystems, such as communications, defense, forestry, international trade, energy, and agricultural development.

Economic Factors Stimulating Overexploitation of Biological Resources

The many factors working to lead species to extinction and habitats to destruction are building in force and combine to result in what some have considered an ''impending extinction spasm'' (Myers, 1987b). The information quoted above suggests that considerable alarm is justified. But species and habitat loss are just the painful symptoms of the problem. The real causes are built on economic foundations.

A growing number of economists have recognized that current economic systems have stimulated the major threats to biological resources (see Clark, 1973a; Dasgupta, 1982; Fisher, 1981b; Norgaard, 1984; Pearce, 1976; and Randall, 1979 for more detailed discussions). These problems are exacerbated by the fact that the tropical countries (including China) have 75 percent of the world's population (rapidly growing) but only about 15 percent of the money. Clearly, different types of biological resources suffer from different problems; open-access fisheries, tropical forests, and land

Table 5: Wildlife Habitat Loss in Tropical Asia.

Country	Original Wildlife Habitat (1000 hectares)	Amount Remaining (1000 hectares)	Habitat Loss (percent)
Bangladesh	14,278	857	94
Bhutan	3,450	2,277	34
Brunei	576	438	24
Burma	77,482	22,598	71
China (a)	42,307	16,500	61
Hong Kong	107	3	97
India	301,701	61,509	80
Indonesia	144,643	74,686	49
Japan (b)	32	14	57
Kampuchea	18,088	4,341	76
Laos	23,675	6,866	71
Malaysia & Singapore	35,625	21,019	41
Nepal	11,707	5,385	54
Pakistan	16,590	3,982	76
Philippines	30,821	6,472	79
Sri Lanka	6,470	1,100	83
Taiwan	3,696	1,072	71
Thailand	50,727	13,004	74
Viet Nam	33,212	6,642	80
TOTAL	**815,186**	**248,765**	**67**

Notes:

a. Tropical portion only (i.e., area south of Yunnan high hills, including the southern coastal strip and the island of Hainan).

b. Tropical portion only (i.e., southern Ryukyu archipelago).

Source: IUCN/UNEP, 1986c.

Table 6: Range Loss and Habitat Protected for Selected Primates in Southeast Asia.

Species	Original Range (1000 hectares)	Remaining Range (1000 hectares)	Percent Loss	Percent Protected
Orangutan	55,300	20,700	63	2.1
Siamang	46,511	16,980	63	6.8
Agile gibbon	53,227	18,435	65	3.7
White-handed gibbon	28,070	10,024	64	13.5
Bornean gibbon	39,500	25,300	36	5.1
Mentawai gibbon	650	450	31	22.9
Javan gibbon	4,327	161	96	1.3
Indochinese gibbon	34,933	8,753	75	3.1
Burmese gibbon	16,835	5,638	67	5.1
Pileated gibbon	7,000	1,120	84	9.9
Long-tailed macaque	38,318	12,332	68	3.4
Pig-tailed macaque	156,862	48,169	69	4.1
Stump-tailed macaque	154,696	55,647	64	3.7
Assamese macaque	80,219	33,500	59	2.5
Rhesus macaque	173,227	56,864	67	2.8
Proboscis monkey	2,950	1,775	40	4.1
Snub-nosed langur	2,969	906	70	1.5
Douc langur	29,600	7,227	76	3.1
Javan lutong	4,327	161	96	1.6
Silvered langur	41,217	16,997	59	3.9
Francois' leaf monkey	9,740	1,411	86	1.2
Phayre's leaf monkey	70,857	19,317	73	3.8

Source: IUCN, 1986.

suitable for agriculture have different economic characteristics and need to be treated in different ways. However, six major issues are of particular concern here (adapted from McNeely, 1988).

First, biological resources are often not given appropriate prices in the marketplace. Even where a biological resource is traded directly in the market, it may have associated values that are not reflected in its price. Further, the benefits of the existence of any given level of biological diversity are conferred on all who value them, and the diversity enjoyed in a non-consumptive way by one individual does not reduce the amount available to others. Biological diversity is therefore a "public good," and individuals and industries can often gain its benefits without paying for them (the "free rider" problem). The often-intangible and widespread costs of depleting biological diversity usually provide ineffectual justification for conservation when balanced against projected monetary benefits of exploitation (which typically accrue to relatively few individuals).

Second, the benefits of protecting natural areas are in practice seldom fully represented in cost-benefit analyses because the social benefits of conserving biological resources are often intangible, widely spread, and not fully reflected in market prices. In contrast, the benefits of exploiting the resources supported by natural areas are often easily measured. Hence, cost-benefit analyses usually underestimate the net benefits of conservation or, equivalently, overestimate the net benefits of the exploitation alternative. As Oldfield (1984) puts it, "Developments are proposed, the development alternatives are evaluated, the social costs of habitat losses or extinction are ignored or casually considered, and the decision to develop is given the go-ahead, actually on the basis of incomplete economic information. It is by this gradual process of land conversion that entire ecosystems and wildlife species have disappeared." In short, today's land use patterns are determined primarily by the rent-producing capacity of the area in question, irrespective of its total value to society in a more natural state, counting all the values discussed above.

Third, those who benefit from exploiting a forest, wetland, or coral reef seldom pay the full social and economic costs of their exploitation; instead, these costs (to be paid either now or in the future) are transferred to society as a whole, or to individuals and institutions who had gained little benefit from the original exploitation. Such "external costs" are often accidental side-effects of development projects, so the loss is not recognized in either private or social cost-benefit analyses. Timber concessionaires, for example, do not need to concern themselves with the downstream siltation they are causing, or the species they are depleting, because they do not pay the full cost of these effects. Once they have logged "their" forest, they will leave, and the downstream farmer will have to pay for the siltation damage, and the nation or world at large for the reduction in biological diversity. It may well be that the greatest cause of the reduction in global biological diversity is inadvertence, an external cost of the more direct financial justification for harvesting certain biological resources.

Fourth, the species, ecosystems, and ecosystem services that are most overexploited tend to be the ones with the weakest ownership. Many of these are open-access resources for which the traditional control mechanisms have failed in the face of growing demands of centralized government, national development, international trade, and population growth. Within modern and centralized systems of administration, the forests and the wildlife they contain are often publicly owned resources that are not valued at market rates, but rather are treated as free commodities for exploitation by concessionaires. Generally speaking, the more well-defined, secure, and exclusive (whether held by individuals, communities, or corporate entities) the property rights to biological resources are, the more effectively can the use of these resources be allocated by markets. When ownership rights are weakly enforced (either by the government or by a private owner), exploitation is allocated not to those who value the resource most, but rather to those who can pay the most for the exploitation rights. In a market situation characterized by central government control over resource use and high consumer demand, the costs of protecting species and ecosystems from exploitation are often prohibitive for government "owners" that usually lack sufficient resources and local knowledge of management needs to control overexploitation through the mechanism of enforcing regulations or other restrictions.

Fifth, the discount rates applied by current economic planning tend to encourage depletion of biological resources rather than conservation. While conservation seeks optimum current benefits and broadly equal access to the same stock of resources for future generations, economic analysis usually discounts future benefits and costs because society tends to value benefits sooner rather than later, to consider future costs as being of less significance than costs today, and to assign value to capital in terms of its opportunity cost in the national economy. The higher the discount rate, the greater the likelihood that a biological resource will be mined. Clark (1976) has shown that when discount rates are high and biological growth rates are low (as in whales or tropical forests), the economically efficient use of a resource may be to deplete it, even to extinction; economic activity would be devoted entirely to the interests of the present generation, at the expense of future generations. Further, the higher the discount rate, the lower the priority that the planning process will give to investments in conservation (Perrings, 1988); very simply, the returns from such investments may sometimes be so distant in the future that, when discounted, they add little by way of current net benefit. However, a low discount rate may make the future better off than the present, but the gain to the future may be in the form of either greater biological diversity or greater consumption (Barrett, 1988).

Sixth, and finally, as Warford (1987b) has observed, conventional measures of national income (such as per capita gross national product) "do not recognize the drawing down of the stock of natural capital, and instead consider the depletion of resources, i.e., the loss of wealth, as net income." Many of the national economies of the tropics are based on biological resources, especially forests, that are being depleted at a rate faster than the net formation of capital. As a result, the total assets of the economy are declining even if per capita gross national product (GNP) is growing (Repetto *et al.*, 1989). Warford estimates that the economic costs of unsustainable forest depletion in major tropical hardwood-exporting countries range between 4 and 6 percent of GNP, offsetting any economic growth that may otherwise have been achieved. Growth built on resource depletion is clearly very different from that obtained from productive efforts, and may be quite unsustainable.

Social Factors that Threaten Biological Resources

Biological resources need protection against inappropriate uses and overexploitation, not against people. Unfortunately, conservation programs have often treated local people as opponents rather than partners. Little distinction has been made between recent migrants into wildlands who lack applicable cultural and technical practices for the particular ecosystems and those peoples with a long tradition of sustainable resource use. The former may require assistance and support to locate and manage their farms adequately on suitable soils and perhaps away from key sites of outstanding ecological value. The latter may collaborate in the management of protected areas and support research efforts with unique knowledge and experience. In situations of extremely long habitation in particular areas, often extending to millennia, there may be a case for cooperative management of sites of mutual interest to conservation for society at large, and for the local people.

The official definition of a national park includes words to the effect that it is not materially altered by human exploitation and occupation. The highest competent authority of the country having jurisdiction over the area is expected to have taken steps to prevent or eliminate as soon as possible exploitation or occupation in the area (IUCN, 1985). In one sense, this approach to habitat protection can be viewed as a reflection of our inability to live in harmony with our natural environment: Conceptually, we would not need national parks if we did not have such an exploitative relationship with nature. This has led to two anomalies that have led in turn to both social and ecological challenges for the managers of national parks.

First, national parks take control for resource management away from the people who are most directly concerned with maintaining the productivity of the resources upon which their welfare depends. The central government, in effect, is asserting that the area is of national interest and that the government can control the land better for that national interest than any local authority could. This assertion has often followed on the heels of a central government's proclaiming ownership and use rights over the forests, and is complicated by land tenure systems that are a combination of feudal, colonial, and democratic approaches.

In many tropical countries, the government is responsible for exploiting the forests (often through concessionaires) and it establishes protected areas as part of the national land-use plan. In such situations, the rights or needs of the local people are often overlooked, and it is not surprising that "poaching" and "encroachment" are common problems. In a period of rapid exploitation of the nationalized tropical forests, national parks have sometimes been used explicitly as mechanisms for extending central government influence into the most distant and least secure parts of the nation, often along international boundaries (Thorsell, 1986).

Second, national parks have boundaries. By their very nature, as being legally established units of land management, national parks have limits on the ground, often marked by fences or other physical manifestations of authority. Yet nature knows no boundaries, and recent advances in conservation biology are showing that national parks are usually too small to effectively conserve the large mammals, birds of prey, or trees they are designed to preserve. The boundary post is too often also a psychological boundary, suggesting that since nature is taken care of by the national park, local people can go ahead and abuse the surrounding lands, thereby isolating the national park as an "island" of habitat that is subject to the usual increased threats that go with insularity (see, for example, Soulé, 1986).

Further, virtually all land is already "occupied" in the sense that the local people living in and around the forest consider that it is "theirs" (Box 9). The very considerable problems involving conflicts between native peoples and the government of Malaysia over logging rights in Sarawak provide a dramatic illustration of this issue (Scott, 1988).

Park managers in many parts of the world have therefore developed a "siege mentality," feeling encroachment from all sides. The dilemma of how to conserve wildlands in a sea of hostile local interests is a serious one. While national parks have been one of the most universally adopted mechanisms for protection that has been devised in our era, and governments have often determined that it is necessary to take a centralist approach when questions of the national interest supersede local aspirations, more effective means are required to ensure that conservation and local people can work together as partners rather than antagonists. The instability described above does not bode well for the long-term survival of protected areas if conflicts persist.

Under today's conditions, governments need to think in terms of ecological and economic viability of their nations. In some situations, especially where sustainable utilization of resources is to be a management objective (multiple-use management areas), governments may wish to supplement their national parks though efforts at decentralization of power and responsibility, and a return of more resource management to local communities (Klee, 1980; McNeely and Pitt, 1984; Marks, 1984). In Central America, Houseal *et al.* (1985) have found that ''native peoples have devised sustainable long-term land use practices combining migratory agricultural practices with arboriculture and wildlife management. . . . Their mixed agricultural and forestry systems produce more labor, more commodity per unit of land, are more ecologically sound and result in more equitable income distribution than other practices currently being imposed upon their lands. There are no other land use models for the tropical rain forest that preserve ecological stability or biological diversity as efficiently as those of the indigenous groups presently encountered there.''

Governments also may wish to establish protected areas that are designed specifically to conserve traditional forms of land use that have proven their success over time. For example, traditional shifting cultivation is a system that is well adapted to the tropical forest environment, helps maintain the biological diversity of the forest, and often provides significant benefits to wildlife populations. The maintenance of such systems is of considerable importance to modern forms of development. The wild relatives of a variety of important crop plants occur in the forests, and these and the primitive cultivars grown by the swidden cultivators are valuable sources of genetic material for modern plant breeders. Rice, for example, provides the main staple for all of Asia, and the traditional rice varieties grown in upland swiddens contain great genetic diversity; the swidden farmers have often cross-bred domestic rice with its wild relatives, bringing new pest resistance to their crops (Oka and Chang, 1961). The species grown in the swiddens are in a state of continuous adaptation to the environment, and in many places the crops are enriched by gene exchange with wild or weedy relatives. Altieri and Merrick (1987) contend that ''maintenance of traditional agroecosystems is the only sensible strategy to preserve *in situ* repositories of crop germplasm.'' Further, the management of traditional systems will be maintained only when guided by the local intimate knowledge of the plants and their requirements, and by the local management practices that are likely to be most productive.

Chapter IV suggests an approach to land management that will accommodate the need both to protect habitats from overexploitation and to ensure that the local people are active participants in conservation activities.

Scientists in Brazil's Amazon region are studying the effects of habitat fragmentation on species, and are learning about the dynamics of species loss and changes in the abundances of species and populations. Our landscape is becoming increasingly fragmented, and few, if any, large tracts of primary forest are expected to remain (photo by R.O. Bierregaard).

Box 9: How "Natural" are Natural Habitats?

Many people assume that "natural" means "totally untouched by any human influence." Following such a definition, no natural habitats remain on earth, because modern human influences through pollution and climate change are pervasive. From a longer historical perspective, humans have been influencing habitats in Africa and Asia for hundreds of thousands of years, ever since fire became a major force in human technology (Hough, 1926); most of the world's savannas are maintained through human influences. As discussed earlier, humans have been part of natural ecosystems in the New World and Australia for tens of thousands of years, and part of Oceanian ecosystems for thousands of years (Martin and Klein, 1984).

Pre-industrial people occupied virtually the entire terrestrial land area, and have had very considerable influences on natural habitats. Spencer (1966), for example, suggests that virtually all Asian forests have been cleared at one time or another by people (mostly for shifting cultivation), and Wharton (1968) has shown that the larger Asian animals are all adapted to feeding in forest clearings and therefore greatly benefitted from shifting cultivation.

Similarly, tribal peoples in Central and South America harvest certain plants and animals in ways that significantly alter their ecosystems to provide them with more of the most-desired products of nature (e.g., Warren *et al.*, 1989; Prance *et al.*, 1987; Boom 1985; Gómez-Pompa 1988; Gómez-Pompa *et al.*, 1987). All in all, historical human influences on the environment, especially through the use of fire and shifting cultivation, have been pervasive and even the ecosystems that appear most "natural" have been significantly altered by humans at some point in the past (Thomas, 1956; Elliott, 1964). Efforts to totally exclude human influence from "natural" ecosystems, as in strictly protected national parks, can lead to a situation that has not occurred for thousands of years and will have unknown ecological implications. The devastating fires that hit Yellowstone National Park in 1988 are one dramatic example of what can happen when nature is allowed to take her own course without human intervention. Lugo (1988) concludes that environmental change and disturbance may be required to maintain a species-rich tropical landscape.

Since the human influence on forests and savannas has been a primary determinant of their current structure, any effort to establish a protected area that excludes people will require active management to maintain its "pristine" nature (which in fact was partially created by human activities in historic and prehistoric times).

Major Obstacles to Greater Progress in Conserving Biological Diversity

At its most fundamental level, biological diversity is threatened because people are out of balance with their environment; benefits are being gained from exploiting natural habitats without paying the full costs of such exploitation. Current human populations and standards of living are subsidized by non-renewable resources that have accumulated over hundreds of millions of years, yet are being consumed in a few generations. Age-old cultures based on sustainable use of renewable resources are being quickly replaced by modern cultures based on over-exploitation. The profound changes in human society called for by the World Commission on Environment and Development (WCED, 1987), the World Conservation Strategy (IUCN, 1980), and the World Charter for Nature will come only with levels of innovation and investment that have not yet been seriously considered by governments.

At the risk of over-simplification, and in the realization that different settings suffer from different problems, six main obstacles to greater progress in conserving biological diversity can be identified:

- *National development objectives give insufficient value to living natural resources.* Maintaining a nation's biological diversity is integral to maintaining its wealth, but the importance of species and ecosystems is seldom sufficiently considered in the formulation of national development policies. Development tends to emphasize short-term exploitation to earn income or foreign exchange rather than long-term sustainable utilization of living natural resources. International development organizations focus on the expressed immediate needs of the developing nations, and tend to seek relatively short-term returns on their investments. As a result, land-use policies are often inappropriate for the long-term benefits of society. Further, the international community tends to encourage this trend in order to facilitate the flow of commodities from south to north.
- *Living natural resources are exploited for profit, not for meeting the legitimate needs of local people.* Uncontrolled worldwide use of wildlife products is contributing to species extinction and loss of biological diversity. Where a significant profit can be made, as in the case of African rhinos or tropical forests, the target species or ecosystems can be devastated, with virtually no benefit to local people. Much of the increasing consumer demand is in markets far removed from the habitat or species involved, and the commercial interests bring few benefits to the local people whose long-term welfare may depend on sustainable use of the overexploited species.
- *The species and ecosystems upon which human survival depend are still poorly known.* The number of specialists working to acquire the necessary knowledge about species,

biological diversity, ecosystems, and human aspects of resource management is woefully inadequate to meet our collective needs. Existing expertise is located primarily in industrialized countries, not in the developing ones that depend upon this expertise to make decisions concerning sustainable utilization of their living natural resources; only 6 percent of the world's scientists and technologists live in the tropical countries (NAS, 1980). Few tropical countries (India is an exception) have sufficient research capacity to address current needs in conservation. And research on biological diversity (taxonomy, inventories, etc.) tends to be located very low on the pecking order of science, thereby suffering from neglect.

- *The available science is insufficiently applied to solving management problems.* The considerable scientific research carried out in recent decades has provided a far better basis for managing resources. Ways need to be developed for applying biological and social science to managing species and ecosystems, helping to restore degraded ecosystems, and bringing the benefits of conservation to the people most directly concerned.
- *Conservation activities by most organizations have had to focus too narrowly.* Most conservation efforts have addressed a small number of species such as mammals, birds, major species of plants, or certain tree species, or the establishment of reserves or other protected areas whose inventory of biological diversity is usually not known. Likewise, management has been largely directed toward conservation of so-called flagship species, usually animals, rather than biological diversity as a whole. Conservation activities often must focus on these narrow objectives in order to obtain funding, focus attention, and achieve results. But attention also needs to be focused on the conservation needs of a wider range of species, to assess how far they are included in existing protected areas, and to determine whether the management plans are suitable for conservation of these identified species. Even more important is research into the reasons for the human behavior that leads to unsustainable use of biological resources.
- *Institutions assigned responsibility for conserving biodiversity have lacked sufficient resources to do the job.* In most countries, those responsible for managing wildlife and protected areas are poorly paid, have insufficient opportunities for advancement, lack specialized training, and have low prestige. Those operating in the field are often isolated from their families and from local communities. While lacking sufficient equipment and managerial capacity, the responsible institutions also suffer from a lack of absorptive capacity and the ability to make good use of new inputs.

A biologist conducts an inventory of a tropical forest in the Atlantic forest region of Brazil (photo by A. Young).

Following page, overleaf: A conservationist explains the importance of endemic primate species to local people in the interior of the state of Minas Gerais, Brazil (photo by R.A. Mittermeier).

أهلاً وسهلاً

MINAS
MURIQUI

CHAPTER IV
APPROACHES TO CONSERVING BIOLOGICAL DIVERSITY

The technologies for conserving biological diversity have tended to focus on protected areas and gene banks. While these approaches will continue to be important, conservation must become more innovative and cross-sectoral. Effective application of conservation technology will call on additional sectors and require increased resources, including personnel, finance, and political commitment.

The threats to biological resources are complex and multifarious, calling for a wide range of responses across a large number of private and public sectors. In general, six kinds of action can be taken by the international community and by governments interested in promoting the conservation of biological diversity: policy changes, integrated land-use management, species protection, habitat protection, *ex situ* conservation, and pollution control. But this short list greatly oversimplifies the matter, for each of the approaches depends to some extent on the others for its success, and weaknesses or failures in any one of the approaches is likely to have negative repercussions on the others.

Policy Shifts, Integrated Land Use, and Biodiversity

Since government policies are often responsible for depleting biological resources, it stands to reason that policy amendments are often a necessary first step toward conservation. National policies dealing directly with wildlands management or forestry, or influencing biodiversity indirectly through land tenure, rural development, family planning, or subsidies for food, pesticides, or energy, can have significant impacts on the conservation of biodiversity. For example, removing subsidies for forest clearing in Brazil is a powerful response to deforestation (Binswanger, 1987; Repetto, 1988), and, in some regions, giving land tenure to rural people who have long lived in balance with their resources can encourage new investments required for sustainable use of biological resources.

Earlier discussions have indicated that many policies outside the traditional conservation sector can have fundamental effects on biodiversity. Repetto and Gillis (1988), for example, discussed the many cases where public policies have led to the misuse of forest resources. The World Commission on Environment and Development indicated a number of the cross-sectoral policy shifts that are required if biological resources are to be used sustainably (WCED, 1987). McNeely (1988) discussed the linkages between biological resources and such sectors as agriculture, tourism, water resources development, research, fisheries, and communications.

The close link between rural development and conservation of biological resources demonstrates that action in either area alone will not solve the problem. Instead, conservation needs to be woven together with agriculture, forestry, fisheries, transport, national defense, and other efforts. The following major policy components might be included in such integrated action:

- to promote cross-sectoral collaboration, the various institutions should share information, develop agreed common objectives, and seek to define problems in the same way.
- the many economic and financial benefits of integrated rural development linked with conservation of biological resources need to be quantified and brought to the attention of policymakers.
- conflicts between the various activities in agriculture, fisheries, forestry, conservation, and rehabilitation need to be identified in integrated plans and programs.
- institutional reform and improvement may be required as part of good design and implementation of integrated sectoral development plans and programs.
- new legislation may need to be formulated consonant with the socio-economic patterns of the target group of people or institutions and the natural resource needs, both to institute disincentives and to ensure that incentives carry the power of law.
- policies and legislation in other sectors need to be reviewed for possible application to conservation of biological resources and community involvement in such work.
- effective incentives need to be devised to accelerate integrated development to close any gap between what the individual sees as an investment benefit and what the government considers to be in the national interest.
- the rural population needs to be involved in the design and follow-up of plans and projects, not simply their implementation (de Camino Velozo, 1987).

One means of initiating improved policy coordination is

through preparing a national or sub-national conservation strategy (NCS), basically an extremely broad national environmental management plan. An NCS can form the basis of a new, broader pattern of well-balanced development that depends upon the conservation of natural resources. Great and lasting benefits are to be gained by bringing the processes of conservation and development together. The preparation of national conservation strategies will assist countries to realize this potential by facilitating the definition of policies and actions, including the conservation of biological diversity, upon which sustainable development can be built.

The first requirement for a successful NCS is the participation of the widest possible range of actors in defining the issues and identifying possible courses for action. Preparing an NCS involves government agencies, non-governmental organizations, private interests, and the community at large in analysis of natural resource issues and assessment of priority actions. In this way, sectoral interests can better perceive their interrelationships with other sectors and new potentials for conservation and development will be revealed. No matter how broadly based a government may be, the nature of the public sector (or indeed of any centralization of power) limits the range of issues that can effectively be considered. The NCS process places government in partnership with NGOs, citizens' groups, universities, industry, financial institutions, and many others in seeking to relate the use of biological resources to national development objectives. It therefore provides an important (and generally non-threatening) forum for reaching national consensus about policies on the use of biological resources. Few better mechanisms seem to exist.

In one form or another, the NCS process has been initiated in over 40 countries. Focusing on national planning and the range of decisions taken by the public sector on the use of biological resources (either deliberately or by default), an NCS can address many of the most fundamental policy issues faced by governments seeking to use their biological resources on a sustainable basis.

In an analysis of how national conservation strategies have addressed biological diversity, Prescott-Allen (1986) concluded that no NCS has yet provided a comprehensive description of the socio-economic contributions of biodiversity to the country concerned, or a comprehensive treatment of the priority needs of biodiversity conservation. He called for better treatment of obstacles and opportunities, cross-sectoral coordination, and integration of conservation and development. The design of policies and practices that would enable the achievement of development and conservation at the same time is the most basic need in most NCS work.

Several other tools have been developed to incorporate what once were regarded as external considerations in development policy decisions. Environmental Impact Assessments (EIA) are one such tool, and their application has yielded many benefits (Ahmad and Sammy, 1985). Yet an EIA generally only offers guidance once fundamental choices among available options have been taken. The NCS approach, in developing a framework where environmental concerns can be related to development objectives, offers the possibility to approach a more appropriate balance point through a process of consensus-seeking.

Integrated rural development can draw on an NCS and on other technologies to promote environmentally sound management of large natural ecosystems. While such programs can contribute to conservation of biodiversity to some extent, many of the most important contributions are made through work directed at stabilization of resource use in areas that are not biologically diverse. These activities focus upon maintaining, or restoring, natural ecosystems so that the ecological and hydrological processes that these support are maintained, and the benefits that they provide to human society are made available on a sustainable basis.

By managing these ecosystems sustainably and stabilizing land use, the root cause of many human population movements can be addressed, with biological diversity being a beneficiary. For example, in many parts of the tropics, forests are being lost because of slash-and-burn agriculture. In most areas, this agricultural practice is a consequence of non-sustainable resource use and declining agricultural productivity in other ecosystems that the rural poor have been forced to leave. By focusing attention on restoring formerly productive agroecosystems, and by maintaining the ecological and hydrological processes that support the productivity of these systems, agricultural pressure on the marginal lands can be reduced and the fields can be allocated to activities more conducive to the conservation of biological diversity.

Integrated rural development that incorporates a component dealing with conservation of biological resources can be an attractive activity for development assistance agencies, as it is likely to fall within their established mandates.

Protecting Species and Habitats: The Need for an Integrated Approach

Species are the building blocks of ecosystems, and often the most obvious indicators of ecosystem health. It is not surprising that they have received considerable attention from governments, NGOs, and international agencies. International measures to protect particular species or groups from destructive exploitation include the International Convention for the Regulation of Whaling (Washington, 1946), the Convention on International Trade in Endangered Species of Wild Fauna and Flora, (Washington, 1973), and the Convention on Conservation of Migratory Species of Wild Animals (Bonn, 1979)(see also Annex 3). At the national level, wild species are protected by hunting regulations, protective legislation, and a wide range of other wildlife management activities.

Species and their genetic resources plainly supply benefits to all human beings. While animals dominate the public consciousness, plants are perhaps even more directly important

Germplasm from perennial corn (*Zea diploperennis*), recently discovered in Mexico, can be crossed with cultivated corn to increase resistence to diseases, thereby increasing agricultural productivity (photos by WWF).

for human welfare; plant germplasm is one of the world's key resources and the future of the world's food supplies will depend on the amount of effort and resources society is prepared to put into its responsible collection and management. Wild genetic resources from Mexico and Central America serve the needs of maize growers and consumers globally. Many of the principal cocoa-growing nations are in West Africa, while the genetic resources on which modern cocoa plantations depend for their continued productivity are found in the forests of western Amazonia.

Coffee growers and drinkers depend for the health of the crop on constant supplies of new genetic material from coffee's wild relatives, principally located in Ethiopia and Madagascar. Brazil, which supplies wild rubber germplasm to Southeast Asia's rubber plantations, itself depends on germplasm supplies from diverse parts of the world to sustain its sugar cane, soybeans, and other leading crops. Over 98 percent of the agricultural produce of the USA is derived from non-native species; on a continental scale, half the crop production in the Americas originated in Asia or Africa, fully 70 percent of Africa's crop production came from Asia or the Americas, and 30 percent of Asia's crop production involves species from America or Africa (Wood, 1988). It is apparent that without access to foreign sources of fresh germplasm year by year, virtually all nations would quickly find their agricultural output declining.

In livestock, as with crops and forestry, controlled breeding and the rapid development of varieties suitable for modern high-energy commercial production is eroding genetic diversity. The rate of loss appears worst in developing countries, where local breeds are being replaced by imported ones. Many of the wild relatives of domestic animals — including wild cattle, wild sheep and goats, and wild elephants — are seriously threatened even though they may be important for breeding purposes.

While a number of species protection measures have been effective and emergency species-specific action is often required to prevent extinction, species are best conserved as parts of larger ecosystems where they can continue to adapt to changing conditions as part of their respective communities. Therefore, governments have long focused on measures to protect particular habitats, such as national parks and other kinds of protected areas. This approach is exemplified at the international level by the Convention on Wetlands of International Importance especially as Waterfowl Habitat (Ramsar, 1971), the Convention Concerning the

Protection of the World Cultural and Natural Heritage (Paris, 1972)(see Annex 3), Unesco's Biosphere Reserves Program, and parts of UNEP's Regional Seas Programs.

Most national governments have established legal means for protecting or regulating the use of habitats that are important for conserving biological resources. These can include: national legislation establishing national parks and other categories of reserves; local laws protecting particular forests, reefs, or wetlands; regulations incorporated within concession agreements; planning restrictions on certain types of land; and customary law protecting sacred groves or other special sites. The responsibility for managing such protected areas is often spread widely among public and private institutions.

Areas that have been given legal protection against conversion to other uses should be among those not considered for alteration or conversion; their contribution to development is typically through maintaining their relatively natural state. In fact, the World Bank's policy on wildlands (Annex 5) expressly prohibits the use of Bank funds to convert legally protected areas to any other uses except under the most stringent and exceptional conditions.

Table 7: Protected Areas of the World. The sites included in this table are all those protected areas over 1000 hectares in size, classified in IUCN categories I to V and managed by the highest competent authority in the country. Data of this sort are dynamic, with new areas being established and information being refined, but this table presents the best available information as of 1 May, 1989. Greenland National Park, covering 70,000,000 ha in the Nearctic Realm, has a significant effect on the total and on comparisons with other realms as it is an order of magnitude larger than any other single site.

REALM	Number of sites	Total area (1000 hectares)
Afrotropical	444	86,090
Indomalayan	676	32,280
Palaearctic	1684	73,190
Oceanian	52	4,890
Nearctic	478	172,560
Neotropical	458	76,810
Australian	623	35,690
Antarctic	130	3,120
Totals	**4545**	**484,630** (3.7 percent of land area)

Source: Protected Areas Data Unit, WCMC, May 1989.

As development has accelerated in the past few decades, governments have recognized the importance of legally protected areas as part of the overall pattern of land use. In the Bali Declaration (in McNeely and Miller, 1984), the world's leading authorities on protected areas pointed out that such sites are an indispensable element of conservation because they maintain those essential ecological processes that depend on natural ecosystems; they preserve the diversity of species and the genetic variation within them, thereby preventing irreversible damage to our natural heritage; they maintain the productive capacities of ecosystems and safeguard habitats critical for the sustainable use of species; and they provide opportunities for scientific research, education, training, recreation, and tourism.

Many traditionally protected areas have been managed for hundreds or even thousands of years, but the modern protected area movement began with the establishment of Yellowstone National Park in 1872. Since that time, protected areas have spread steadily throughout the world as the primary means for conserving natural habitats. Today, over 4,500 protected areas of over 1,000 ha each (in IUCN categories I-V) have been established, covering nearly 500 million hectares, roughly equivalent to the size of most of the countries of Western Europe combined, or twice the size of Indonesia. The distribution of these nationally protected conservation areas by region is presented in Table 7. Many new areas are added every year, and over 130 nations have accepted the importance of protected areas as a part of balanced land use. But many more areas need to be recognized for the important contributions they make to sustaining society (Box 10).

Box 10: Where New Protected Areas are Needed.

Since 1970, the area legally protected has expanded by more than 80 percent, with around two-thirds of the growth in the Third World. But a great deal more remains to be done; a consensus of professional opinion suggests that the total expanse of protected areas needs to be increased at least three times if the global system is to achieve long-term environmentally sound management of the earth's biological resources.

Reviews of the protected areas of tropical Asia (IUCN/UNEP, 1986a), Africa (IUCN/UNEP, 1986b), and the South Pacific (IUCN/UNEP, 1986c) have been conducted by IUCN's Commission on National Parks and Protected Areas in collaboration with UNEP and numerous other institutions. While many of the extensive parks and reserves necessary to protect the world's most outstanding natural areas are now in place, a number of large gaps remain to be filled. In Indomalaya, ten biounits (regions with unique assemblages of flora and fauna) were reported greatly under-protected, four more need some adjustments, and only ten are noted as being adequately protected. In Africa, five biogeographical units are in need of substantially greater protection, seven need only minor additions, and four are judged adequate. In Oceania, dozens of sites have been identified where protection measures are needed. While the systems review for Latin America and the Caribbean has not yet begun, it will probably report a roughly similar state of affairs. Worldwide, the coastal and marine habitats remain woefully under-represented in the system and far more work remains if these habitats are to be protected effectively.

The nomenclature for protected areas is extremely varied. The same name can mean quite different things; for example, National Parks in Canada do not feature human habitation, while National Parks in the United Kingdom all contain human communities. Significantly, while uniformity of nomenclature and criteria for establishment and management of protected areas is useful to foster management, international communication, and cooperation, the exact form of protection provided to individual areas can vary greatly from country to country, or even from locality to locality.

Population growth and economic development are threatening many protected areas. Furthermore, the list of demands placed by society upon wildland reserves is expanding. Thus, governments today recognize that strictly protected areas cannot be managed to meet society's growing list of goods and services, involving such diverse activities as genetic resource management, watershed protection, recreation, and education. Additional approaches to the management of wildlands are required to supplement the idea of strictly protected national parks, where some sustainable harvesting of biological resources can be among the objectives of area management.

Following the principles outlined above, new approaches to linking protected areas to surrounding lands are required if the appropriate benefits are to flow to society. While the specifics will vary from case to case, the major generalization as stated in the Bali Action Plan (Annex 4) is that local support for protected areas must be increased through such measures as education, revenue sharing, participation in decisions, complementary development schemes adjacent to the protected area, and, where compatible with the protected area's objectives, access to resources. In short, economic incentives should be used to enable people to behave according to their own enlightened self-interest, and sound government policies should be designed to ensure that conservation is indeed in their self-interest (see McNeely, 1988, for more specific recommendations on how to implement such incentives).

In seeking additional land management mechanisms or technologies that can contribute to conserving biological diversity while contributing to sustainable development, IUCN (1985) has prepared a system of categories of protected areas, each designed to achieve an array of management objectives (Box 11). While national parks by definition need to be protected against human exploitation on a

Box 11: Categories and Management Objectives of Protected Areas.

While all protected areas control human occupancy or use of resources to some extent, considerable latitude is available in the degree of such control. The following categories are arranged in ascending order of degree of direct human use permitted in the area.

I. ***Scientific reserve/strict nature reserve.*** To protect nature and maintain natural processes in an undisturbed state in order to have ecologically representative examples of the natural environment available for scientific study, environmental monitoring, and education, and for the maintenance of genetic resources in a dynamic and evolutionary state.

II. ***National park.*** To protect outstanding natural and scenic areas of national or international significance for scientific, educational, and recreational use. These are relatively large natural areas not materially altered by human activity, and where commercial extractive uses are not permitted.

III. ***Natural monument/natural landmark.*** To protect and preserve nationally significant natural features because of their special interest or unique characteristics. These are relatively small areas focused on protection of specific features.

IV. ***Managed nature reserve/wildlife sanctuary.*** To ensure the natural conditions necessary to protect nationally significant species, groups of species, biotic communities, or physical features of the environment when these require specific human manipulation for their perpetuation. Controlled harvesting of some resources may be permitted.

V. ***Protected landscapes.*** To maintain nationally significant landscapes characteristic of the harmonious interaction of resident people and land while providing opportunities for public enjoyment through recreation and tourism within the normal life-style and economic activity of these areas.

VI. ***Resource reserve.*** To protect the natural resources of the area for future designation and prevent or contain development activities that could affect the resource pending the establishment of objectives based on appropriate knowledge and planning.

VII. ***Natural biotic area/anthropological reserve.*** To foster the way of life of societies living in harmony with the environment to continue little disturbed by modern technology; resource extraction by indigenous people is conducted in a traditional manner.

VIII. ***Multiple-use management area/managed resource area.*** To provide for the sustained production of water, timber, wildlife, pasture, and outdoor recreation, with the conservation of nature primarily oriented to the support of the economic activities (although specific zones can also be designed within these areas to achieve specific conservation objectives).

Source: IUCN, 1985.

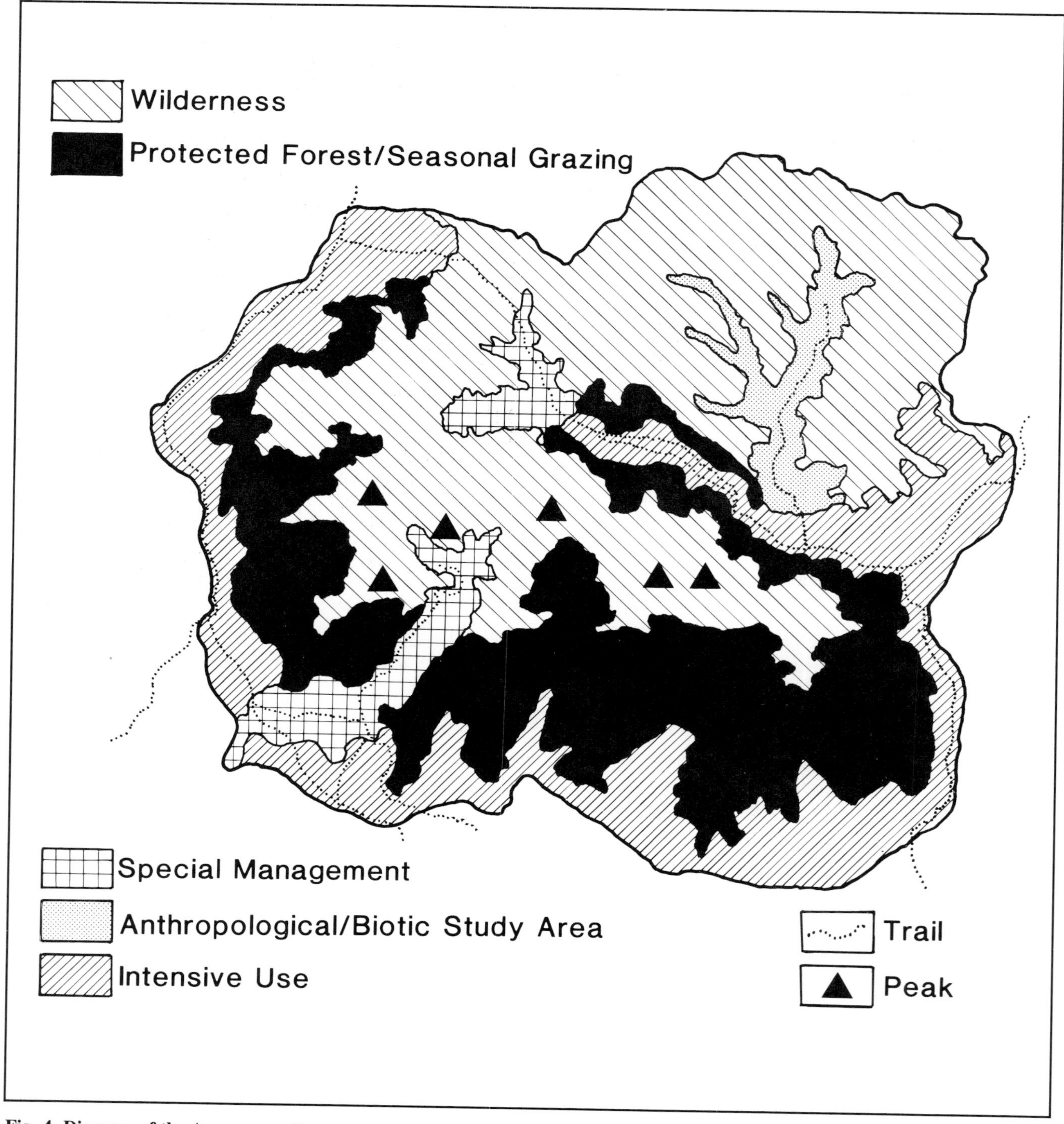

Fig. 4. Diagram of the Annapurna Conservation Area in Nepal, showing different kinds of land-use (Source: National Geographic Society).

commercial scale, other categories of protected areas — such as game reserves, protected landscapes, and multiple-use management areas — can be established around the strictly protected areas to prevent them from becoming biologically impoverished islands, or they can stand by themselves to make important contributions to systems of land management (IUCN, 1985). Several of these categories of protected areas can include sustainable utilization of resources as a management objective, to both conserve biological diversity and provide sustainable benefits to local human communities from the use of those resources. For example, IUCN's Category VI can be used for protecting traditional forms of agriculture, as an integral part of a nation's protected area system.

Recent advances in conservation biology (e.g., Harris,

1984; Diamond and May, 1976; Higgs, 1981; Soulé, 1986) have shown that by themselves the strictly protected categories (Categories I, II and III) will not be able to conserve all — or even most — species, genetic resources, and ecological processes. Far greater expanses are required for conservation than modern societies are willing to remove from direct production. The best answer to this dilemma is to design and manage different types of protected areas — including very large expanses in the categories that permit, and even encourage, compatible human uses of resources — to support among them the overall fabric of social and economic development (Figure 4). *Through a planned mix of national parks and other categories of reserves, amidst productive forests, agriculture, and grazing, protected areas can serve human communities and safeguard the well-being of future generations of people living in balance with their local ecosystems.*

Improvements in conservation over the coming decades will lie primarily in the establishment, implementation, and improved management of those categories of protected areas where some human use will be tolerated or even encouraged, or on new types of protected areas in degraded landscapes that have been restored to productive use for conservation. Strictly protected areas (Categories I, II and III) are unlikely to ever cover more than about 4 percent of the globe. But since permanent agriculture seldom covers more than about a quarter of a nation's land area, ample land exists for forestry, shifting cultivation, grazing, and other uses that may, with proper management, contribute to conservation of biological diversity. Many such areas might surround the more strictly protected national parks, helping to buffer them against the more negative human influences.

In addition, small reserves can also make important contributions to conservation (Simberloff, 1982, 1983). Reserves of less than 10 ha can be effective in conserving viable populations of plants, provided their boundaries can be secured. Numerous plants have survived for as long as botanists have recorded them as populations confined to a hectare or so of land. D.R. Given (pers. comm.) has recorded a case where plants appear to have evolved over millions of years in a site of this size. In Mauritius, an IUCN-WWF project will by the end of 1989 have secured a network of about 10 mini-reserves that will include about 80 percent of the 300 endemic plants, a flora considered essentially doomed by most scientists (H. Synge, pers. comm.). Because they are small, the reserves can be fenced and weeded of damaging introduced plants, yet still contain populations of as many as 240,000 specimens per hectare.

Finally, the protected areas will succeed in realizing their conservation objectives only to the extent that the areas themselves are effectively managed, and to the extent that the management of the land surrounding them is compatible with the objectives of the protected areas. This will typically involve protected areas becoming parts of larger regional schemes to ensure biological and social sustainability, and to deliver appropriate benefits to the rural population. In general, the agencies responsible for managing protected areas need to seek new and more powerful partners in protected area management — local communities; ministries of forestry, agriculture, and foreign affairs; development aid agencies and international banks; and politicians.

In further elaborating the network of protected areas, the following points also need to be taken into account:

- management of a protected area and that of the adjacent land must be planned together, since few protected areas are self-contained entities. The establishment of "transition zones," in which human activities including uses of natural resources in adjacent land are compatible with the conservation of biological diversity within the more strictly protected core area, are often vital to the integrity of the latter.
- the management context and likely ecological resilience of the area in the face of climatic trends and human pressures need critical review, taking into account the likely trend in human numbers in and surrounding the area in question.
- certain "keystone" and critical species, especially of vertebrates, may be used as diagnostic indicators of the adequacy of the protected area system, it being assumed that if habitats capable of assuring the survival of viable populations are protected, the lesser-known species will also be safeguarded (though this approach has some shortcomings — see Landres *et al.*, 1988).
- a conscious relationship needs to be established between *in situ* and *ex situ* approaches to conservation and these methods need to be integrated within overall regional development (see following section).
- the acceptance of protection depends on putting a sufficient economic value on natural resources and biological diversity and, often, on demonstrating that such areas bring a positive benefit to the local communities around them (see Chapter II).
- the national infrastructure needs to be so designed as to ensure that the protected area system is properly evaluated as a national asset and that adequate resources are deployed in its management.
- management policy and practice must be reviewed, especially since these may not be best suited to the conservation of biological diversity. For example, the excessive restrictions on collection of material for study purposes that have been instituted in some national parks hampers the evaluation of their biological diversity and also of certain management interventions that may be required to manage populations; furthermore, the approaches used to address other objectives, such as tourism, may not always be compatible with the requirements of conserving particular life forms.
- much greater efforts must be made to ensure that research in both the natural and the social sciences is made available to protected area managers, and that managers consider

all ecosystem management procedures as scientific experiments to be monitored continuously as their effects become apparent. Since such management needs to be based on the best available information, many protected areas will find it useful — even essential — to institute their own long-term research programs to assess basic ecological relationships, dynamics of change, possible results of manipulation, effects of tourism, and so forth (Thorsell, 1989).

- a major effort is needed to raise public consciousness, to enlist the aid of professionals in the field (e.g., in universities, museums, and professional networks), and to educate local communities about the value of the region.

Contributions of Ex Situ Mechanisms to the Conservation of Biodiversity

While it is universally agreed that the most effective and efficient mechanism for the conservation of biodiversity is habitat protection, it is also acknowledged that off-site (*ex situ*) facilities can be critical components of a comprehensive conservation program (Conway, 1988; Ashton, 1988). Measures to promote *ex situ* conservation of species can include botanic gardens, game farms, captive breeding programs in zoos, and gene banks. The most extensive efforts in *ex situ* conservation have been applied during the past 20 years to crop species (mainly food plants), to some trees, and to pasture plants by FAO, the Consultative Group on International Agricultural Research (CGIAR)(which includes the International Board on Plant Genetic Resources — IBPGR — and 12 other international agricultural research centers throughout the world), 150 or so genebanks around the world, and other crop genetic resource centers. Together these cover about 500 species of plants (including wild relatives of crops), or about 0.2 percent of the total. For the majority of wild species, most *ex situ* germplasm is maintained by the 1,300 botanic gardens in the world. International coordination of the *ex situ* plant efforts is maintained by the Botanic Gardens Conservation Secretariat (operated under the auspices of IUCN), which holds records of 20,000 species of which material is cultivated in botanic gardens (roughly 8 percent of the world's plants). A major expansion of this program is being planned.

Ex situ conservation programs supplement *in situ* conservation by providing for the long-term storage, analysis, testing, and propagation of threatened and rare species of plants and animals and their propagules. They are particularly important for wild species whose populations are highly reduced in numbers, serving as a backup to *in situ* conservation, as a source of material for reintroductions, and as a major repository of genetic material for future breeding programs of domestic species. Some *ex situ* facilities — notably zoos and botanic gardens — also provide important opportunities for public education.

Even for wild species that are not threatened, *ex situ* collections are needed to make the material readily available for breeding — breeders do not normally go out into the field for their material, though regular infusions from wild sources are required.

Ex situ methods suffer from some limitations. First, it is not feasible economically to keep more than a limited sample of the genetic diversity of a species in a zoo, seed bank, or botanic garden. Conway (1988) concludes that because of limitations of space and the numbers of individuals required to maintain viable populations, it is impractical for zoos to sustain in the long term more than about 900 species of vertebrates and probably far fewer in conventional breeding programs. Second, little directional habitat-responsive evolution can take place *ex situ*, so the captive populations are not adapting to changing environmental conditions. Third, the *ex situ* population is likely to have a narrow genetic base, and is unlikely to have been collected so as to ensure the representation of a wide range of genotypes. Fourth, *ex situ* conservation depends on continuity in policy and funding, which is far from assured, especially in the tropics (Ledig, 1988).

In conclusion, *ex situ* contributions to conservation are essential for ensuring the survival of crop plants dependent on humans and can provide an extremely valuable insurance policy against the extinction of wild species of plants and animals that have been reduced to very low levels in the wild. Yet *ex situ* approaches depend on *in situ* approaches to enable their genetic stocks to be replenished. Therefore, the two approaches should be seen as opposite ends of the total spectrum required for effective conservation.

Zoological gardens

Over 3,000 vertebrate species are being bred in zoos and other captive animal facilities, totalling some 540,000 individuals (Conway, 1988). Despite the fact that this number is trivial in terms of the total wild population and is roughly equal to 1 percent of the domestic cats in the USA, criticism is sometimes levelled at zoos, aquaria, and similar institutions for holding and breeding endangered animals. Zoos were indeed once a net drain on wildlife, but today most modern zoological gardens are largely self-sufficient in terms of animal production and some are working to reintroduce various species, many of them endangered, into the wild. The Arabian oryx, addax, Przewalski horse, European bison, giant panda, black-footed ferret, golden lion tamarin, Hawaiian goose, Bali starling, Guam rail, peregrine falcon, and whooping crane have all benefitted from captive breeding. Considerable work is still required to ensure that species such as gorillas, giant pandas, elephants, and chimpanzees can be maintained as viable captive populations without needing to draw on wild populations.

A wealth of experience is available in modern zoos, including husbandry, veterinary medicine, reproductive biology, behavior, and genetics. These facilities offer space for supporting populations of many threatened taxa, drawing on resources that do not compete with those for *in situ*

An enclosure of the Rio Primate Center in Brazil where endangered primates are bred in captivity (photo by R.A. Mittermeier).

conservation. Indeed, a number of major zoos, including those of San Diego, Chicago, New York, Washington D.C., and Frankfurt, have major field research and conservation programs that support *in situ* management of species and protected areas.

When a serious attempt is made, most species can be bred in captivity and viable populations can be maintained over the long term. U.S. Seal (1988), the Chairman of the IUCN/SSC Captive Breeding Specialist Group, lists the following major contributions to successful captive management programs over the past 20 years:

- improved nutrition and prepared diets;
- chemical immobilization and anesthesia;
- vaccinations and antibiotics;
- individual animal identification and records combined with a central database (the International Species Inventory System) (ISIS);
- reproductive control (contraception) and enhancement;
- population biology and molecular genetics;
- information technology and microcomputers; and
- decision analysis and crisis management.

But the establishment of captive breeding populations has often come too late, when the species is perilously near extinction. Instead, management to best reduce the risk of extinction requires the establishment of supporting captive populations when the wild population is still in the thousands. Vertebrate species with populations below one thousand individuals in the wild require close and swift cooperation between field conservationists and captive breeding specialists (IUCN, 1987b). This principle is well illustrated by the Kouprey Action Plan, which involves government agencies in the range States, a number of zoos, and field scientists in a major conservation effort (MacKinnon and Stuart, 1989).

Through their research on captive populations, zoos are also learning lessons about wildlife management that can be applied to protected areas that contain relatively small populations of certain species of particular concern. Methodologies and management techniques such as induced ovulation, transplanting of certain individuals between populations to ensure gene flow, and various veterinary procedures developed in zoos will often need to be applied to protected areas that become isolated islands of habitats (and therefore large, semi-natural zoos).

The zoos are very well organized to contribute to conservation, with a number of national and international associations. Both the International Union of Directors of Zoological Parks (IUDZG) and the American Association of Zoological Parks and Aquaria (AAZPA) have significant conservation programs, and provide considerable support to IUCN's Captive Breeding Specialist Group. Studbooks are kept for many of the most important species, masterplans are compiled for key species, breeding stock is freely exchanged, and a series of regular publications ensures that zoo professionals are well informed of current progress.

The zoo community is generally well funded and zoos worldwide receive hundreds of millions of visitors every year. For many people, a visit to the zoo is the only chance they will ever have to experience much of the world's most spectacular biological resources. Today's modern zoos are educating visitors about conservation of biodiversity, supporting field conservation, providing training opportunities for wildlife managers, and holding in captivity breeding populations of species that are critically endangered but that may one day be reintroduced into their historical natural habitats (Kleiman, 1989). These contributions earn zoos an important role as part of the world conservation movement.

Botanic gardens

Some 1,300 botanic gardens and arboreta have been established to hold and exhibit plants. Many devote considerable attention to investigating those aspects of plant biology that require growing a variety of large or long-lived plants of wild origin, or that involve growing plants over long periods or in large experimental plantings. They are potentially well equipped for research into aspects of plant propagation, including seed physiology, germination and establishment, and vegetative reproduction. They are also well placed to conduct research on breeding systems, pathology and herbivory, symbiotic relationships, and minimum viable population sizes for conservation (both *in situ* and *ex situ*).

Botanic gardens are already playing an increasing role in conservation and the maintenance of genetic diversity. Current activities include:

- documenting the local flora, including sending expeditions to explore new areas and conducting systematic studies, and investigation of hitherto unrecognized species in the flora;
- maintaining specimen collections so that records of the distribution, abundance, and habitats of species may be available for research, including assessing species of potential agronomic, horticultural, medicinal, or other economic interest;
- building up expertise among botanic garden staff through research and field investigations (often leading to recommendations on areas for *in situ* conservation or on management policies for sustained survival of plants in reserves);
- maintaining and monitoring nature reserves either within or associated with the garden or arboretum (over 250

Only one individual of this species of palm (*Pritchardia munroii*) remains in the wild. These plants are being grown at Hawaii's Waimea Arboretum and Botanical Garden (photo by Waimea Arboretum and Botanical Garden).

botanic gardens and arboreta maintain natural vegetation areas or reserves either inside their area or separately, ranging from a hectare or less to over 100,000 ha); and

- preserving samples of rare or endangered species in cultivation, multiplying and producing rare and endangered species for reintroduction into the wild and for use in restoration or rehabilitation of habitats, and maintaining special conservation collections (such collections are grown by over 350 gardens and often include rare or endangered species).

In recognizing the important conservation role of botanic gardens, IUCN in 1987 established a Botanic Gardens Conservation Secretariat (BGCS) to mobilize the world's botanic gardens into an effective force for conservation. Its objectives include: to promote the implementation of the Botanic Gardens Conservation Strategy; to monitor and coordinate *ex situ* collections of conservation-worthy plants; to develop a program for liaison and training; to arrange the Botanic Gardens Conservation Congress every three years; and to develop an education program (BGCS, 1987).

Today botanic gardens should be viewed, and should view themselves, as resource centers for conservation, research, and development. Their value in conservation should not be seen in a narrow sense but as linked with various aspects of applied science. In the words of Ashton (1984): "Botanic gardens have an opportunity, indeed an obligation which is open to them alone, to bridge between the traditional concerns of systematic biology and the returning needs of agriculture, forestry, and medicine for the exploration and conservation of biological diversity."

It is apparent, however, that the supply of botanic gardens in the tropical countries falls far short of the needs. While a number of outstanding gardens have been established in the tropics — as in Java, Sri Lanka, and Colombia — a considerable expansion of such areas must be a very high priority as a means of augmenting the *in situ* efforts to conserve natural habitats (Table 8). Greatly increased international support for the tropical botanic gardens will enable them to participate fully in the international effort to conserve biological diversity.

Seed banks

The storage of conservation material in the form of seeds is one of the most widespread and valuable *ex situ* approaches. Extensive expertise has been developed in this field by the agencies and institutions involved with plant genetic resources over the past 20 years. Seed storage has considerable advantages over other methods of *ex situ* conservation, including ease of storage, economy of space, relatively low labor demands, and consequently the ability to maintain large samples at an economically viable cost.

The disadvantages of seed storage (apart from those inherent in all *ex situ* methods) include their dependence on secure power supplies, the need to monitor the viability of the seeds, and the need for periodic regeneration. In addition, many species have seeds that are "recalcitrant" in that they are easily killed by the usual techniques of storing under reduced temperature and humidity. It is estimated that up to 15 percent of the world's flora possess recalcitrant seeds (i.e., some 37,500 species) and therefore cannot be conserved in seed banks with current technology (BGCS, 1989). Far greater research efforts are required to determine how such species can be maintained *ex situ*, as the most effective way to make their genetic contributions available for research.

Over the past few decades, significant investments have been made to develop seed banks for the major world food crops, often using the CGIAR network. Today, more than 50 seed banks have been established worldwide, over half of them in developing countries. Most of these operate under a set of guidelines or procedures developed through IBPGR, based on three main principles: germplasm shall be available to all *bona fide* scientists and researchers, regardless of political or institutional affiliation; collections made in a particular country will be carried out in partnership with the country concerned, and half of all samples collected will be left in the country of origin; and all germplasm collections will be duplicated elsewhere.

For many of the major staple food crops — plants of global economic value such as wheat, maize, oats, and potatoes — more than 90 percent of the variation in landraces has now been preserved in *ex situ* collections (Plucknett *et al.*, 1987). For other species, such as rice, sorghum, and millet, it is estimated that the major part of the work involved in preserving landraces will be completed by 1990 (Williams, 1984). This sounds promising, but Peeters and Williams (1984) estimated that of the 2 million accessions of plant germplasm in seed banks worldwide, 65 percent lack even basic data on source; 80 percent lack data on useful characteristics, including methods of propagation; and 95 percent lack any evaluation data such as responses to germinability tests. Extensive data are held on only 1 percent of the specimens, and it is feared that a substantial proportion of the accessions not tested for germinability may be dead.

Despite the important achievements of seed banks, far more work needs to be done in securing crop genetic diversity. First, international efforts have focused on crops of widespread importance, so numerous species that may be of low global importance but of high priority for particular regions or countries, or for specific purposes such as medicinal plants, are poorly represented in seed banks (Baskin and Baskin, 1978).

Second, many plants of economic importance are poorly represented in *ex situ* collections because of the difficulty of storing their seed or because the species are normally propagated vegetatively. For example, the seeds of many tropical forest species cannot withstand drying or freezing (Plucknett *et al.*, 1987). Crops like rubber, cacao, palms, and many tropical fruits can only be conserved in field gene banks akin to botanic gardens. Many root crops that are propagated vegetatively must be planted each year to maintain the strain;

because of the expense associated with this procedure, only potatoes among the root crops have received coverage of more than 50 percent of landraces, even though cassava (manioc) is the fourth most important dietary source of calories in tropical developing countries after rice, maize, and sugar cane (Cock, 1982; Gulick *et al.*, 1983).

Finally, with only two exceptions — wheat and tomatoes — the wild relatives of crops are extremely poorly represented in *ex situ* collections, constituting only about 2 percent of the varieties stored in seed banks (Table 8). While wild relatives of domesticated plants have traditionally been considered a last resort of plant breeders, they have nonetheless played an important role in sustaining agricultural productivity. Over 20 wild species, for example, have contributed genes to potatoes. The difficulty of interspecific crosses has restricted the role of wild relatives, but developments in biotechnology may substantially increase their importance in the future.

Table 8: Wild Relatives of Major Crops Held in Seed Banks.

Only the wild relatives of a few crops such as wheat, potato and tomato have been widely collected and preserved in seed banks. In most cases, wild germplasm represents less than 2 percent of the seed bank holdings for each crop and most wild relatives of crops still thrive only in the wild.

Crop	Wild species held in all seed banks as percent of total holdings	Estimated percent of wild species still to be collected
CEREALS		
Rice	2	70
Wheat	10	20-25
Sorghum	0.5	9
Pearl millet	10	50
Barley	5	0-10
Corn (maize)	5	50
Minor millets	0.5	90
ROOT CROPS		
Potato	40	30
Cassava	2	80
Sweet potato	10	40
LEGUMES		
Beans	1.2	50
Chickpea	0.1	50
Cowpea	0.5	70
Groundnut	6	30
Pigeonpea	0.5	40

Source: Hoyt, 1988.

Management Action in Response to Pollution and Climate Change

Measures to curb the contamination of the biosphere with pollutants are perhaps the most widespread conservation measures, are the most expensive, and have attracted the greatest attention from both the public and government. When selenium accumulating in water draining from irrigated fields killed or deformed hundreds of aquatic birds at the Kesterson National Wildlife Refuge in California, a massive clean-up was ordered, with a final bill that could reach $50 million (Anderson, 1987). International measures include the Convention on the Prevention of Marine Pollution by Dumping of Wastes and Other Matter (London, 1972, as amended), the Convention for the Protection of the Ozone Layer (Vienna, 1985) and the Montreal Protocol on Substances that Deplete the Ozone Layer (1987). Action against pollution generally began with national measures to remedy acute contamination of rivers and urban air, but in recent years has extended to regional problems (like long-range transboundary air pollution creating acid deposition in areas remote from the source of the gases concerned) or global problems like those of stratospheric ozone depletion.

Biological diversity is threatened by various forms of chemical pollution. Sufficient evidence has been presented to convince governments that the depletion of stratospheric ozone over Antarctica in springtime is linked with chlorofluorocarbons released to the atmosphere through their use as aerosol propellants, refrigerants, and generators of plastic foams. New indications suggest that stratospheric ozone in middle latitudes may have been depleted generally by some 3 percent (McElroy and Salawitch, 1989), thus permitting more damaging ultraviolet radiation to reach the earth's surface, with consequences ranging from reduced production of algae in the surface waters of the sea to increased skin cancer in humans with fair skin. The deposition of sulphate and nitrate produced as sulphur and nitrogen oxides in fossil fuel combustion has acidified lakes, rivers, and soils over considerable areas of northern Europe and North America and, in conjunction with oxidants produced by reactions involving hydrocarbons and nitrogen oxide originating especially from motor vehicle emissions, is incriminated in forest dieback in these regions.

But the gravest threat — or at least the straw that breaks the camel's back — may be climate change brought about by air pollution and the increase in atmospheric carbon dioxide due to deforestation and burning fossil fuels. While the earth has benefitted from a greenhouse effect for hundreds of millions of years — it is what makes the planet habitable — the effect is now becoming intensified to the extent that some habitats may become unsuitable for the species currently living there at a time when those habitats are so isolated by surrounding agricultural lands that the wildlife has no other place to go (Strain, 1987).

The greenhouse effect due to the accumulation of carbon dioxide, methane, nitrous oxide, and chlorofluorocarbons in the atmosphere is likely to raise mean world temperatures by about 2 °C by 2030 and mean sea levels by around 30-50 centimeters on a comparable time scale (Warrick *et al.*, 1988). By the end of the next century, global average surface temperatures are expected to increase by 2-6 °C, with an attendant rise of sea level of 0.5 -1.5 meters (Schneider, 1989).

These effects threaten biological diversity because the combination of the magnitude and the rate of the changes involved lies outside the range of variation to which living organisms have been exposed over the past hundreds of thousands — or even millions — of years (Holdgate, 1989). Rising sea level could outstrip the rate of growth of coral reefs, and compress zones of coastal mangroves so much that coastlines are no longer adequately protected from waves and storm surges. Coral reefs are showing signs of dieback for unknown reasons in many parts of the world. Recent studies have detected a very alarming trend in the accumulation of polychlorinated biphenyls (PCBs) in the oceans and freshwater systems and their biomagnification to elevated levels in the tissues of such marine mammals as whales, dolphins, and seals (Cummins, 1988). Major international rivers like the Rhine, and major freshwater systems like the North American Great Lakes, have been biologically impoverished by chemical pollution. Action to ensure that this pollution does not increase in proportion to growing human populations and global industrialization is of the highest importance. Action to reduce pollutants should include the earliest possible phasing out of the chlorofluorocarbons involved in ozone depletion, reduction of the release of other "greenhouse gases" to a minimum level, and a stringent precautionary approach that minimizes the discharges of harmful substances into the world's oceans.

The impacts of changes in atmospheric carbon dioxide levels and attendant climate change on species and ecosystems are likely to be dramatic. Strain (1987) has shown that increasing the atmospheric carbon dioxide concentration under experimental conditions alters the growth rate and reproductive potential of plants, and must ultimately affect interactions at the community level and beyond. Crowley and North (1988) have found that rapid climate change may have contributed to major extinction events in the earth's history. Looking more specifically at the effects of climate change on nature reserves, Peters and Darling (1985) conclude that because many reserves are now "islands" of habitat to which species are closely adapted, climate change could well cause extinctions among reserve species without being compensated by immigrating "new" species. MacArthur (1972) has derived some broad rules about how ecosystems respond to climate change, and suggested that a change of 3 °C can lead to a shift in habitat type of roughly 250 kilometers in latitude or 500 meters in elevation. This is not to suggest that all species will migrate together, like soldiers marching off to war; different species can be expected to react differently to climate change, so the characteristic species content of ecosystems will also change.

Nor can ecosystems be expected to react quickly to climate change, except when the change is accompanied by other ecological factors such as fire or disease. Soil types change very slowly, and many trees are very long-lived and will survive for hundreds of years even if they do not reproduce. Further, the species that exist today are already adapted to the fairly rapid climate changes that have characterized the past 2 million years, and the ranges of many species appear to be affected more significantly by factors such as competition than by climate change. MacArthur (1972) has discussed this point in some detail, pointing out that many species tend to be persistent once they have become established. Which plant or animal species will become established in "new" communities will be greatly influenced by the ones that have survived from the "old" communities. The only safe conclusion seems to be that under conditions of changing climates, variable responses by the resident plants and animals are to be expected and these are likely to be highly unpredictable given our current state of knowledge.

But it is unrealistic to expect the boundaries of existing protected areas to change very much, because they are usually surrounded by human land uses that will not allow much change. Instead, more intensive forms of management intervention — such as habitat enrichment, artificial insemination, and boreholes to provide drinking water — will be required to maintain systems deemed desirable; the alternative, which might be more attractive in some cases, is to allow nature to take her own course and for existing protected areas to be allowed to develop their own "new" ecosystems.

Of particular interest in this regard are the protected areas that have relatively large altitudinal gradients, thereby containing numerous and diverse ecosystems. As Peters and Darling (1985) pointed out, many reserves have been placed in mountainous land because such areas are generally less suitable for agriculture. In attempting to assess how the world's protected areas are distributed by altitudinal range, McNeely and Harrison (in prep.) reviewed all protected areas of over 1,000 ha in size in IUCN categories I to V and having altitudinal range data. Of the 4,518 sites meeting the first two criteria, altitudinal range data were available for 2,290 sites (51 percent). The results are presented in Table 9. More than half of these protected areas have altitudinal ranges of less than 1,000 meters.

Table 9: Altitudinal Range of Protected Areas.

Biogeographic Realm	**Altitudinal Range (in meters)** 0-999	1000-1999	2000-2999	3000-3999	4000-4999	5000-5999	6000-6999	**Totals**
Nearctic	171	41	27	4	6	2	2	253
Palaearctic	423	146	49	25	6	4	4	657
Afrotropical	319	50	14	2	2	-	-	387
Indomalaya	346	92	21	7	-	-	-	466
Oceania	23	8	1	2	2	1	-	37
Australia	85	26	-	-	-	-	-	111
Antarctic	76	21	4	3	-	-	-	104
Neotropical	161	51	34	21	5	3	-	275
Totals	**1,604**	**435**	**150**	**64**	**21**	**10**	**6**	**2,290**

The implications of anthropogenic climate change for biological diversity are profound, and detailed studies are required to prescribe steps that can be taken by governments and the international community to adapt to the changes that seem almost certain to come (even though these changes are not predictable at the local level with a degree of certainty that will support detailed plans). Such studies should build on three principles:

- First, maintaining maximum biological diversity assumes far greater urgency as the world becomes increasingly threatened by rapid climatic change. Diversity in species provides the raw materials with which different communities will adapt to these changes, and the loss of each additional species reduces the options for nature — and people — to respond to changing conditions.
- Second, global generalizations are unlikely to be sufficient as a basis for response to the problems. While broad patterns of climatic change can be predicted, the real impacts will be felt locally, and these impacts are unlikely to be predictable with much precision. Recommending action in the face of great uncertainty is a risky business, but it is surely sensible to provide local communities with the capacity to adapt to these changes, based (among other things) on traditional knowledge about local ecosystems and their management.
- Third, all indications are that climatic change is a continuing phenomenon that follows a number of inscrutable cycles. Considerably greater scientific attention needs to be given to the implications of climate change for all ecosystems — terrestrial, marine, and freshwater — and to possible management steps that could be taken to maintain biological diversity in the face of it.

A New Global Convention on the Conservation of Biological Diversity

The international legislation cited above and summarized in Annex 3 has proved an important means of promoting international cooperation in the conservation of biological diversity. The World Charter for Nature (Annex 2) has provided "soft law" as further support for this cooperation. However, species and ecosystems are still being exploited at rates that far exceed their sustainable yield. Far more international cooperation is required to reverse this trend.

Recognizing the growing severity of threats to biological diversity and the increasingly international nature of the actions required to address the threats, IUCN and UNEP have embarked on the preparation of an international convention on the conservation of biological diversity. This effort has gained the broad support of governments, including a joint resolution from the U.S. Congress (H.R. Res. 648, 27 September, 1988), which called upon the President to promote efforts "to achieve the earliest possible negotiation of an international convention to conserve the Earth's biological diversity, including the protection of a representative system of ecosystems adequate to conserve biological diversity."

A high-level group of experts advised the Executive Director of the United Nations Environment Programme that a new convention is required, and that the IUCN draft convention provided a useful starting point for such a convention. Meeting in Nairobi, Kenya, in August 1988, the experts advised that the aim of a global convention for conserving biological diversity should be to engage action to conserve as much of the world's biological diversity as possible. It should provide a forum for international identification of priorities. The obligations it imposes should relate to results, leaving the contracting parties to adopt whatever specific means their national legal and administrative systems find most appropriate. The primary approach should be through the protection of habitats, but within national frameworks for land-use planning and species protection that safeguard biological diversity to the maximum extent practicable outside as well as within designated protected sites. The convention should also commit its parties to the adoption of measures to minimize the threats to biological diversity like unsustainable exploitation, pollution of the biosphere, introduction of undesirable alien and genetically modified species, and other factors. It should be recognized that the motivation for global action should be both the value of biological diversity to humanity and the intrinsic and ethical value of species themselves. It should be further recognized that as stewards of biological diversity, States should not only safeguard their own natural heritage but refrain from actions that threaten that of other States.

The central question that needs to be addressed is how a global approach to the conservation of biological diversity should be financed. Some governments are already investing considerable sums in national conservation programs, but there are ample grounds for considering these to be insufficient. The new convention might be financed directly from the uses made of biological resources, perhaps through the mechanism of imposing several small new taxes on certain uses of biological diversity, such as the exploitation of germplasm (for breeding programs, or for the development of new drugs), harvesting (in forestry, fishery, or via authorized and licensed wildlife products), recreational use (in tourism), and for the disposal and recycling of wastes. (See Chapter VIII for a further discussion of funding mechanisms.)

Certain other features of the proposed international convention need emphasis. Clearly, its efficacy depends on the soundness of its scientific foundation, without which national inventories of key areas, international assessments of priority, and specific needs for action will not readily be established. The contracting parties to a convention need to meet periodically in conference to review the working of the convention, and to this end they need a strong, professional, and independent scientific advisory committee (or the services of a body like IUCN in this capacity). These are matters that demand further attention.

The urgency of the problem demands action. The dilemma is over the readiness of the world community to accept a major new measure with financial obligations. Without the provision of new resources, and/or the substantial redistribution of existing resources, the issues considered in this report will not be addressed effectively. A global convention would be a powerful catalyst drawing together the efforts of the various sectoral and regional conventions in this field, by giving overall shape and strategic direction to the whole world effort. But a global convention must be more than a series of noble aspirations. It must do more than state on paper needs for action that cannot be fulfilled for want of resources, or that will not be fulfilled for lack of political will among governments. A convention must not be adopted as a substitute for action, or it will blunt and deflect the efforts the world needs. Accordingly, any convention should only proceed if it:

- has a sound basis in science;
- is truly comprehensive in scope, covering *in situ* and *ex situ* conservation and the protection of the biosphere from all significant damaging impacts, and is in harmony with and supplements existing conventions, agreements, and programs in this broad field;
- is practical in defining obligations and goals, leaving the contracting parties the responsibility of achieving them;
- has the commitment of governments to funding at a realistic level;
- provides realistically for the transfer of resources to allow implementation of the convention by the poorer countries that are the custodians of much of the biological heritage of the earth; and
- is capable of catalyzing and coordinating the efforts of governments and other agencies under other conventions in this field.

The time is coming when the earth's endowment of species and natural ecosystems will be broadly appreciated as assets to be conserved and managed for the benefit of all humanity. This will necessarily add the challenge of species conservation to the international political agenda. It implies two forms of responsibility. First, all nations have the duty to safeguard species within their territories, on behalf of everyone. Second, all humanity has the duty to offer whatever support is required — finance, skills, and so on — to enable individual nations to discharge their duties.

Following page: The rare thin-spined porcupine (*Chaetomys subspinosus*) was only recently rediscovered in Brazil's Atlantic forest region (photo by I. Santos).

CHAPTER V
THE INFORMATION REQUIRED TO CONSERVE BIOLOGICAL DIVERSITY

Many kinds of information are required to conserve biological diversity, including human uses, basic taxonomy, distribution, status and trends, and ecological relationships. Recent advances in data management technology make such information more accessible than ever.

Effective action must be based on accurate information. And the more widely shared the information, the more likely it is that individuals and institutions will agree on the definition of problems and solutions. Developing and using information is therefore an essential part of conservation at all levels, from the local to the global community.

Earlier chapters have indicated how diverse nature is, and suggested how diversity can be conserved to better serve the processes of development. But the current state of knowledge about species and ecosystems is woefully inadequate; detailed knowledge is still lacking on the distribution and population sizes of even such large and well-studied animals as African primates (Oates, 1985). Given the current rates of extinction, the next few years will provide the only opportunity to collect information about issues of major importance to human welfare. For many kinds of tropical organisms, the specimens that are collected in these few years may be the only samples available for future study, providing posterity with an indication of how rich the earth was before people claimed priority for living space.

Insufficient knowledge results from two main factors. First, species and ecosystems both on land and in the sea are so diverse that major efforts are required to collect systematic information. It has taken about 230 years to describe the world's current 1.4 million species; based on the figure of 10 million species currently alive in the world, which we are using for convenience (and realizing that the figure could be far higher), it would take 1,643 years to describe the world's species if we continue at this same rate. Second, basic taxonomic research is no longer fashionable and few major research institutions — especially in the tropics — are engaged in describing the diversity of species and ecosystems. It seems self-evident that increasing knowledge about the kind and variety of organisms that inhabit the earth — and the ways that these organisms relate to each other and to humans — must be a foundation of conservation action. Therefore, a major effort is required to:

- document the wealth of the world's species of plants and animals, involving museums, botanical gardens, universities, and research stations — this work needs to assess the genetic diversity of the population of especially important species, as well as the size and nature of the gene pool represented there;
- carry out ecological fieldwork to show how the various pieces fit together, discover the population dynamics of species of particular concern, assess the effects of fragmentation of natural habitats, and determine what management steps are required to enable ecosystems to flourish with their full complements of species;
- develop new mechanisms for *ex situ* conservation, including both captive propagation and eventual release into "natural" ecosystems;
- monitor the changes in ecosystem diversity and function as the influences of humans become more pervasive, including climate change, deforestation, and various forms of pollution;
- assess the ecological differences between relatively large but minimally disturbed ecosystems and ecosystems that have been heavily affected by humans, as a basis for enhancing productivity and restoring degraded ecosystems to a more productive state;
- carry out research in the social sciences to determine how local people manage their resources, how changes in resource availability and land use affect human behavior, and how people decide how to use their biological resources.

Such basic inventory and fundamental research work should be carried out simultaneously with field action, with the two forms of activity reinforcing each other. More detailed discussions of research priorities are available in Committee on Research Priorities in Tropical Biology (NAS, 1980) and in Soulé and Kohm (1989).

Types of Information Needed

Scientific knowledge

As more taxonomic and survey work is done, scientific knowledge grows but so does awareness of ignorance; the more new discoveries are made, the more new gaps are found in the data. Action — surveys, inventory, taxonomy, and

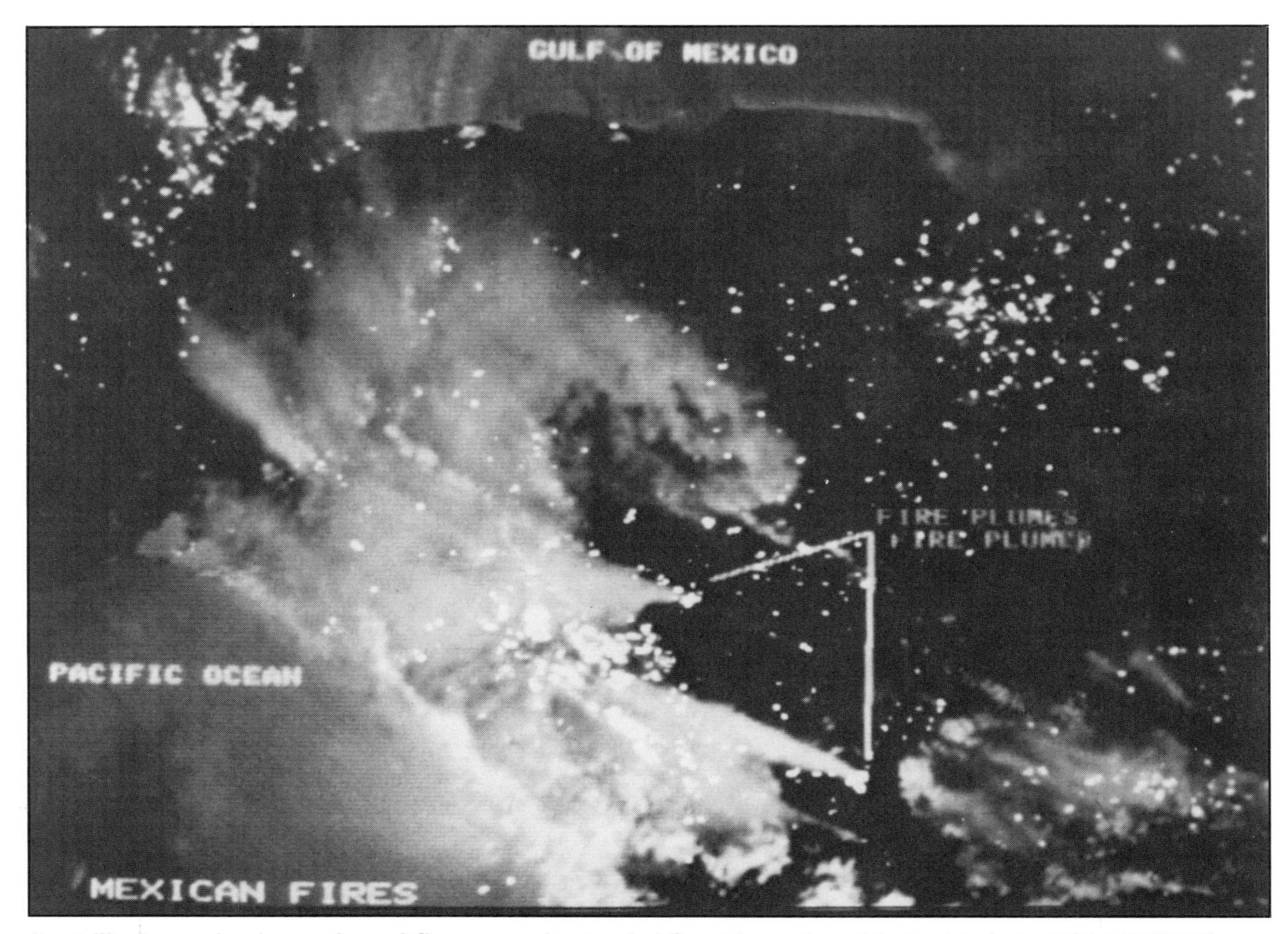

A satellite image showing smoke and fires consuming tropical forest in southern Mexico (photo by NOAA/NESDIS).

analysis — is required at both national and international levels, and especially in the most species-rich habitats.

This is a daunting challenge, but it must be met. Far less than 5 percent of the species in the tropics (and the figure could be an order of magnitude less) have been described as yet, and the number of taxonomists able to identify tropical organisms is shrinking as the urgency of the problem becomes greater. The Committee on Research Priorities in Tropical Biology (NAS, 1980) concluded that at least a five-fold increase in the number of systematists is necessary to deal with a significant proportion of the estimated diversity while it is still available for study. Approaches to addressing this problem could include:

- closer cooperation between major taxonomic institutions (museums and universities) in temperate and tropical countries;
- better use of the Biosphere Reserve network, which includes some 260 areas managed at least partly to enable research to be carried out on basic questions of conservation (Unesco, 1985); and
- training a large number of parataxonomists to collect and document specimens, and to make initial identifications, along the lines that are currently being adopted in Costa Rica (Janzen, 1988).

A major effort is required to establish tropical research centers, train the personnel to carry out the required research, and provide the incentives necessary to give the work the high prestige it deserves. Such research centers are describing the basis of long-term human welfare, and warrant major investments from society. Without the knowledge that comes from ecological field research, it will be impossible to develop the ecologically sound systems of resource management required to support the people now living in the tropics, to say nothing of improving their condition in the future.

For all of these reasons, very greatly increased efforts are required to enable the taxonomic institutions to form close symbiotic relationships with the conservation agencies, which in turn need to work far more closely with the more applied fields of plant and animal genetic resources for agriculture and forestry.

On a global scale, modern technology is available but is insufficiently used. Given the importance to the world of knowing how much tropical forest remains and given the capability of carrying out a reasonably accurate inventory by satellite imagery, it is of considerable concern that the world is still relying on highly questionable data on tropical forest coverage produced by FAO (1981) on the basis of information from the late 1970s.

However, availability of information does not mean just carrying out surveys and publishing results. The information must be used. This may involve creating a network of centers at local, national, and international levels that know what information is where and that can tap into available information and present it to planners and decision-makers in useful forms.

New technology makes data management more productive than ever before (Box 12). The development of computerized Geographic Information Systems (GIS) has greatly simplified the preparation of integrated biodiversity conservation strategies. Recent advances have made it possible to store and analyze multiple layers of geographic data on relatively inexpensive micro-computer systems. Computers can respond to search commands to identify gaps in the system of protected areas from a variety of perspectives, or to demonstrate how various management or development options are likely to affect environmentally sensitive areas (Chalk *et al.*, 1984), or even to define the relative sensitivity of areas.

The new technology can make it easier to determine which species and communities are currently protected and to identify alternative conservation strategies to achieve various levels of protection of other areas of high biological diversity. It is therefore not surprising that development agencies are increasingly turning to GIS as important planning tools (Bailey and Hogg, 1986).

Local Knowledge

One extremely rich source of information about resource management is usually ignored by decision-makers and even scientists: the knowledge of local people whose livelihoods depend on their management of biological resources (McNeely and Pitt, 1984; Geertz, 1983; Warren *et al.*, 1989).

Rural communities often have profound and detailed knowledge of the ecosystems and species with which they are in contact and have developed effective ways of ensuring they are used sustainably (IUCN, 1980), so information should be collected — especially in tropical countries — about the use that indigenous peoples make of biological resources, and the management approaches they have developed.

Since local cooperation is essential for the long-term success of conservation efforts, it is usually advisable to undertake a socio-economic survey of the communities affected by projects that involve controlling use of biological resources in order to determine what resources are used, how they are harvested, the degree of awareness about controlling regulations, and possible alternative sources of income.

Such surveys can also provide the necessary raw material for determining the sorts of incentives required to bring about the desired changes in behavior, as well as the best means of providing incentives and ensuring that they are perceived as fair, equitable, and fairly earned. Information collected might cover the ethnic diversity of the communities and their social structure, including the traditional location and proximity of householder and kin groups for ritual, labor exchange, and other important community activities. This in-

Box 12: Remote Sensing.

The technology that is used in remote sensing was developed as early as 1972, but only recently became a tool in conservation. In remote sensing, maps are created from numerical data collected by satellites that measure the amount of reflected energy from different land-cover types. These data are then translated into an image by assigning visible colors to the numerical values. Images generated from this procedure reveal distinctions in habitats, such as forests, savannas, rivers, roads, cities, etc. Depending upon what a technician may want to show, different land types may be highlighted according to the chosen color scheme.

Satellite data can supply information to a Geographical Information System (GIS), which in combination with other data sources can produce an analysis of land-use and habitat modification. The information provided by a map produced by remote sensing is verified in the field by comparing the image with on-the-ground observations — a procedure known as "ground truthing." The final image reveals important information on human activity and the natural composition of the area under study. This provides conservationists with a picture of what is happening in a given area, and improves their ability to make sound management decisions. New data collected periodically — possible with the satellite's ability to make frequent passes over an area — can also generate an image of change in large areas over time.

In the tropics, remote sensing is being applied to the challenges of conserving biodiversity. For example, deforestation can be monitored over areas too large to be monitored by ground techniques. Satellite images are also valuable tools in local campaigns to preserve natural areas, supplying individuals with a visual proof of environmental disruption, and a map of natural resources. However, because conservation has only recently focused on the tropics special obstacles have arisen, such as the difficulty of distinguishing separate elements in a rich, diverse habitat like tropical rain forest. But the spectral characteristics of satellite images are improving, and remote sensing technology is becoming increasingly sensitive in distinguishing different vegetation types. Moreover, remote sensing is more cost-effective than on-the-ground mapping. The former costs about $5 for each square kilometer, where field methods cost about $80 for an area of the same size (Conservation International, unpublished data, 1989). Remote sensing, and other technologies that comprise integrated information systems are continually improving, and will undoubtedly make important contributions to conservation methods.

formation can provide managers of biological resources with the necessary insights into the needs and desires of the local people, and can avoid misunderstandings and disruptions when implementing incentive systems.

These efforts at assessing the relevant biological and human resources will help governments to recognize the consequences of their development activities on the biological resources of the nation, and help ensure that external effects of development projects on biological diversity are clearly identified.

Information Needs At National And Local Levels

To develop informed policies on resource depletion rates, rates of sustainable yield, national accounting systems, and land-use planning, governments require reliable information on the current status and trends of biological resources in their countries. A careful analysis of existing information is therefore required before any significant decisions are made that might affect those resources.

Governments need to determine their own needs for information that would improve conservation at the national level. They might consider three major questions:

- What type of information is needed to support changes in policies (e.g., information on the economic importance of biological resources, or on how traditional peoples have depended on biological resources)?
- What information is needed to help identify sites important for conservation (e.g., ecologically sensitive areas), and to assign priorities for investment?
- What information is needed to manage these sites (e.g., information on resource distribution and use within the area, or on the social and economic needs of the communities living in and around the sites)?

All governments should build the capacity to assess the status, trends, and utility of their biological resources as an essential foundation for planning and implementing development action. This capacity, which should build on existing knowledge and form a permanent part of the management enterprise, should include:

- national compilations of the flora and fauna contained within the nation, in addition to the more usual assessment of stocks of timber, fish, and minerals;
- institutionalized biological surveys, perhaps carried out by university departments of biology, to determine what

A woman drying fish in Sarawak, Malaysia (photo by R.A. Mittermeier).

species occur where and in what numbers, or perhaps involving innovative approaches including "barefoot taxonomists" (D. Janzen, University of Pennsylvania, pers. comm., 1988) and local screening of organisms for useful chemicals (T. Eisner, in press);
- a national system for monitoring the status and trends of biological resources, linked to international systems such as UNEP's Global Environmental Monitoring System and the World Conservation Monitoring Centre; and
- regular publication of the available information on status and trends of biological resources, and the various forces affecting these trends.

All natural resources are managed in some way, whether through protection, production, or consumption. In order to manage resources in a sustainable manner, it is essential that the effects of different management practices are monitored and understood, and that any lessons learned are applied in future management. The types of questions that need answering include:
- What type of management is being carried out in each area, and to achieve what objectives?
- What staff resources and infrastructure does this involve?
- What is the effect of management on the natural resources and their value?
- What further information is required to improve management of these sites and resources?

An important part of assessing biological resources is estimating the economic contribution that they make to the national economy. This requires:
- developing methodologies for assigning values to non-marketed biological resources, appropriate to the needs of the country;
- ensuring that national accounting systems make explicit the tradeoffs and value judgments regarding impacts on biological resources that may not be measured in monetary terms;
- conducting research on methodologies for assessing the cross-sectoral impacts of resource utilization;
- documenting the past, present, and potential value of wild species, their products and derivatives to human societies for medicinal, nutritional, and other socially valuable uses;
- collecting information on the physical properties of resources in specific environments and for specific uses; and
- evaluating the true economic productivity of various ecosystems.

Managing the flow of information

Significant differences are apparent between countries and regions, and even between sectors within countries, in institutional arrangements, technological development, and availability of data. Information is often scattered, and sometimes difficult to obtain, and some is not even directly available in the country from which it was collected. In some countries, good databases exist for certain regions or sectors, but are conspicuously absent in others.

In a detailed study of the users of information regarding resource management in Botswana, Zambia, and Zimbabwe, Rennie and Convis (1989) determined that the main need was not to obtain information from a single source, but rather to identify the sources, to obtain the information efficiently and in useable form, and to integrate this information into other work that is being carried out. The primary frustrations of the users were difficulties in finding the information, obtaining it, and integrating it into a planning or management framework.

Another potential problem is variability in data quality, age, and presentation, and in the ways data sets (on computer or otherwise) are maintained. This affects two main sets of users: first, where information is used to plan management action, it needs to be provided to planners and policymakers in a standard and useable format; second, the subsets of information provided to other information centers (such as national agencies) need to be comparable with information for other regions or sectors.

Planners and policymakers should not be expected to deal with incoming information in a wide range of formats and with a wide range of variability when careful planning and coordination by those managing information could ensure a more focused and coordinated approach. Standard methods of presenting information should be devised, indicating its accuracy and currency, based on the levels of information available and on the technology used.

Regarding compatability of information between databases (or information centers), clearly standardization is most desirable and most achieveable at the point of data exchange (BGCS, 1987). Information centers cannot be expected to use methods, classification systems, or technology (software of hardware) prescribed by others, but efforts can be made to ensure that what they hold can be converted into standard (and properly documented) data sets that can then be of wider utility.

Key issues in improvement of information flow (and hence the better use of available information in management decisions) are therefore:
- development of databases on what information is available and where (probably including extensive bibliographic servicing);
- development of standard methods of presentation (which covers both user involvement in information development and the education of users); and
- development of standard transfer formats.

Local Information Management

Most day-to-day resource management decisions are made at the local level, so information must be managed there to provide managers and planners with what they require. Provincial planners thus need their information to be integrated on a geographical basis, which will enable local land-use plans to be prepared, appropriate permits to be issued, and planning restrictions to be enforced. At this level, site-specific information is highly important, and it is crucial that this information be integrated with material from a range of sectors so that appropriate decisions can be made. Efficient information management is therefore essential at the local level (be this a provincial government or a national park), and the data management systems must be designed with the full involvement of those who will use them. Constant consultation with the users will ensure that the technical and institutional design of the database will enable it to fulfill the function for which it is created, provide information to those who need it, and provide the information in a form that can be used (Rennie and Convis, 1989).

However, biological resources are also managed in a more general way at the regional, national, and international level; consequently, information on resources within the country as a whole is also required and this involves interactions with other information centers and wide exchange of information. But resource management at the local level requires far more detail than at the national level; the latter is usually only a subset of the former.

National Conservation Information Management

While the most detailed information will be held by the sectoral and local planning agencies, national conservation databases can maintain a detailed overview of what natural resources data sets are held where, can maintain a single database on the more important data aggregated at national level to provide a general overview, can identify which sites are important at the national level, can interact with other national or international databases, and can indicate where major gaps in national data need to be filled. A national database need not necessarily be one central office (even though this might be the ideal in some cases), and might be distributed between agencies, building on existing initiatives.

The main consideration in establishing a national data center is to identify, support, and develop a national institution (or institutions) already active in the data management business. This bottom-up approach must involve from the outset the likely data users, especially the government agencies, for unless they participate in the planning of the center, its outputs may be politically unacceptable.

With the widespread availability of micro-computers and the growing sophistication and ease of use of software, national conservation databases are becoming more prevalent. The Nature Conservancy (TNC) has long been a leader in establishing such data centers, especially in Latin America, working in close collaboration with various national institutions. IUCN is working with the governments of Saudi Arabia, Bangladesh, and the Sahelian and SADCC countries of Africa to develop national databases. WWF has also supported work in establishing databases, notably in Thailand, Brazil, and China. UNEP has helped establish an environmental database for Uganda and is collaborating with the Costa Rican Conservation Data Center.

Perhaps the most effective efforts to date to help ensure that conservation and development decisions are based on good information, particularly about species, are the national "conservation data centers" (CDCs) started in Peru, Bolivia, Colombia, Costa Rica, Panama, Netherlands Antilles, Puerto Rico, Venezuela, Belize, and Paraguay. At these centers, financed initially with funds from The Nature Conservancy (U.S.A.) and Conservation International, a small staff of biologists and ecologists continually assess the current status of species and ecosystems in the country, thus putting a sharper edge on recommendations for conservation. New and more detailed vegetation maps are being produced to analyze the protection of vegetation types in established protected areas at an ever finer level of resolution.

The next step is to transfer this information regularly to government departments and development agencies so that it can be used in planning natural resources development (Jenkins, 1985). But the data remain just a tool, and the CDCs are also building the human expertise for turning these data into information that can be applied to solving management problems. In the long run, CDCs may become the most authoritative and up-to-date centers of conservation information at the national level, making data available for planning conservation systems, monitoring the status of wildlife and critical ecosystems, and reviewing the environmental impacts of development projects.

To have sufficient impact on resource decisions, the national data center should integrate not only conservation data but also data covering the whole spectrum of natural resources management. The conservation data must be integrated with (or capable of being integrated with) agricultural, forestry, fisheries, land-use, soil, climate, human settlement, and other data sets if they are to be of practical value to the resource planner. The national database being developed in China (Box 13) illustrates such a system. The outputs must also be produced in mapped form, with GIS analysis producing details such as ecologically sensitive areas, areas suitable for sustainable use of their resources, and particularly important areas for the conservation of biological diversity. Such GIS software can now be run on personal computers, and does not require a level of training inappropriate for use in most developing countries. An outstanding example of the application of GIS is the assessment produced for Costa Rica (Backus *et al.*, 1988).

The importance of data exchange between databases was mentioned above, along with the need for consistency. While the ideal way to achieve this might be for all databases to use one design and set of programs, this would not work in reality, and existing initiatives need to be developed or supported rather than outside solutions imposed. While the development of data transfer formats and accepted standards is one answer to ensuring consistency, provision of data management tools can help ensure better communication. These need not be complete database programs, but tools to assist in management of parts of a database, which can be added to existing programs and facilities.

Box 13: A National Database for China.

The database proposed by WWF for China is particularly relevant as an illustration of how national databases might be structured, as it has the primary purpose of conserving biological diversity. The project will involve collating a number of different types of information to produce six primary resource classifications for each of the 27 provinces and autonomous regions of China. It will include:

- A mapped land classification of species and genetic value, to be based on consideration of species richness, levels of local endemism and degree of threat, species of particular economic (medicinal) importance, and Pleistocene refugia.
- A mapped land classification of habitat threat, based on consideration of the rarity, rate of loss, and degree of protection of the country's various natural vegetation types and its freshwater and coastal habitats.
- A map showing the location of all existing and proposed protected areas scored for their natural importance, and a map of sites of high scenic and landscape value, rated for relative importance.
- A mapped watershed classification system, based on rainfall intensity, slope, soil type, and levels of hazard (potential for flooding, dependence of irrigation system, reservoirs, fisheries, etc).
- A mapped land classification of relative human pressure on the environment, based on population pressure for agricultural expansion (current density x regional growth x potential for agricultural expansion), forestry pressures (standing crop x value x accessibility), and economic pressures (economic incentive (e.g., special economic zones) x potential (mineral and energy resources + access)).

These types of information will be brought together by overlaying maps using a Geographic Information System to give two intermediate maps for each of the provinces and autonomous regions of China. The first will be a classification summarizing "genepool sensitivity" (combining the first three maps listed above), while the second will be a classification of physical environmental sensitivity (combining the last two maps above). These two maps will finally be combined to give an overall environmental classification, indicating areas where development should not be permitted to further disturb the environment and areas where development should only be permitted if various environmental safeguards are taken. These areas will be coded for the type of safeguards needed (e.g., water catchment protection, pollution controls, wildlife issues, etc).

These data are handled using software developed for WWF for use on personal computers. Regional databases of this type will be set up in each of the seven biogeographic divisions of China. The data will be collated together with the data on the physical environment in Beijing, where the Commission for Integrated Survey has a mainframe computer with GIS capability.

Information Management at the International Level

In addition to the need for information at the national level, a number of users — particularly international agencies — require a global center for information on biodiversity, with a global overview database on biodiversity. An international database is also required for some national applications, such as dealing with migratory species or animal trade.

The World Conservation Monitoring Centre (WCMC) is the primary clearinghouse for data on species and ecosystems. WCMC is a joint venture between the three partners in the World Conservation Strategy (IUCN, UNEP, and WWF) established to provide a central repository of information on the world's biological diversity. In trying to keep the data management exercise within reasonable bounds, WCMC has focused on assessing:

- the status and distribution of species of conservation concern (some 18,000 animal species are held in the database, of which some 4,500 are listed as globally threatened (IUCN, 1988a), together with some 52,000 species of plants, of which 19,500 are categorized as threatened (Davis *et al.*, 1986));
- critical sites for biological diversity (particularly tropical forest sites, but also wetlands and coral reefs, (IUCN/UNEP 1988));
- the protected areas of the world (details of some 17,000 sites are now included in the database, along with more detailed information on all protected area systems and many individual sites); and
- international trade in threatened species and their derivative products (some 9 million trade transaction records are held on CITES-listed species).

These data are collected from a variety of sources — published literature, unpublished reports and government documents, conservation organizations, and a wide network of contacts and correspondents throughout the developing world. Protected areas and species data are also provided by members of IUCN's Commission on National Parks and Protected Areas and the Species Survival Commission.

WCMC Data on Species

While the databases can be tapped to provide a wide range of integrated outputs, the best known products from the species database are the Red Data Books, which provide the standard for assessing the status of threatened species (including extinct, endangered, vulnerable, and rare). Data on such species are held in a computerized database at WCMC, in the data fields listed in Box 14; information held in text files is listed in Box 15. Additional information that needs to be added so that conservation needs can be assessed on a sound biological basis includes known range, breeding season, age at first reproduction, litter or clutch size, mean longevity, and primary and secondary food.

Box 14: WCMC Data Fields on Species.

1. Higher taxonomic names (family)
2. Taxon name (genus, species)
3. ISIS code
4. Authority for the taxon name
5. Common name
6. Basic habitat (under development)
7. IUCN conservation category
8. Wild population size
9. Wild population trend
10. Captivity status
11. Captive population size
12. Exploitation by man
13. Threats to the taxon
14. Text file identification
15. CITES data
16. Status of distribution info
17. Country where taxon occurs
18. Occurrence in protected areas
19. Introduction status
20. Area population size, trend
21. Legal coverage in area
22. Geographical qualifier

Box 15: WCMC Text Files on Species.

1. Summary
2. Common and scientific name
3. Authority for name
4. Order and Family
5. World IUCN category
6. Distribution
7. Population
8. Habitat and Ecology
9. Threats to Survival
10. Conservation measures taken
11. Conservation measures proposed
12. Captive breeding
13. References
14. Summary of Information

On the basis of such information, IUCN publishes every two years the *Red List of Threatened Animals* (e.g., IUCN, 1988a), as well as geographically or taxonomically based Red Data Books (see bibliography for examples). The importance of this quantitative approach to species conservation is shown by the great number of national and international Red Data Books that have been prepared using the IUCN threat categories as their model; Burton (1984) listed some 154 Red Data Books on animal species alone published until the end of 1982. Much of this work has been sponsored by WWF, UNEP, and other international conservation organizations.

The conservation status of plants has proved more difficult to assess, partly because there are far more of them and partly because they are much less known; botanists estimate that 10-20,000 flowering plants are still unknown to science (A. Gentry, pers. comm.) and many countries do not have even a reasonably complete description of their respective vascular floras (Davis *et al.*, 1986). The lack of a standard nomenclature and the rate of taxonomic change create problems for maintaining a global database. Even so, WCMC's Threatened Plants Unit now has documented 18,000 globally threatened plant species, and national experts have prepared numerous Plant Red Data Books.

About two-thirds of the world's plants are found in the tropics, but assessing the conservation status of any single species is often extremely difficult because of the lack of data. Instead, WCMC has decided that its resources would be more effectively used to promote plant conservation by identifying a relatively limited number of key sites of high plant diversity that, if protected, would ensure the survival of a large percentage of the world's plant species. In addition, since animal diversity tends to reflect plant diversity, protection of these sites will also conserve a high percentage of animal diversity. While the Centers of Plant Diversity program is just beginning, approaching conservation of plants through the mechanisms of sites instead of species has a number of advantages. Botanists are well able to make an assessment of plant diversity in a country or region without knowing about each species; certain soil or geological substrata — such as limestone or ultra-basic rocks — tend to have distinct floras; mountains with rich soils (such as some volcanoes) often have highly diverse floras spread over their altitudinal gradients, and older mountains often contain relict floras in their higher elevations.

Climatic changes are likely to affect natural vegetation much more slowly than they affect agricultural crops, so areas established to protect high plant diversity could serve as "Holocene refugia," which will provide gene centers containing plants that could be used to repopulate abandoned lands. Furthermore, identifying important conservation areas on the basis of plants will enable an independent cross-check to be made on areas identified on the basis of birds or mammals, and will facilitate greater cooperation between zoologists and botanists.

WCMC Data on Protected Habitats

The Protected Areas Data Unit of WCMC works closely with the IUCN Commission on National Parks and Protected Areas in managing an information center on protected habitats. To assist its work, PADU maintains basic information on all major protected areas of the world (see Box 16 for a list of the data fields held on the database, and Box 17 for information held in text files).

Box 16: WCMC Data Fields on Protected Areas.

1. Name of protected area
2. Country it lies within
3. Size of area
4. Year of establishment
5. IUCN management category
6. Category within country
7. Biogeographic code
8. Latitude and longitude
9. Altitude
10. Text file identification
11. Types of maps on file
12. If management plan on file

Box 17: WCMC Text Files on Protected Areas.

1. Name of protected area
2. Country
3. Management category
4. Biogeographical province
5. Legal protection
6. Date, history of establishment
7. Geographical location
8. Altitude
9. Area
10. Land tenure
11. Cultural heritage
12. Local human population
13. Physical features
14. Vegetation
15. Fauna
16. Management problems
17. Conservation management
18. Visitors and visitor facilities
19. Scientific research
20. Principal References
21. Staff and Budget
22. Local administration

On the basis of this information, IUCN publishes periodic editions of the *United Nations List of Protected Areas* (which lists the areas, their management category, their size, and their biogeographic province, plus lists of World Heritage Sites, Biosphere Reserves, and Wetlands of International Importance). IUCN and WCMC, together with a number of other agencies, have also produced a number of more detailed directories of protected areas. These include directories of protected areas of given regions, such as the Afrotropical Realm (IUCN, 1987a), of all Biosphere Reserves (IUCN/Unesco, 1987), and of all sites listed under an international convention, such as the Ramsar Convention (IUCN, 1986). Virtually any other configuration is also possible.

The information for these directories is collected from the management agencies in each country, often through the mechanism of meetings held by CNPPA in the respective regions, and from a range of other contacts familiar with the protected area systems. The information is kept up to date through correspondence, reviews of published and unpublished literature, and regular meetings in the countries or regions involved. The process of maintaining this information helps encourage national protected area management authorities to maintain their own databases, and facilitates the assessment of international conservation priorities.

WCMC is also working to identify critical sites for the conservation of biological diversity outside the protected areas network. This program commenced with the Protected Areas Reviews (IUCN/UNEP, 1986a, b, c), which on a continental basis analyzed the representation of the main vegetation types within the protected areas system. This work has now been extended by the introduction of a GIS that enables geo-referenced biological diversity data to be integrated with other biogeographical and ecological data sets to produce a variety of mapped outputs of practical value for the assessment and management of biological resources. The technique is now being used by WCMC to digitize the protected areas network and then to overlay this on the distribution of main tropical forest types to show the adequacy of their protection on a global and regional level.

A study recently completed in seven central African countries — Cameroon, Central African Republic, Congo, Gabon, Equatorial Guinea, São Tomé and Principe, and Zaire — has produced a database containing information on 104 sites known to be of critical importance for forest conservation; 41 already have some sort of protected status (IUCN, 1989a). It is likely that as more forest is converted to other uses or is brought under intensive management, and as knowledge of biological resources improves, managers will be increasingly preoccupied with the problems of conserving smaller forest sites that are subject to varying degrees of extractive use. WCMC's "critical sites database" is designed to monitor such sites. Results of initial inventories will be made available in a series of Tropical Forest Resource Atlases, the first of which will be published in 1990, covering Asia. The digitized GIS resulting from this work will be constantly updated, and national and regional critical sites maps will be published periodically.

Finally, IUCN has just published a three-volume compilation of coral reefs of the world, which identifies the most important coral reefs for protection and includes detailed information on all important corals (IUCN/UNEP, 1988).

WCMC Data on Wildlife Trade

The excessive harvesting of wildlife species for commercial gain is one of the main threats to species diversity. The international treaty established to regulate this trade is the Convention on the International Trade in Endangered Species (CITES), to which some 96 countries are now Parties. WCMC maintains the database that includes all records of international trade in species listed in the CITES Appendices involving a Party state. The Centre conducts analyses of significant trade in selected listed species (IUCN, 1988b), and monitors the trade in unlisted species to recommend their inclusion. It also carries out major surveys of the status of species of conservation interest and the threat imposed by illegal trade. This monitoring of the wildlife trade and its impact is essential for the effective operation of CITES.

WCMC also manages the ivory trade database that monitors the transactions of African elephant ivory, regulated under the CITES elephant quota system. Because of the threat created by the illegal trade in ivory, the African elephant is the focus of a major conservation initiative backed by governments and NGOs, including IUCN. To support this program, WCMC is developing an elephant database that will link the ivory trade statistics to the current elephant populations on a country-by-country basis. Other trade-related organizations with which WCMC works include the TRAFFIC Network, the international coordination of which is managed by the Centre, and the International Tropical Timber Organization (ITTO) with which the Centre is undertaking a major analysis of tropical timber species in trade.

The economic benefits derived from wildlife utilization can be a powerful argument to promote conservation. WCMC is now developing a database on sustainable wildlife utilization programs that can be analyzed to look for common features of success or failure. The Centre is also active in the investigation of sectoral utilization programs, such as crocodile ranching, butterfly farming, or game meat production, and in assessing their significance for conservation.

Information on Legislation

Much law — both international and national — exists today throughout the world on the subject of biological diversity and particularly on species of plants and animals. Development of national law has been stimulated by the adoption of many multilateral, bilateral, and regional conventions (Annex 3), which in turn have evolved as States have become increasingly aware of the need to cooperate in maintaining biological diversity.

Scientists, administrators, and lawyers throughout the world need to be able to find at a glance information on the law relating to species in other regions or States, thereby enabling national laws to be enforced more effectively. Convention secretariats need to obtain information on implementation of conventions through the development of national law, and researchers who map and monitor the status of species require information on the applicable law in order to be able to make recommendations for future conservation measures. Finally, information on species law is required by legislators who may wish to see the precise steps that have been taken elsewhere before deciding on measures to be taken in their own country or region.

For all these reasons, an index to species mentioned in legislation is being compiled by IUCN's Environmental Law Centre (ELC), located in Bonn, Federal Republic of Germany. Over 1,500 international and national legal instruments form part of the index, and 10,000 taxa of mammals, birds, reptiles, amphibians, fish, and invertebrates are included in the database. Access to the data can be gained by the taxa concerned, the type of legal protection provided, the jurisdiction, or any combination of these three.

The IUCN/ELC index to the species data bank provides an overview of the legal status of species and higher taxa in legislation, provided as accurately and concisely as possible through symbols, notes, and abstracts. Such an overview has its limitations, and the index is not a substitute for the full texts; copies of the texts of all legislation mentioned in the index are available from ELC (IUCN/ELC, 1984).

Conclusions

Government agencies, local communities, and conservation organizations all need information to enable them to manage their biological resources more effectively. Information tools that can help meet this need include basic descriptions of fauna and flora, practical handbooks for field identification, rapid inventory techniques, and basic computer programs for use with micro-computers.

The information needs in the tropics are particularly important, because these areas hold the majority of the world's biological diversity and they are losing species at rates that far exceed the world's capacity to record them. Highest priority for basic inventory work should be given to the sites of greatest diversity and endemism coupled with the greatest threat, for the information contained by the species held in these areas could disappear before humanity even knows what it is losing (see Chapter VI).

Development assistance agencies should provide support for national efforts to establish local, sectoral, and national information management systems, through demonstrating methodologies, providing training opportunities for taxonomists and biologists, and subsidizing publication of status reports. Universities, research institutions, and NGOs need to be strengthened so that they can help governments assess their biological resources. Closer working relationships should be established between museums and other taxonomic-oriented institutions and those concerned with conservation of biological diversity.

Information centers should be developed at appropriate levels to ensure that the information is available where it is needed, whether in a single area (such as a national park), in a country or region, or at the international level. Land-use decisions, which can radically affect local biological resources, are often made at the regional or provincial level; this must be reflected in the structuring of the national information system to ensure that regional resource planners have access to regional data. In particular, national databases managing information on the resources of the country should be implemented as part of a full National Conservation Strategy.

All agencies managing information on natural resources at the supranational level should work together to ensure a minimum of duplication of effort, and to ensure that national and local sources are not asked by more than one agency for the same information. The agencies should also work together to ensure that information from the various sectors can be brought effectively together, in particular through the UNEP GEMS and GRID programs.

Agencies involved at the international level in natural resource database design and function should collaborate to assist in the establishment and development of local and national databases, building on the needs and expectations of the local users, and building on the experience gained and initiatives undertaken in the country already. Networks of such information centers and databases should be developed to improve standardization and to facilitate exchange of experience and methodologies. The decision of WCMC and The Nature Conservancy to pool their expertise for the development of database systems appropriate for developing countries is an encouraging start.

The international agencies with an interest in the conservation of biological resources, including development aid agencies, governments, the UN system, and various NGOs, should collaborate to prepare global overviews on the status and management of biological resources. These overviews can be an incentive to action by these agencies, stimulating greatly increased flows of funds and other kinds of support. The objective of their collaboration must be to ensure that land-use decisions affecting biological resources are based upon reliable, useable information: ignorance must no longer be an excuse for environmental degradation.

Following page: An opening in the canopy of a rain forest in Costa Rica (photo by C. Isenhart/J. Birmingham).

CHAPTER VI
ESTABLISHING PRIORITIES FOR CONSERVING BIOLOGICAL DIVERSITY

Problems and opportunities far exceed the resources available for conserving biological diversity. What methods are available for deciding what to do first? The question will be answered in different ways for achieving different objectives.

When governments approved the World Charter for Nature in the United Nations in 1982, they agreed that all species and habitats should be safeguarded to the extent that it is technically, economically, and politically feasible. But resources for conservation are always limited, so efforts spent in deciding what to do first are usually well repaid in savings of time, finances, and personnel.

Determining priorities is a complex task. The genetic landscape is constantly changing through evolutionary processes, and the world contains more variability than can be expected to be protected by explicit conservation programs. Further, the capacity of governments or private organizations to deal with environmental problems is limited and many urgent demands compete for their attention. So governments, international organizations, and conservation agencies seeking to conserve biological diversity must be selective, and ask which species and habitats most merit a public involvement in protective measures. In the case of the species and habitats not given high priority for such treatment, governments should enact national laws and public policies that encourage individual, community, or corporate responsibility as appropriate.

Further, some habitats may be well conserved with virtually no investment by governments, merely by enabling local communities to continue to manage their resources in a sustainable way. ''Benign neglect'' may be the best strategy in such cases, though this can involve an opportunity cost, such as when a government would earn foreign exchange from selling a timber concession in an area occupied by a tribal group that harvests non-timber products from the forest.

No generally accepted scheme exists for establishing priorities for the conservation of biological diversity, nor is it either possible or advisable for such a scheme to be devised. Different organizations and institutions can be expected to have different ways of establishing priorities because of their differing goals.

The various methods of establishing priorities that have been considered suggest different types of conservation action and will result in the conservation of different subsets of the world's biological resources. Each system has its own strengths and weaknesses, with the major point of difference being the objective for which the system was devised. Chapter V discussed the information that is required as a basis for determining objectives and priorities, and Chapter VII will discuss how to apply the priorities. This chapter discusses several approaches to establishing priorities, concluding with suggestions on how to determine priorities so that resource allocations can be based on credible criteria.

Establishing Priorities within a Nation

The biological resources contained within each nation need to be managed in ways that provide sustainable benefits. With rising populations, this may require that some natural habitats are converted into agricultural systems, forest plantations, and other uses that are heavily affected by people. But the natural value of some areas is so significant that they need to be converted with great care, or even left in their natural state.

Areas of outstanding natural value for hydrological, geological, scenic, wildlife, or vegetation reasons should be converted with great care or not at all, and can be termed Ecologically Sensitive Areas (ESAs). (A similar concept — Environmentally Sensitive Areas — has been enshrined in legislation in the United Kingdom and enables local farmers access to grant aid to conduct traditional farming methods that are more favorable to conservation than the modern intensive approach is.) They may contain unique features and processes, such as large aquifers and lakes, cave systems, headwaters, steep or unstable slopes, rare plants or animals and their habitats, or important breeding areas for wildlife. Some ESAs are natural, while others have been significantly altered by certain human activities. Nations with tribal peoples may wish to include tribal homelands as a category within the ESA framework in order to ensure that the relationship between culture and nature already discussed is appropriately managed. In terms of management, some ESAs will prosper with minimal inputs while others will require intensive management to restore or maintain their natural values.

Habitats can be considered ecologically sensitive areas if they:

- provide protection of steep slopes, especially in watershed areas, against erosion;
- support important natural vegetation on soils of inherently low productivity that would yield little of value to human communities if transformed;
- regulate and purify water flow (as valley forests and wetlands often do);
- provide conditions essential for the perpetuation of species of medicinal and genetic conservation value;
- maintain conditions vital for the perpetuation of species that enhance the attractiveness of the landscape or the viability of protected areas; or
- provide critical habitat that threatened species use for breeding, feeding, or staging.

The last two points traditionally have been used to select areas protected for conservation purposes. Current networks of protected areas will seldom address all criteria for ESAs however, and such situations as those covered by the first three points will require additional approaches to management. Because ESAs are often proposed for conversion to other purposes, or are subject to ineffective management, criteria are required in each nation to guide decisions on whether an ESA can be converted to alternative uses, and, if so, under what conditions. Such criteria will help ensure that ESAs contribute to the development process in a carefully considered manner, and are not needlessly destroyed through ignorance or inadvertence.

The productivity of many ESAs has already been reduced through inappropriate uses, and many others face very real threats. People have shown the capacity to convert almost any piece of natural habitat into agricultural land that can produce a crop for at least a few growing seasons. Indeed, Spencer (1966), in his classic work on shifting cultivation in Southeast Asia, concludes that virtually all of mainland tropical Asia has been cleared at one time or another by shifting cultivators over the past 10,000 years. But areas that are inappropriate for such use quickly degrade into wastelands, such as the great expanses of *Imperata* grasslands found in much of tropical Asia. Worse, inappropriate conversions of watersheds, for example by illegal logging, can contribute to very high human costs downstream through floods, erosion, siltation, and other external factors.

A useful scale for guiding decision-making is the ecosystem, a community of organisms interacting with the local living and non-living elements of the environment and forming a system in which life-sustaining processes are maintained. The functioning of an ecosystem involves the accumulation, circulation, and transformation of matter and energy through such biological processes as photosynthesis and decomposition. The ecological processes often work though the means of water, which provides a medium of transfer and storage of energy and materials used by living organisms within the ecosystem (Siegfried and Davies, 1982).

The boundaries of an ecosystem are often identified by changes in vegetation, soil, or landscape form. The scale of the ecosystem depends on the purpose of analysis; a small mountain pond is an ecosystem, and so is the mountain on which it is located. Almost all ecosystems are connected with others of various scales. Protected areas with artificial boundaries may be whole or partial ecosystems, depending on the size of the area, the scale of analysis, and the form of the boundary; a protected area surrounded by forest is a much different ecosystem than one surrounded by agricultural land.

Some ecosystems are relatively robust and resist permanent damage, but others are very sensitive to disturbance and may require long periods to recover from disruptions. Grasslands naturally subject to periodic fires, for example, are robust, while mature tropical rain forests may be easily disturbed and require decades or even centuries to recover (e.g., Gomez-Pompa *et al.*, 1972). Particularly sensitive ecosystems include those that lie on geologically unstable substrata, such as steep slopes subject to landslides, and those dependent on influences from outside the system, such as estuaries and deltas.

How to Identify Ecologically Sensitive Areas

Ecologically Sensitive Areas already exist in every nation. Some are well known, others are known only to local communities, and a few may not be recognized by anyone. In order to determine how ESAs are to be managed for their adequate protection and appropriate use, the logical first step is a survey of all remaining natural habitats to determine which have the highest value in their present state and those with the least value if transformed (i.e., the least opportunity cost for conservation); other biological and social parameters should also be included in the survey, often based on existing data. Criteria for identifying ESAs need to be established on the basis of this information, and the most important areas should be designated for special treatment (Box 18).

The identification of ESAs outside of existing protected areas ideally requires considerable research and information, but the pace of development is such that it will inevitably prove necessary to take some relatively arbitrary decisions on the limited information available. In practical terms, and as a working rule of thumb where complete information is not yet available, it may be best to accept the rationale behind the current processes used by resource management authorities to select ESAs for particular protection. These vary considerably from country to country, and from management authority to management authority; even within a country, a national parks department may use selection criteria that are quite different from those used by a watershed protection department. In general, this would mean that first-priority ESAs should include all areas that are given legal protection (though recognizing that not all areas given legal protection really qualify as ESAs), second-priority

Box 18: Four Steps Toward a National System of Protected Ecologically Sensitive Areas.

Each country will need to design its own approach to identifying the Ecologically Sensitive Areas (ESAs) in its country, as a basis for deciding how to manage the ESAs. The following general steps may provide a useful foundation.

Step 1. Evaluate patterns of habitats and vegetation, soils, mineral resources, topography, rivers and other hydrological features, climate, current land use, ethnic groups, and population density.

Step 2. Establish criteria for identifying ESAs and for providing objective guidelines on appropriate management regimes.

Step 3. Based on the criteria established, identify especially vulnerable locations, areas of high biological diversity, and areas of high economic value in the natural state.

Step 4. Prepare a national strategy for conserving ESAs, including establishing national objectives, identifying economic relationships, designing any necessary legislation, and assigning institutional responsibility for the ESAs.

ESAs should include all areas proposed for protection or rehabilitation, and third-priority ESAs should be all other areas where natural habitat remains.

Criteria for Designating and Managing Ecologically Sensitive Areas

In planning a system to protect ESAs for supporting national development goals, criteria for selection and management are essential. A system of criteria will enable a relatively organized comparison of different sites to be made, help tell decision-makers why certain areas or policy initiatives are important, help focus research on the most important questions, facilitate the drawing of boundaries for the ESA by specifying the features that need special management, and promote public information programs.

The sorts of protective regimes that are most appropriate for each major ESA (including, but not limited to, designation as a protected area) will be determined by the local social, political, and economic factors that need to be considered along with the ecological ones. The following set of model criteria is presented in roughly descending order of importance, though modifications will be required for adapting to each particular situation. Each criterion is presented as an ideal against which a given site can be considered. No one site can be expected to meet the ideal, and different criteria will be relevant to different sectors. In some cases, it may be appropriate for planning purposes to assign numerical scores to the various criteria.

1) *Criteria that determine the importance of the site to human society:*

Economic benefit. The site provides obvious long-term economic benefits, such as watershed protection or tourism, and does not involve great opportunity costs. This will often be the most important criterion for the production-oriented sectors.

Diversity. The site has a great variety of species and ecosystems, and is sufficiently large to contain viable populations of most species; it contains a variety of geomorphological features, soils, water regimes, and microhabitats.

Critical habitat, international. The site is essential to the survival of one or more threatened species that occur in no other country, contains the only example of certain types of ecosystems, or contains landscapes of outstanding universal value.

Critical habitat, national. The site is essential to the survival of one or more species threatened nationally, or contains the nation's only example of certain types of ecosystems. The ecological functioning of the area is vital to the healthy maintenance of a natural system beyond its boundaries (such as habitat for migratory species, an important catchment area for lowland irrigation systems, protection of the coast against typhoons, etc.).

Cultural diversity. The site supports populations of indigenous people who have developed mechanisms for living in sustainable balance with the natural ecosystems, and whose continued presence in the ESA would help ensure that the diversity of the area is maintained.

Urgency. Action is required quickly at the site in order to avert an immediate threat (though it should be realized that this is often a "damage control" action; it is usually best to protect far in advance of threat).

2) *Criteria to determine additional elements that enhance the value of the site:*

Demonstration. The site demonstrates the benefits, values, or methods of protection, and can show how to resolve conflicts between natural resource values and human activities.

Representativeness. The site is representative of a habitat type, ecological process, biological community, physiographic feature, or other natural characteristic.

Tourism. The site lends itself to forms of tourism compatible with the aims of conservation; this criterion is often related to those of economic benefit and social acceptance.

Landscape. The site has features of outstanding natural beauty; these are usually also unique, easily destroyed, and attractive to tourists, and any alteration would significantly reduce the area's amenity value.

Recreation. The site provides local communities with opportunities to use, enjoy, and learn about their natural environment.

Research and monitoring. The site can serve as a non-manipulated area against which to measure changes occurring elsewhere; it can form the basis for assessing any ecological change. Research has been carried out over a long period in the site, and major field studies have been carried out to provide a strong foundation on which new research can build. The site represents ecological characteristics of regional value so research can yield arguments that can have impacts far beyond the protected area.

Awareness. Education and training within the site can contribute knowledge and appreciation of regional values. The site can serve to exemplify techniques or scientific methods, making it particularly important for education purposes.

3) *Criteria to help determine the management feasibility of a site:*

Social acceptance. The site is already protected by local people, or official protection by the government (particularly against outside exploitation) would be welcomed.

Opportunism. Existing conditions or actions at the site lend themselves to further action (such as the extension of an existing protected area or establishment of a buffer zone around an existing park).

Availability. The site can be acquired easily, through inter-departmental transfer, easements, or other legal forms of control.

Convenience. The site is accessible to researchers or students for scientific and educational uses.

International Approaches to Determining Priorities

Just as some ecosystems within a nation have more species than others, so do some nations have more species than others (usually because they contain more ecological diversity). It is not suggested that biological diversity should be the only criterion to guide conservation investments; criteria such as degree of human need and opportunity for success, also need to be considered. However, it still seems worthwhile to identify which parts of the world contain the greatest diversity. This problem can be approached on at least three different levels: the region, the nation, and the site.

The Regional Approach: Critical Areas in Tropical Forests and Temperate Areas

It is generally accepted that the greatest threat of species loss is in the tropical forests, which are thought to contain at least half the world's species on just 7 percent of the world's land surface (Wilson, 1988a). But within this richest of the world's biomes, a relatively small number of particularly rich areas harbor an inordinately large share of the earth's biodiversity, featuring exceptional concentrations of species with exceptional levels of endemism.

The Committee on Research Priorities in Tropical Biology (NAS, 1980), drawing on very wide consultation with experts in various fields throughout the world, identified 11 areas in the tropics that, because of their great biological diversity, high levels of endemism, and the rate with which their forests are being converted to other purposes, seem to demand special attention. These are:

- Coastal forests of Ecuador,
- The "cocoa region" of Brazil,
- Eastern and southern Brazilian Amazon,
- Cameroon,
- Mountains of Tanzania,
- Madagascar,
- Sri Lanka,
- Borneo,
- Sulawesi,
- New Caledonia, and
- Hawaii.

Insects comprise the largest group of organisms on earth, with estimates as high as 30 million species (photos by A. Young).

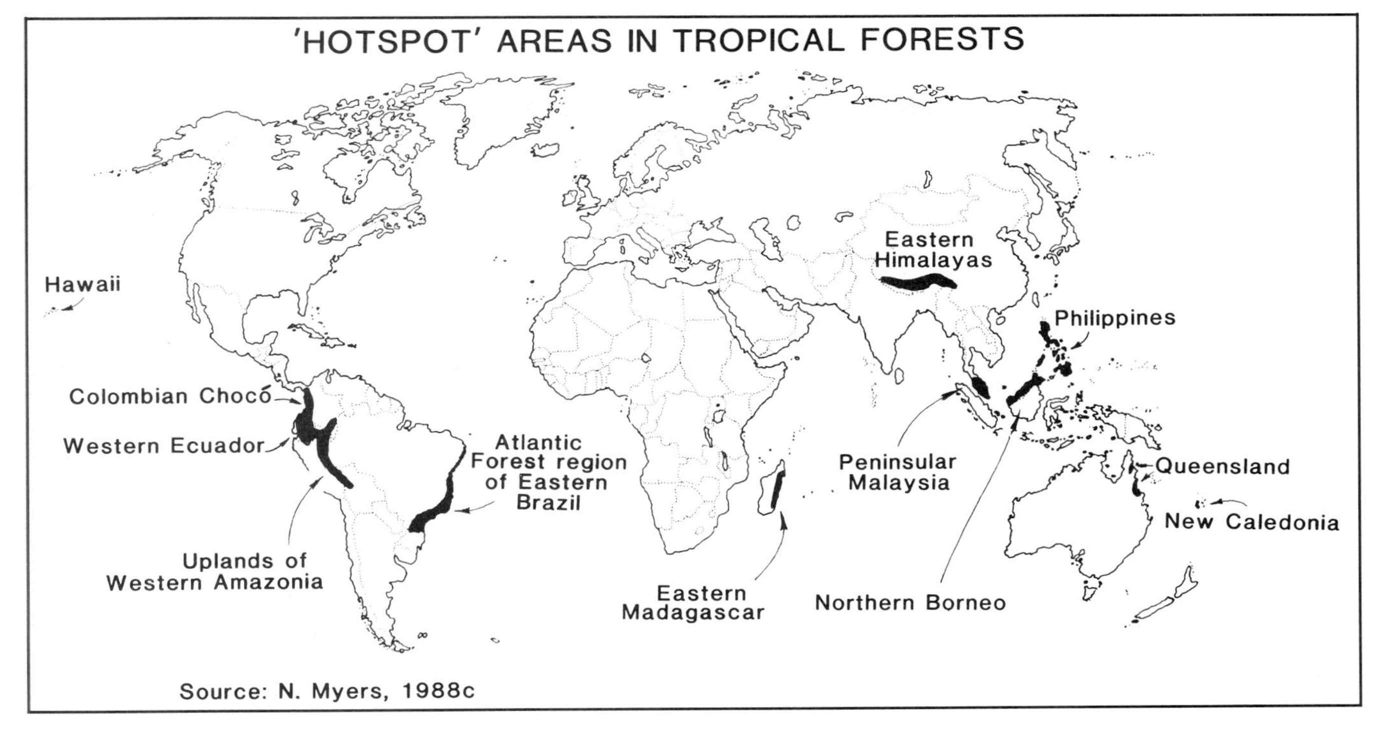

Fig. 5.

Myers (1988c) developed the approach to critical areas in a somewhat different way, identifying 10 tropical forest "hotspots," (plus 2 in the developed world — Hawaii and Queensland) (Fig. 5) totalling about 3.5 percent of the primary tropical forest remaining (and only 0.2 percent of the land surface of the planet) but containing at least 27 percent of the higher plant species found in the tropics; no less than 13.8 percent of the world's plants are found only in these hotspots (Table 10).

Myers was meticulous in pointing out the limitations of data that bedevil the conservation biologist attempting to compile data to enable such assessments to be made. While some figures can be taken as accurate to within five percent or better, others are little more than best guesses of specialists who have worked in the areas involved for many years. But Myers concludes that the overall approach, uneven as it is, is justified as an analytical exercise that seeks to delineate the conservation challenge facing the tropical forests.

The focus on tropical countries, however, can lead to insufficient attention being given to extremely important temperate areas. For example, of the 23,200 species of plants estimated to occur in southern Africa (South Africa, Lesotho, Swaziland, Namibia, and Botswana), 18,560 (i.e., 80 percent) are endemic to the region (Davis *et al.*, 1986). This gives the area the highest species richness (calculated as species/area ratio) in the world, 1.7 times greater than that of Brazil. Of these, 2,373 have been reported as threatened, 1,621 of which are in the Cape Floristic Kingdom, which gives this region the highest concentration of threatened plants of any temperate region (Heywood, 1989).

Table 10: "Hotspot" Areas in Tropical Forests.

The last figure in each line is the percentage of the flora of each region that is endemic to that region, the total figure of 13.8 percent is the percentage of the world's flora endemic to these ten regions.

Area	Original extent of forest (1000 ha)	Present primary forest (1000 ha)	Plant species in original forest	Number of endemics in original forests (percentage)	
Madagascar	6,200	1,000	6,000	4,900	(82)
Atlantic Forest, Brazil	100,000	2,000	10,000	5,000	(50)
Western Ecuador	2,700	250	10,000	2,500	(25)
Colombian Chocó	10,000	7,200	10,000	2,500	(25)
W. Amazonian Uplands	10,000	3,500	20,000	5,000	(25)
Eastern Himalayas	34,000	5,300	9,000	3,500	(39)
Peninsular Malaysia	12,000	2,600	8,500	2,400	(28)
Northern Borneo	19,000	6,400	9,000	3,500	(39)
Philippines	25,000	800	8,500	3,700	(44)
New Caledonia	1,500	150	1,580	1,400	(89)
	220,400	**29,200**	*	**34,400**	**(13.8)**

* It is not meaningful to total these figures because of overlap between some areas, notably in Borneo, Peninsular Malaysia, and the Philippines.

Source: N. Myers, 1988c.

The Regional Approach: Diversity in the Seas

While the tropical forests are thought to still contain millions of undescribed species, the world's oceans are also poorly known and regularly yield major new discoveries: a totally new phylum, *Loricifera,* was described only in 1986 (Baker *et al.*, 1986); a shark over 5 meters long (known as "the megamouth shark") has been discovered in the past decade (Ray, 1988); a species of mussel living near hydrocarbon seeps in the Gulf of Mexico was found to be feeding on methane (Childress *et al.*, 1986); deep-sea communities have been found to be far richer than suspected, with seafloor sediments at depths of 1,500 to 2,500 meters off the coast of New Jersey found to contain 898 species in more than a hundred families and a dozen phyla (Grassle, 1989); and entirely new habitats — hydrothermal ocean vents — have been discovered in the past decade to consist of at least 16 previously unknown families of invertebrates (Grassle, 1985). At the higher taxonomic level of phylum, marine ecosystems are actually more diverse than either terrestrial or freshwater biomes, with more phyla and endemic phyla (Table 11). Within this great diversity, Vermeij (1978) has shown that some areas are considerably richer than others (Table 12).

Table 11: Distribution of Animal Phyla by Habitat.

SYMBIOTIC	MARINE	
Orthonectida	*Placozoa*	Porifera
Dicyemida	*Ctenophora*	Cnidaria
Nematomorpha	*Gnathostomulida*	Platyhelminthes
Acanthocephala	*Kinorhyncha*	Nemertea
	Loricifera	Nematoda
Porifera	*Priapula*	Rotifera
Cnidaria	*Pogonophora*	Gastrotricha
Platyhelminthes	*Echiura*	Tardigrada
Nemertea	*Chaetognatha*	Mollusca
Nematoda	*Phoronida*	Kamptozoa
Rotifera	*Brachiopoda*	Sipuncula
Mollusca	*Echinodermata*	Annelida
Kamptozoa	*Hemichordata*	Arthropoda
Annelida		Bryozoa
Arthropoda		Chordata
Chordata		
= 15 (4 endemic)	= 28 (13 endemic)	
= 14 (0 endemic)	= 11 (1 endemic)	
Porifera	*Onychopohora*	
Cnidaria		
Platyhelminthes	Platyhelminthes	
Nemertea	Nemertea	
Nematoda	Nematoda	
Rotifera	Rotifera	
Gastrotricha	Tardigrada	
Tardigrada	Mollusca	
Mollusca	Sipuncula	
Kamptozoa	Annelida	
Annelida	Arthropoda	
Arthropoda	Chordata	
Bryozoa		
Chordata		
FRESHWATER	TERRESTRIAL	

The oceans are a great new frontier whose productivity is just beginning to be harnessed by humans (though already some areas show disturbing signs of overexploitation). As society seeks to exploit this productivity, far greater efforts are required to ensure that the exploitation is based on a more complete knowledge of how marine ecosystems function, how marine biodiversity contributes to productivity, and what management activities are required to ensure that the characteristic diversity of the seas is maintained. Establishing priorities within this context is a daunting task.

Table 12: Species Richness in Tropical Waters.

Group	Number of Species			
	Indo-West Pacific	Eastern Pacific	Western Atlantic	Eastern Atlantic
Molluscs	6,000+	2,100	1,200	500
Crustaceans				
Stomatopods	150+	40	60	10
Brachyura	700+	390	385	200
Fishes	1,500	650	900	280

Source: Vermeij, 1978.

The National Approach: "Megadiversity Countries"

As developed by Mittermeier (1988, Mittermeier and Werner, 1989), the megadiversity country concept recognizes that:

- although basic scientific information on biodiversity and endangered ecosystems should be our first step in assessing international conservation priorities, conservation programs are developed with and by the governments of sovereign nations; that
- biodiversity is by no means evenly distributed among the world's 168 countries; and that
- a very small number of countries, lying partly or entirely within the tropics, accounts for a very high percentage of the world's biodiversity (including marine, freshwater and terrestrial diversity), and that these countries require very special international conservation attention.

The megadiversity concept integrates biological information of many different kinds, but the two main criteria for inclusion in this category are total species numbers and levels of endemism both at the species level and at higher taxonomic categories (e.g., genus, family).

Although data are still being gathered on this topic, preliminary indications are that about a dozen countries belong on the megadiversity list, including Brazil, Colombia, Ecuador, Peru, Mexico, Zaire, Madagascar, Australia, China, India, Indonesia and Malaysia, and that these countries by themselves account for 60 to 70 percent (and perhaps even more) of all the world's biodiversity. Of these, Brazil,

Colombia, Indonesia, and Mexico are especially rich in species numbers (and often have high endemism as well) for most groups of organisms on which information is available. Madagascar and Australia, though usually not as high in total species numbers (but see Box 19 for reptile diversity in Australia), belong on the megadiversity list because of their very high degrees of endemism, both at the species level and also at higher taxonomic categories like the genus and family (Table 13). For example, Brazil, though highest in the world in total plant numbers, has no endemic plant families, whereas Madagascar has five and Australia 12. Although the megadiversity countries (e.g., Brazil, China) are among the world's largest and would be expected to have high diversity simply because of their size, their diversity far exceeds that of other countries of similar size (e.g., Canada, U.S.A, U.S.S.R.). Furthermore, several of the megadiversity countries are quite small (e.g. Ecuador, Madagascar, Malaysia), with the diversity being due more to topography, climate and/or long isolation than to surface area. In any case, large, small or medium-sized, these countries have great strategic importance in conservation of biological diversity worldwide. They themselves have a great responsibility to the world to conserve their biological wealth, and the international community has a special responsibility to them to provide any assistance that might be required to achieve their conservation goals.

It should be emphasized that focus on these megadiversity countries is in no way intended to imply a triage approach in which the focus is exclusively on a limited number of countries to the exclusion of all others (*sensu* Myers, 1979). Obviously, the biological resources of each and every country are of critical importance, at least to their own survival and well-being (even if not particularly important in the global picture), and therefore worthy of national and international attention for that reason alone. Rather, the megadiversity approach is aimed at focusing attention on these highly diverse, strategically critical megadiversity countries roughly in proportion to the biological wealth that they harbor and regardless of how difficult it might be to achieve conservation within them. Furthermore, it recognizes that if we do not pay sufficient attention to these megadiversity

Table 13: Number of Endemic Families in Australia and Madagascar.

	Australia	Madagascar
Birds	4	3
Mammals	7	5
Angiosperms	12	8

Source: Conservation International, numerous sources.

Table 14: Countries with the Highest Numbers of Species for Selected Organisms.

	MAMMALS	BIRDS	AMPHIBIANS
1.	Indonesia . . . (515)	Colombia . (1721)	Brazil . . . . . (516)
2.	Mexico . . . . (449)	Peru . . . . . . (1701)	Colombia . . (407)
3.	Brazil . . . . . . (428)	Brazil . . . . . (1622)	Ecuador . . . (358)
4.	Zaire . . . . . . (409)	Indonesia . . (1519)	Mexico . . . . (282)
5.	China . . . . . . (394)	Ecuador . . . (1447)	Indonesia . . (270)
6.	Peru . . . . (361)	Venezuela . . (1275)	China . . . . . (265)
7.	Colombia . . . (359)	Bolivia . . (±1250)	Peru . . . . . . (251)
8.	India . . . . (350)	India . . . . . . (1200)	Zaire . . . . . . (216)
9.	Uganda . . . . (311)	Malaysia . (±1200)	U.S.A. . . . . (205)
10.	Tanzania . . . (310)	China . . . . . (1195)	Venezuela Australia . . (197)

	REPTILES	SWALLOWTAIL BUTTERFLIES[1]	ANGIOSPERMS[2] (est.)
1.	Mexico . . . . . (717)	Indonesia . . . (121)	Brazil . . (55,000)
2.	Australia . . . (686)	China . . . (99-104)	Colombia . (45,000)
3.	Indonesia . (±600)	India . . . . (77)	China . . (27,000)
4.	Brazil . . . . . . (467)	Brazil . . . . (74)	Mexico . (25,000)
5.	India . . . . (453)	Burma . . . . (68)	Australia . . (23,000)
6.	Colombia . . . (383)	Ecuador . . . . . (64)	S. Africa(21,000)
7.	Ecuador . . . . (345)	Colombia . . . . (59)	Indonesia(20,000)
8.	Peru . . . . (297)	Peru . . . . . . (58-59)	Venezuela(20,000)
9.	Malaysia . . . (294)	Malaysia . . (54-56)	Peru . . . . (20,000)
10.	Thailand Papua NG . . (282)	Mexico . . . . . . (52)	U.S.S.R. (20,000)

Source: Conservation International, numerous sources.
[1]Collins and Morris, 1985.
[2]Davis *et. al.*, 1986.

Table 15: Asian Countries with the Highest Numbers of Species for Selected Organisms.

	MAMMALS	BIRDS	AMPHIBIANS
1.	Indonesia . . . (515)	Indonesia . . (1519)	Indonesia . . (270)
2.	China . . . . . . (394)	India . . . . . . (1200)	China . . . . . (265)
3.	India . . . . (350)	Malaysia . (±1200)	Australia . . . (197)
4.	Burma . . . . . (300)	China . . . . . (1195)	Papua NG . (183)
5.	Malaysia . . . (293)	Burma . . . . . . (967)	India . . . . . . (182)
6.	U.S.S.R. . . . (276)	Nepal . . . . . . (835)	Malaysia . . . (171)
7.	Thailand . . . . (263)	Thailand . . . . (800)	Thailand . . . (101)
8.	Australia . . . (255)	U.S.S.R. . . . (728)	Philippines . . (77)
9.	Vietnam . . . . (201)	Pakistan . . . . (612)	Burma . . . . . . (75)
10.	Philippines . (165)	Philippines . (541)	Vietnam . . . . (72)

	REPTILES	SWALLOWTAIL BUTTERFLIES[1]	ANGIOSPERMS[2] (est.)
1.	Australia . . . (686)	Indonesia . . . (121)	China . . (27,000)
2.	Indonesia (>600)	China . . . (99-104)	Australia (23,500)
3.	India . . . . (453)	India . . . . (77)	Indonesia(20,000)
4.	Malaysia . . . (294)	Burma . . . . (68)	U.S.S.R. (20,000)
5.	Thailand Papua NG . . (282)	(Indochina) (66-69)	Malaysia (15,000)
6.	China . . . . . . (278)	Malaysia . . (54-56)	India . . . (14,500)
7.	Burma . . . . . (241)	Philippines . . . (49)	Thailand (11,500)
8.	Vietnam Philippines . (212)	Nepal . . (37 or 38)	Vietnam (11,500)
9.	Bangladesh . (129)	Papua NG . . . (37)	Philippines (8000)
10.	U.S.S.R. . . . (125)	Brunei . . . . (35-37)	Burma . . . . (7000)

Source: Conservation International, numerous sources.
[1]Collins and Morris, 1985.
[2]Davis *et. al.*, 1986.

Table 16: African Countries with the Highest Numbers of Species for Selected Organisms.

	MAMMALS	BIRDS	AMPHIBIANS
1.	Zaire . . . . . . (409)	Zaire . . . . . . (1086)	Zaire . . . . . . (216)
2.	Uganda . . . . (311)	Kenya . . . . . (1046)	Cameroon . . (190)
3.	Tanzania . . . (310)	Uganda . . . . . (973)	Madagascar (144)
4.	Kenya . . . . . . (308)	Tanzania . . . . (969)	Tanzania . . . (127)
5.	Cameroon . . (297)	Cameroon . . . (849)	Nigeria . . . . . (96)
6.	S. Africa . . (279)	Ethiopia . . . . (827)	S. Africa . . . (93)
7.	Angola . . . . . (275)	Nigeria . . . . . (824)	Congo . . . . . . (88)
8.	Nigeria . . . . . (274)	Zambia . . . . . (728)	Angola Gabon . . . . . . (86)
9.	Sudan . . . . . . (266)	S. Africa . . . (725)	Côte d'Ivoire(80)
10.	Ethiopia . . . . (256)	Ghana . . . . . . (721)	Kenya . . . . . . (79)
	REPTILES	**SWALLOWTAIL BUTTERFLIES[1]**	**ANGIO-SPERMS[2] (est.)**
1.	S. Africa . . . (281)	Zaire . . . . (48)	S. Africa(21,000)
2.	Zaire . . . . . . (280)	Cameroon . . . . (39)	Zaire . . . (10,000)
3.	Madagascar (269)	Congo . . . . (37-38)	Madagascar(10,000)
4.	Tanzania . . . (244)	Tanzania . . . . . (34)	Tanzania (10,000)
5.	Angola . . . . . (217)	Uganda (W. Africa)(31-32)	Cameroon . (9000)
6.	Cameroon . . (183)	Kenya . . . . (30)	Gabon . . . . (7900)
7.	Namibia Somalia . . . . (166)	Angola . . . . . . (27)	Kenya . . . . (6750)
8.	Mozambique (159)	Gabon . . . . (25-31)	Ethiopia . . (6200)
9.	Nigeria . . . . . (147)	C.A.R. . . . (24-29)	Mozambique (5000)
10.	Uganda . . . . (143)	Zambia . . . . . . (23)	Uganda . . . (4500)

Source: Conservation International, numerous sources.
[1]Collins and Morris, 1985.
[2]Davis *et. al.*, 1986.

countries, we will lose a major percentage of the world's biodiversity regardless of how successful we are in the other less diverse countries.

As examples of the critical importance of these megadiversity countries for different groups of organisms, four of them (Brazil, Zaire, Madagascar, and Indonesia) by themselves account for two-thirds of all primate species; four (Mexico, Brazil, Indonesia and Australia) are home to more than a third of all reptiles; seven (Brazil, Colombia, Mexico, Zaire, China, Indonesia and Australia) have more than half of all flowering plants; and three (Brazil, Zaire, Indonesia) have within their borders roughly half of all the world's tropical rainforest. Table 14 presents a summary of the top 10 countries for mammal, bird, reptile, amphibian and swallowtail butterfly diversity, (taxa selected because they are among the best known and most conspicuous forms of life). Tables 15 to 17 present the top 10 in each of the major tropical regions, and Boxes 19 to 25 profile in more detail seven of the most important megadiversity countries: Brazil, Colombia, Mexico, Zaire, Madagascar, Indonesia and Australia.

The megadiversity approach is an important catalyst that has already helped generate interest and mobilize funds from new sources, both from within the megadiversity countries themselves and from the international community, and it has given these countries a new awareness of and pride in the importance of their biological heritage. As a result of this approach the World Bank, for example, has increased its attention to biodiversity in countries like Brazil and Madagascar (without overlooking less diverse countries like Mauritius and Botswana), and megadiversity thinking has also influenced the investments of international conservation organizations like CI and WWF. CI, in particular, is incorporating the megadiversity concept into its Rain Forest Imperative Campaign for the critical decade of the 1990s, and is already making major investments in Brazil, Mexico and Madagascar.

Of course, to be truly effective the megadiversity approach cannot stand alone, but should be recognized as a way of packaging biodiversity priorities in terms of political boundaries and realities. In all cases, it needs to be followed up by site-specific efforts to determine the highest priority ecosystems, endangered species, etc. within the country in question and to consider questions that transcend national boundaries, like conservation of migratory species. Furthermore, we emphasize once again that it should not supplant efforts to conserve the biological wealth of other less diverse countries that also play a role in the global strategies to conserve biological diversity.

Table 17: Neotropical Countries with the Highest Numbers of Species for Selected Organisms.

	MAMMALS	BIRDS	AMPHIBIANS
1.	Mexico . . . . . (449)	Colombia . . (1721)	Brazil . . . . . (516)
2.	Brazil . . . . . . (428)	Peru . . . . . . (1701)	Colombia . . (407)
3.	Peru . . . . (361)	Brazil . . . . . (1622)	Ecuador . . . (358)
4.	Colombia . . . (359)	Ecuador . . . (1447)	Mexico . . . . (282)
5.	Venezuela . . (305)	Venezuela . . (1275)	Peru . . . . . . (251)
6.	Ecuador . . . . (280)	Bolivia . . (±1250)	Venezuela . . (197)
7.	Bolivia . . . . . (267)	Mexico . . . . (1010)	Panama . . . . (159)
8.	Argentina . . . (255)	Argentina . . . (942)	Costa Rica . (150)
9.	Panama . . . . (217)	Panama . . . . . (907)	Argentina . . (130)
10.	Costa Rica . . (203)	Costa Rica . . (796)	Guyana . . . . (100)
	REPTILES	**SWALLOWTAIL BUTTERFLIES[1]**	**ANGIO-SPERMS[2] (est.)**
1.	Mexico . . . . . (717)	Brazil . . . . (74)	Brazil . . (55,000)
2.	Brazil . . . . . . (467)	Ecuador . . . . . (64)	Colombia (45,000)
3.	Colombia . . . (383)	Colombia . . . . (59)	Mexico . (25,000)
4.	Ecuador . . . . (345)	Peru . . . . . . (58-59)	Venezuela(20,000)
5.	Peru . . . . (297)	(C. America)(57-58)	Peru . . . . (20,000)
6.	Venezuela . . (246)	Mexico . . . . . . (52)	Ecuador . (15,000)
7.	Costa Rica . . (218)	Bolivia . . . (43-44)	Bolivia . (15,000)
8.	Panama . . . . (212)	Argentina . (36-37)	Argentina . (8500)
9.	Argentina Guatemala . . (204)	Venezuela . (35-39)	Costa Rica(8000)
10.	Bolivia . . . . . (180)	Guyana Suriname . . (30-31)	Panama . . . (7750)

Source: Conservation International, numerous sources.
[1]Collins and Morris, 1985.
[2]Davis *et. al.*, 1986.

Box 19: Biological Diversity in Australia.

The country of Australia is not only an entire continent but along with New Guinea and nearby islands is sometimes considered to be one of the eight major biogeographical realms on earth. The distinctiveness of its flora and fauna, with very high levels of endemism, results from the country's long isolation. It covers an area of 7,682,300 km² (Times, 1988), and has a human population of 16,820,000 (PRB, 1989).

Although Australia is the driest continent, with a good portion of its territory consisting of arid zones, it also has considerable habitat diversity, ranging from the Great Barrier Reef to tropical rainforests of the northeast. Australia has the planet's second highest number of reptile species (686), is fifth in flowering plants (23,000) and tenth in amphibians (197). More significant, however, is the high percentage of organisms that occur only in Australia, and that this endemism extends up to the higher taxonomic categories of genus and family. Seven families of mammals, including that of the platypus and that of the koala, four of birds, and twelve of flowering plants are endemic — far more endemic families than any other country. At the generic level, 45% of birds and 37% of mammals are endemic. At the species level, the mean percentage of endemism for terrestrial vertebrates and flowering plants is 81%, the same figure for Madagascar.

The quokka (*Setonix brachyurus*), one of the world's smallest wallabies, is an endemic species found only in a small area of southwestern Australia (photo by S.D. Nash).

The koala (*Phascolarctos cinereus*) is the only species of the Phascolarctidae, a family endemic to Australia (photo by R.A. Mittermeier).

The northern part of Queensland, the state in the country's northeast, is one of the two areas in the developed world identified by Norman Myers as a "threatened hotspot." In the past 50 years, approximately half of the tropical rain forest has been removed from this species-rich habitat, and logging pressures still imperil its survival. Off the coast to the east and southeast of this forest lies the Great Barrier Reef, the world's largest coral reef system, and one of the most diverse in marine organisms. Although tourism is on the increase in this ecosystem, most of the region is still pristine, so that well-managed tourism should have minimal impact.

Ninety-five species of Australia's vertebrates are listed by IUCN as under some degree of threat. The highest number of species at risk for any group of vertebrate is for mammals, 32 of which are listed. This figure excludes the 16 that are believed to have gone extinct within historical times, mainly from the effects of human settlement and introduced species.

Source: Conservation International, unpublished data, 1989.

Box 20: Biological Diversity in Brazil.

Brazil, with an area of 8,511,965 km^2 (Times, 1988) and a human population of 147,393,000 (PRB, 1989), is perhaps the single richest country in the world in overall species diversity. It tops the world list in diversity for many different groups of organisms, among them primates (55 species; 24 percent of the world total), amphibians (516 species), terrestrial vertebrates (3,010 species), endangered and vulnerable vertebrates (310 species), flowering plants (55,000; 22 percent of the world total), freshwater fish (more than 3,000 species, three times more than any other country), and insects (estimated at 10-15 million species, most of them still undescribed by scientists). When not the single richest country, Brazil is usually not far behind, ranking fourth in reptile diversity (467 species), third in birds (1,622 species) and palms (387 species), and fourth in mammals (428 species).

In addition, Brazil has by far the most closed tropical forest in the world, with its 357 million ha accounting for almost 30 percent of the world total and exceeding that of the second richest country (Indonesia) by three times. Indeed, Brazil has more tropical rain forest than the rest of South and Central America combined, more than all of Asia, and more than all of Africa. Taking all kinds of forest into consideration, Brazil is second only to the U.S.S.R. in total forest cover.

The largest portion of Amazonian forest (62 percent) is found within Brazil, and covers 42 percent of the country, comprising the most extensive tropical forest region falling within the borders of any one nation. About 80 percent of it is still intact, but forest destruction over the past decade has been especially heavy in certain regions (e.g., Rondônia, southern Pará), and the trends are not promising. For example, during 1987 alone, it is estimated (from satellite imagery analysis by Brazilian specialists) that some 8 million ha of primary forest were destroyed.

The Atlantic forest region of eastern Brazil runs from the states of Rio Grande do Norte and Ceará in northeastern Brazil in a narrow belt south as far as Rio Grande do Sul, in the southernmost Brazilian state. It once covered 100 to 120 million ha, or about 12 percent of the country. However, it was the first part of Brazil to be colonized, is now the major agricultural and industrial center of the country, and has been largely deforested — to the point that only 1 to 5 percent of the original forest cover remains.

The Pantanal region of Mato Grosso and Mato Grosso do Sul is a vast, low-lying swampland in the center of the continent, and has some of the most spectacular and visible concentrations of wildlife in all of South America. Though still largely intact, it is being increasingly threatened by various kinds of pollution, mining, siltation resulting from deforestation at the headwaters of rivers feeding into the region, overfishing, poaching, and ill-conceived development projects.

Source: Conservation International, unpublished data, 1989.

The muriqui (*Brachyteles arachnoides*), a monospecific genus of primate found only in Brazil's Atlantic forest (photo by A. Young).

Box 21: Biological Diversity in Colombia.

The cotton-top tamarin (*Saguinus oedipus*) is an endemic primate of Colombia (photo by R. Mast).

Exceedingly high diversity coupled with the high degree of threat posed by unchecked human (pop. 31,192,000) (PRB, 1989) development and commercial resource exploitation places Colombia among the highest conservation priorities on earth. Colombia is one of the world's richest countries in terms of species diversity per unit area and is second only to Brazil in overall species numbers.

Though Colombia's land area of 1,138,915 km^2 (Times, 1988) accounts for only 0.77 percent of the earth's surface, it is home to approximately 10 percent of the earth's species of terrestrial plants and animals. Colombia contains 45,000 to 50,000 higher plant species, thus nearly reaching Brazil's total in an area less than one-seventh the size. In comparison, only 30,000 species of plants are found in all of sub-Saharan Africa. Colombia tops the world list for numbers of orchid species (about 3,500), amounting to a full 15 percent of the world's total. More species of birds live in Colombia than in any other country (1,721 species, or nearly 20 percent of the world total). Colombia contains the third highest number of terrestrial vertebrates for any country on earth (2,890 species), and more than one-third of all neotropical primates (27 species). Colombia's documented diversity will undoubtedly increase substantially with further biological inventory, especially in groups such as invertebrates and plants.

In addition to its biotic diversity, Colombia is home to numerous indigenous tribes who still utilize native vegetation for medicinal and other purposes. Many of the economically important South American plants currently being used in the industrial countries of the world originated in the forests of Colombia, and ethnobotanical research indicates that yet untapped knowledge of indigenous peoples of Colombia could yield invaluable data regarding medicinal and industrially important plants (Plotkin, 1988).

Colombia's high species diversity and endemism in many groups of organisms (e.g., plants, amphibians) can be attributed largely to the country's geographic position bridging North and South America, its high precipitation (reaching 13 meters/year in the Chocó region, the highest rainfall on earth) and its mountainous aspect (with three Andean chains reaching almost 6,000 meters above sea level). The effects of altitude and climate have combined to create a myriad of microhabitats along their slopes, each home to its own distinct and unique flora and fauna. In addition to the high-altitude páramos, superpáramos, evergreen, and cloud forests of the Andes, Colombia harbors a wide variety of lowland habitats including the arid deserts of the Guajira peninsula, the rich littoral zones and mangrove forests on both Atlantic and Pacific coasts, pristine coral reefs, island environments, Amazonian forest, and vast savannas (the llanos) to the east.

Source: Conservation International, unpublished data, 1989.

The plate-billed mountain-toucan (*Andigena laminirostris*), a species found within Colombia and northwestern Ecuador (photo by R. Mast).

The world's largest lizard, the komodo dragon (*Varanus komodoensis*), only occurs on a small island of Indonesia (photo by R.A. Mittermeier).

The endangered Sumatran rhinoceros (*Dicerorhinus sumatrensis*) (photo by B. Bunting).

Box 22: Biological Diversity in Indonesia.

Indonesia is by far the most biologically diverse country in Asia in almost every faunal and floral category, and is at or near the top of the world list in several categories. This is largely due to the unique biogeography of the country and its large amount of tropical rain forest. Indonesia bridges two of the earth's biogeographic realms (Indomalaya and Oceania) and numerous biogeographic provinces, and has within its borders a major transition zone between these two realms, Wallacea, that includes a large number of endemics. With 114 million hectares of closed forest, Indonesia possesses more tropical forest than any other single African or Asian country and is second only to Brazil worldwide in tropical forest area. Within Asia, Indonesia has more than double the forest of the next most forested country.

These species-rich forests are home to the world's greatest diversity of palm species and an estimated 20,000 species of flowering plants. Many plant species of global and national economic importance originated in Indonesia, including citrus fruit, black pepper, and sugar cane.

Indonesia also harbors a rich fauna, including the greatest mammal diversity on earth (515 species, of which 36 percent are endemic), more psittacine birds (parrots and macaws) (78 species, 40 percent endemic), the highest number of threatened bird taxa in the world (126 of Indonesia's 1,519 avian species are threatened), more swallowtail butterfly species than any other country (121 species, 44 percent endemic), and more species of primates than any other Asian nation (33 species, 18 endemic).

A country of 1,919,445 km² (Times, 1988) with over 13,000 islands, Indonesia covers more marine than terrestrial territory, and its largely unstudied marine fauna is certainly among the earth's most diverse. Indonesia possesses the most extensive reef areas in the Indo-Pacific Ocean and more total marine coastline than any other tropical country (about 5,500,000 ha). Approximately 7,000 species of marine fishes are described from Indonesia, and in just one small area (the Sangkarang Archipelago), a recent survey described 262 species of hard corals, more than is known from anywhere else in the Indo-Pacific region.

Indonesia has the highest human population of all the profiled megadiversity countries, with 184,583,000 inhabitants (PRB, 1989).

Source: Conservation International, unpublished data, 1989.

Box 23: Biological Diversity in Madagascar.

Although this country is only 594,180 km² in area (Times, 1988), it is a mini-continent with a wide variety of species found nowhere else on earth. Located only about 400 km off the east coast of Africa, it has been isolated for a very long time, perhaps as long as 200 million years. As a result, it has become a unique evolutionary experiment, a living laboratory where evolution has followed a course different from anywhere else on the planet. As a result of its long isolation, it has some of the highest levels of species endemism anywhere on earth, and because of its climate and position it also has very high species diversity in certain groups of organisms. Twenty-eight of Madagascar's 30 species of primates are restricted to the island, the highest level of primate endemism found anywhere, and four of the five families of primates occurring on the island are found nowhere else. Eight of the nine species of carnivores, 29 of the 30 species of tenrecs, 237 of the 269 reptiles, 142 of the 144 amphibians, and 128 of 133 palms are restricted to the island. Though Madagascar's birds are less endemic than other groups at the species level (106 of 250), they include a total of three endemic families — an extremely high level of familial endemism matched only by Australia. Madagascar's approximately 7,900 species of flowering plants account for about 20 percent of all the plants in the African region, and 80 percent of them are endemic, including 5 endemic families. Madagascar has more species of orchids than all of mainland Africa, inspite of the fact that it occupies only 1.9 percent of the region.

Madagascar has more species of baobab (*Adansonia* spp.) than any other country (photo by R.A. Mittermeier).

Madagascar's remaining forest ecosystems can be divided into three broad categories: the southern spiny desert, the western dry deciduous forest, and the eastern rain forest, each with its own complement of species and each with very high levels of endemism as well. For instance, 48 percent of the plant genera and 95 percent of the species found in the southern spiny desert are endemic not only to Madagascar, but to just this small region of the country.

Madagascar's unique species and ecosystems are highly endangered. The central plateau of the country, a region once home to many now-extinct species, is almost entirely deforested, and trends in the other ecosystems indicate that action is needed as soon as possible. At least 80 percent and perhaps as much as 90 percent of Madagascar's forests are already gone, and the remainder are being chipped away for firewood and charcoal, and being cut down for slash-and-burn agriculture. In a region of such diverse species and ecosystems that often occupy only a small area, the effects of deforestation and other forms of habitat conversion are usually devastating to biodiversity. The arrival of man some 1500-2000 years ago and the pursuant environmental modifications have already resulted in the extinction of species including a pygmy hippopotamus, an ardvark, at least six genera of lemurs, huge elephant birds, and at least two giant tortoises. Human population growth is high at about 3.1 percent per year, and inhabitants already number 11,602,000 (PRB, 1989). Pressures on the island's natural habitats persist, and, if uncontrolled, are expected to cause further eradication of many of the planet's most spectacular organisms.

A conservation strategy has been developed for this country, entitled, "An Action Plan for Conservation of Biological Diversity in Madagascar," and the World Bank and USAID, among others, are collaborating in efforts to conserve this country's unmatched natural heritage.

Source: Conservation International, unpublished data, 1989.

Box 24: Biological Diversity in Mexico.

Mexico covers 1,972,545 km² (Times, 1988) and has a human population of 86,740,000 (PRB, 1989). It has the highest reptile (717 species, of which 53 percent are endemic) diversity in the world, is only exceeded by Indonesia in numbers of mammals (449 species, with 33 percent endemic), and ranks fourth globally in numbers of amphibians (282 species, with 63 percent endemic). It has nearly 30 percent more bird species (1010) than the U.S.A. and Canada together, and is by far the most important wintering area for many U.S. and Canadian migratory bird species. For example, Mexico hosts 51 percent of all migratory bird species from North America every year, and these birds spend from six to nine months of their lives in the country. Migratory species of butterflies, fishes, whales, bats, and turtles are additional examples of these internationally shared resources.

The Mexican flora is also very rich in species diversity and endemism, with more than 2,000 genera of flowering plants alone, and 22,000 known and 30,000 expected species. More than 15 percent of plant genera and approximately 50 to 60 percent of plant species are endemic to the country. These endemics include 50 percent of the world's *Phaseolus* species (bean family), 82 percent of the *Agave* species, 88 percent of the *Salvia* species, and 75 percent of the *Scutellaria* species, some of them with medicinal characteristics (e.g., probably anticancer properties), to name a few.

This biological richness is due to the great habitat variation and diverse ecological regions, complex topography, heterogeneity of soils and climate, geological history, and geographic location. Like Indonesia, Mexico bridges two major biogeographic realms of the world — the Nearctic and the Neotropical — that facilitates the exchange between elements of northern boreal and tropical origins. This great array of interacting species and organisms creates unique ecosystems of international importance.

Mexico has important marine habitats on both the Atlantic and Pacific coasts. The Sea of Cortes, dividing the Baja peninsula from the rest of Mexico, contains the only breeding ground of the Guadelupe fur seal (*Arctocephalus townsendii*), and also the endemic harbor porpoise (cochito) (*Phocoena sinus*), an animal only recently described with an unknown number of individuals.

Threats to Mexico's biodiversity mainly come from conversion of land to agriculture and cattle-ranching. Less than 40% of the country is still considered natural habitat. Also, there is an active trade in wildlife, legal and illegal, with a huge demand in the United States for cacti and birds. Fortunately, despite many pressures on Mexico's habitats, there is a rising interest in conservation and a good number of professionals, many with backgrounds in natural history, that can lead the protection of the country's natural heritage.

Source: Conservation International/WWF-US, unpublished data, 1989.

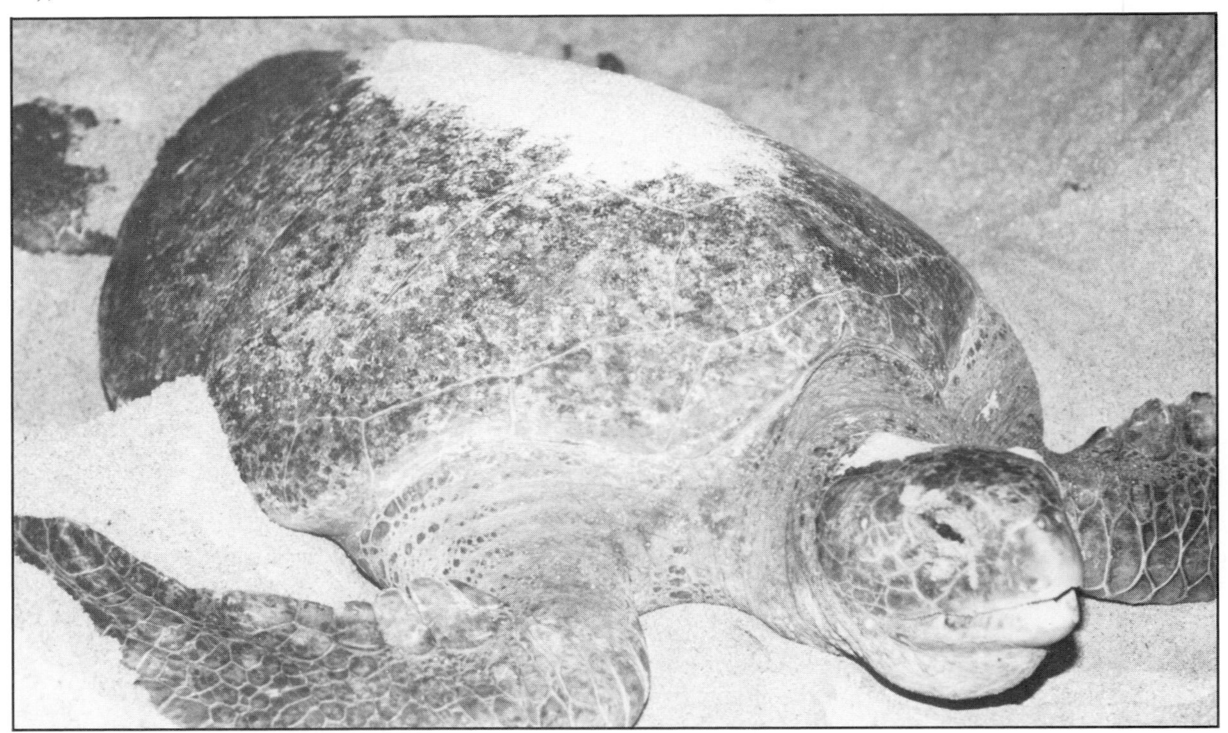

Seven of the world's eight species of sea turtle nest on the coasts of Mexico, including this Pacific green turtle (*Chelonia agassizii*) (photo by R. Mast).

Box 25: Biological Diversity in Zaire.

Zaire is one of the most important countries in Africa for the conservation of biological diversity, with more species of vertebrates than any country on the continent. Indeed, it is the only megadiversity country on the African mainland, and has some of the most important major tropical forest wilderness areas left on earth. It is also wealthy in natural resources, being rich in minerals and ranking third in the world for its amount of closed tropical forest. Zaire covers an area of 2,345,410 km^2 (Times, 1988), has a human population of 34,853,000 (PRB, 1989), and contains the world's second largest river system (the Zaire River), the second deepest lake (Lake Tanganyika), and the largest tropical forest park (Salonga National Park).

Even though the Africotropical realm is generally less diverse in overall species than the Neotropical and Indomalayan realms, Zaire still appears high on the global list, being the fourth highest in mammals (409 species) and probably second in freshwater fish diversity (with an estimated 70% endemism). Within Africa, Zaire has more species of mammals, primates (29-32), birds (1086), amphibians (216), fishes, and swallowtail butterflies (48) than any other country. Plant diversity (11,000 species) is second on the continent after South Africa, and the number is expected to increase as more exploration takes place.

So far, low human population and a low deforestation rate (.15-.5%/year) has kept the pressures on Zaire's forests and resident wildlife to a minimum. However, poaching is a serious problem, as exemplified by the current plight of the African elephant. Zaire has some of the largest populations of both the forest and savanna elephant; they have been under siege for many years, and prospects for their survival are bleak.

The huge patches of pristine African rain forest in Zaire deserve special conservation attention. Twenty-one of the 36 mammals of Zaire on IUCN's List of Threatened Animals occur in forest, as do 18 of its 28 threatened birds. Eastern Zaire contains several forest refugia, areas of the planet with remnant populations of more ancient species, many of which are found nowhere else. Among the wildlife of these refugia are globally important species like the okapi (*Okapia johnstoni*) and the mountain gorilla (*Gorilla gorilla beringei*). Another forest species, the pygmy chimp (*Pan paniscus*), man's closest living relative, is endemic to central Zaire.

The pygmy chimpanzee (*Pan paniscus*) (photo by R.A. Mittermeier).

As the earth's resources continue to dwindle, human commercial and subsistence activities will enter those areas rich in products like timber, minerals, and game, that previously were unexploited because of their remoteness. The rainforests of Zaire are one of the few remaining areas on the planet where human impact on the land has not yet been severe, though steps are needed to ensure the survival of its unique and rich biological diversity.

Sources: Conservation International, unpublished data, 1989.
Goodson, 1988.

Box 26: Major Tropical Wilderness Areas.

The southern Guianas: Including the southern portion of the countries of Guyana and Suriname, and also the French department of French Guiana. Perhaps the most important of the three is Suriname, which has a total human population of 397,000 (PRB, 1989) in an area of 163,820 km² (Times, 1989). Fully 95 percent of the population lives along the coast, with the vast interior, making up 80 percent of the country, inhabited only by scattered groups of Amerindians and Bushnegroes. An excellent nature protection system is in place in Suriname and requires only modest investment to continue and expand. A five-year Action Plan for Conservation of Biological Diversity in Suriname has also been prepared (Malone *et al.*, 1990).

Southern Venezuela: Venezuela south of the Orinoco is also a very important wilderness area. A vast zone that is only sparsely inhabited by several Amerindian groups, the terrain is largely undisturbed rain forest, with some savanna regions as well, and its mountainous nature makes road construction very difficult. Indeed, one attempt to colonize the area has already failed, and the Venezuelan government does not seem to have any immediate plans to exploit the southern wilderness. The largest protected area in South America, Canaima National Park (3 million ha) also falls within this region.

Northernmost Brazil Amazonia: The parts of Brazilian Amazonia adjacent to the Guianas and southern Venezuela are also in largely pristine condition, including the state of Amapá, and northernmost Amazonas and Roraima. It is uncertain how long this area will remain undisturbed, however, since plans for the northern perimeter road (Calha Norte), abandoned in the 1970s, are again under serious discussion.

Parts of the western Amazonian lowlands of Brazil, Colombia, Ecuador, Peru and Bolivia: Upper Amazonia is a vast region that still includes enormous areas of primary rain forest. As colonists spill down from the Andes and move up from southern and northeastern Brazil through Acre and Rondônia, it is unlikely that this area will remain pristine for much longer. Nonetheless, in its present state, much of the region still can be considered a major wilderness.

The central Zaire basin, Gabon, and the Congo Republic: Much of this portion of equatorial Africa is still relatively undisturbed, and has a low human population density. For instance, Gabon covers an area of 267,665,000 km² (Times, 1988) and has a human population of only 1.1 million (PRB, 1989), a population density comparable to that of Suriname. Central Zaire and the Congo Republic also contain major forests with low human populations (IUCN, 1989a).

The island of New Guinea, including Papua New Guinea and Irian Jaya: The entire island of New Guinea is still largely unaffected by modern forms of human exploitation. The Indonesian portion of the island, Irian Jaya, has a human population of only 1.4 million and covers an area of 345,670 km² (Times, 1988) while the larger eastern portion, belonging to the country of Papua New Guinea, has 3.9 million people (PRB, 1989) in 462,840 km² (Times, 1988). Although transmigration schemes intended to move large numbers of ethnic Javans to Irian Jaya threaten the future of this part of the island, and although some development is taking place in Papua New Guinea, this island still has the largest tracts of mature rain forest in the Asian/Pacific region.

Source: Conservation International, unpublished data, 1989.

The National Approach: Major Tropical Wilderness Areas

Major tropical wilderness areas are becoming increasingly rare with each passing day. Simply stated, these are the few remaining parts of the world where very large tracts of primary forest still exist and, because of low human population and little or no development pressure, are likely to continue to exist well into the next century (Fig. 6). As discussed in Chapter III, much of the remaining natural habitat that will persist into the 21st century will be in the form of forest "islands" protected by law but surrounded by a mosaic of degraded lands, agriculture, pasture lands, and urban development. Such small representative tracts of the major forest regions that once existed will need careful management in order to ensure that the species they were created to protect do not disappear. In effect, this will mean that even the evolutionary processes in these will be to a considerable extent controlled by the hand of our own species.

However, in the few major tropical wilderness areas (Box 26) remaining, the situation will be somewhat different, and these areas will become increasingly important for a variety of resons. They will:

- be the last areas where major evolutionary processes can continue to take place with only limited impacts by humans (though the threats of pollution and climate change are increasingly pervasive);
- serve as controls against which the success or failure of the managed ecosystems — the forest "islands" — can be measured;
- be major storehouses of biological diversity, where large numbers of individuals of many different plant and animal species will continue to exist;

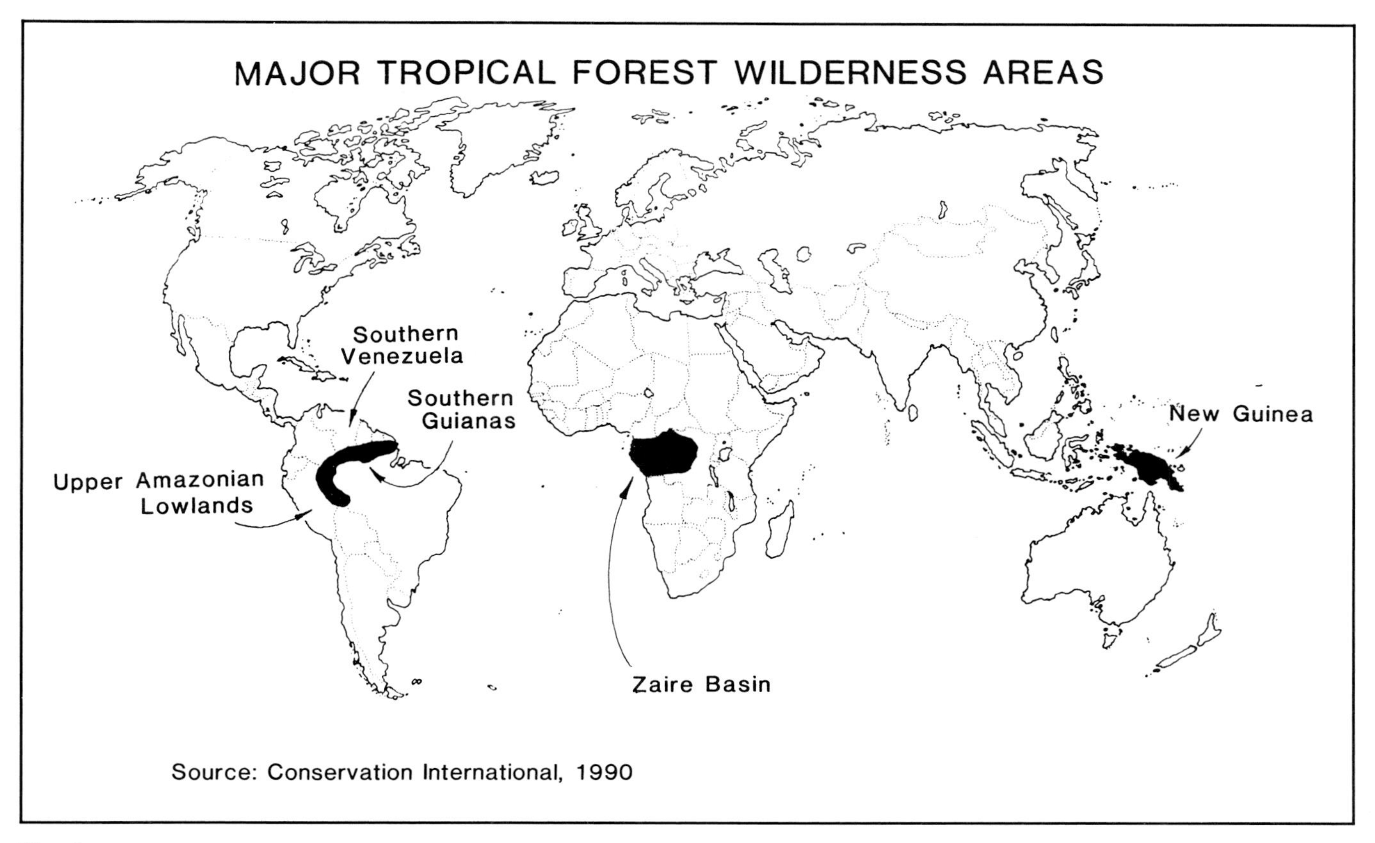

Fig. 6.

- play a key role in maintaining local and, because of their size, global climate patterns (Bunyard, 1987);
- be the last areas where tropical aboriginal human groups can continue to live their traditional lifestyles; and they will
- have ever-increasing aesthetic, spiritual, and scientific value on an increasingly overcrowded, urbanized planet.

Although these areas are clearly not in urgent need of attention and do not require the level of funding that must go immediately into more threatened areas, they should not be overlooked. Those governments fortunate enough to possess them should be convinced of the value the world gives to such areas, and supported and encouraged to maintain them as an investment in the future.

Sites of Outstanding Diversity: Habitats of Threatened Birds in Africa

In their study of threatened birds in Africa and surrounding islands, Collar and Stuart (1985) highlighted the areas where more than one threatened bird occurred, recognizing the economies that are to be had when action aimed at saving a single species becomes a component of action to save an entire ecosystem. They identified five main regions in continental Africa as critically important for threatened birds: the Upper Guinea lowland rain forest block of West Africa; the montane and adjacent lowland forests of Cameroon and adjacent areas in Nigeria, Gabon, and Equatorial Guinea; the forests along and adjacent to the Albertine Rift; the East African coastal and montane relict forests extending south as far as the Eastern Highlands of Zimbabwe; and the forest patches of the Angolan escarpment.

Carrying this approach a step further, they identified those specific forests of Africa, Madagascar, and surrounding islands that would be most important for conserving threatened birds (Collar and Stuart, 1988) (Figure 7). A forest was considered "important" if it fulfilled one of the following criteria:

- it held (or very probably holds) more than one threatened bird species;
- it held only one threatened species, but one that occurs nowhere else or only in one or a few much less significant (or less studied) localities; or
- it held one threatened and one or more near-threatened species.

A scoring system was adopted on the basis of the Red Data Book category of the bird species involved, ranging from 5 points for an Endangered Species to 1 point for a Near-threatened Species. If a species is endemic to a forest, or effectively so for practical conservation purposes, its score was doubled. No weighting was given to species with higher taxonomic distinctiveness (i.e., endemic genera or families), the chief effect of which would be merely to expand the score of the island forests.

Based on these criteria, they identified a total of 75 forests

Key to map of forests in Africa and Madagascar important for the conservation of threatened birds (opposite page):

1. Gola Forest (Sierra Leone)
2. Lofa-Mano proposed national park (Liberia)
3. Mount Nimba (Liberia)
4. Sapo National Park (Liberia)
5. Grand Gedeh County/Grebo National Forest (Liberia)
6. Tai National Park (Côte d'Ivoire)
7. Bia National Park (Ghana)
8. Obudu Plateau (Nigeria)
9. Korup National Park and Mamfe region
10. Rumpi Hills
11. Mount Cameroon
12. Mount Kupe
13. Mount Manenguba
14. Mount Nlonako
15. Mount Oku
16. Dja Game Reserve
17. Forests in Gabon
18. Lendu Plateau (Zaire)
19. Ituri Forest (Zaire)
20. Semliki (Bwamba) Forest (Uganda)
21. Kibale Forest (Uganda)
22. Kakamega and Nandi Forests (Kenya)
23. Forest west of Lake Edward (Zaire)
24. Impenetrable (Bwindi) Forest (Uganda)
25. Nyungwe (Rugege) Forest (Rwanda)
26. Forest west of Lake Kivu (Zaire)
27. Itombwe Mountains (Zaire)
28. Mount Kabobo (Zaire)
29. Marungu Highlands (Zaire)
30. Lower Tana riverine forests (Kenya)
31. Sokoke Forest (Kenya)
32. Taita Hills (Kenya)
33. Coastal forests in south-east Kenya
34. Usambara Mountains (Tanzania)
35. Nguru Mountains (Tanzania)
36. Ukaguru Mountains (Tanzania)
37. Pugu Hills (Tanzania)
38. Uluguru Mountains (Tanzania)
39. Uzungwa escarpment (Tanzania)
40. Southern Highlands (Tanzania)
41. Mount Namuli (Mozambique)
42. Mount Chiradzulu (Malawi)
43. Mount Soche (Malawi)
44. Mout Mulanje (Malawi)
45. Mount Thyolo (Malawi)
46. Mount Chiperone (Mozambique)
47. Gorongosa Mountain (Mozambique)
48. Vumba Highlands (Zimbabwe and Mozambique)
49. Chirinda Forest (Zimbabwe)
50. Coastal forests in Sofala, Mozambique
51. Amboim and adjacent forests, Gabela region (Angola)
52. Bailundu Highlands (Mount Moco) (Angola)
53. Forests of northern Angola and western Zaire
54. Day Forest (Djibouti)
55. Forests around Neghelli (Ethiopia)
56. Forests of south-western Nigeria
57. Daloh Forest reserve (Somalia)
58. Ngoye Forest (South Africa)
59. Bush forest north of Tuléar (Madagascar)
60. Zombitse Forest (Madagascar)
61. Ankarafantsika Réserve Naturelle Intégrale (Madagascar)
62. Andohahela R.N.I. (Parcel 1) (Madagascar)
63. Tsarafidy and Ankazomivady Forests (Madagascar)
64. Ranomafana (Madagascar)
65. Périnet-Analamazaotra Special Reserve (Madagascar)
66. "Sihanaka Forest" (Madagascar)
67. Tsaratanana Massif (Madagascar)
68. Forests around Maroantsetra (Madagascar)
69. Marojejy Réserve Naturelle Intégrale and Andapa region (Madagascar)
70. Forests in south-west São Tomé (São Tomé e Principe)
71. Mount Malabo on Bioko (Equatorial Guinea)
72. Mount Karthala on Grand Comoro (Comoro Islands)
73. Central highland rainforest, Mahé (Seychelles)
74. Plaine des Chicots, Réunion (to France)
75. Macchabé/Bel Ombre Nature Reserve (Mauritius)

as being important for conserving threatened birds. Collar and Stuart (1988) recognize that other forests might well qualify if more were known, but conclude that, in the absence of full threatened-species analysis for any other class of vertebrate, birds can serve very effectively as practical first indicators of sites of general biological importance (notably in terms of endemism). The 75 forests they identify are not considered to represent the minimum number that need conservation in Africa, but they are proposed to be part of whatever that number might be.

The International Council for Bird Preservation (ICBP) is continuing its efforts to identify the most important sites in the world for birds; combined with the work of SSC to identify the top priority sites for a number of other taxa, a very real possibility exists for identifying the sites of great significance for key species at an international level.

Sites of Outstanding Diversity: Plants

The conservation of plant species requires a somewhat different approach from that used to manage animal species. Plants differ from animals in five important ways:

- longevity, either in the case of long-lived perennials, such as trees that can live orders of magnitude longer than mammals, or in the case of annuals or other short-lived species which may possess seed banks in the soil that can survive for decades or even centuries.
- a sedentary nature, as opposed to more motile animals. Plants are rooted in place, while many animals require a range of habitats for breeding, feeding, or for different parts of their life cycle. In some cases a plant species exists as a single population with only one habitat.
- reproduction, that does not always require two sexes. Vegetative reproduction can enable many plant species to recover even if the population has been reduced to a single individual.

Fig. 7: Forests in Africa and Madagascar identified by the International Council for Bird Preservation as important for the conservation of threatened birds (Source: Collar and Stuart, 1988).

- tolerance to inbreeding, so that the disastrous consequences of inbreeding observed in animals are normally avoided. A high level of homozygosity does not militate against the vigor or health of populations of inbreeding plant species.
- a very low ratio of habitat requirement to body size. Many plants can survive for several centuries in a forest niche scarcely larger than the diameter of its leaf rosette.

These key differences between plants and animals in their habitat requirements clearly affect policies in terms of reserve selection and size. While it is generally preferable to establish large reserves, small reserves focusing on plants are a viable alternative in many cases where large reserve establishment is not possible.

Specific action to conserve large numbers of plant species in the tropics requires identification of specific sites where the action can take place, and some sites have extraordinary levels of diversity. For example, Mount Kinabalu in Sabah, Malaysia, has been described as containing "the richest and most remarkable assemblage of plants in the world" (Corner, 1978), including 78 species of figs, 15 species of pitcher plants, and about 1,000 species of orchids. IUCN (1987d) is preparing a book on such centers of plant diversity that will identify some 150 of the world's most important plant habitats (Figure 8). In choosing the sites, it was recognized that it is not always possible to compare individual sites using a set of rigidly applied objective criteria because the amount of information about different sites is so variable. Furthermore, one site may be high in diversity but low in endemism, while another may have lower species diversity but may contain many endemics (as on islands).

With those provisos, two broad criteria have been established: first, the site is evidently species-rich, even though the total number of species present may not be known with great accuracy; second, the site is known to contain a large number of species endemic to it. Sites to be included in the list must meet at least one of these two criteria. Four additional characteristics will also be considered in the selection, and may provide criteria for more detailed site selection at the local level or within a vegetation type: the site is

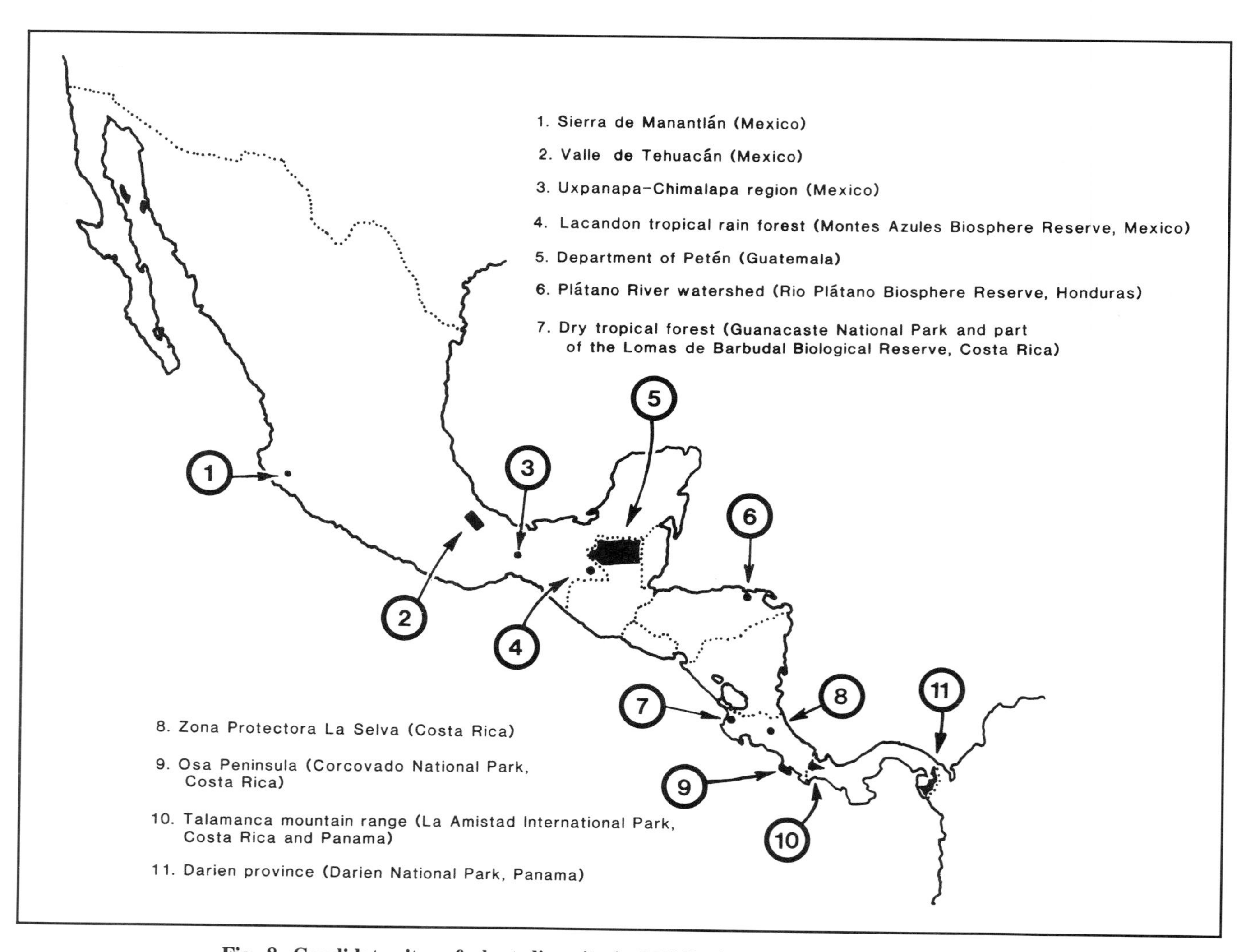

Fig. 8. Candidate sites of plant diversity in Middle America as identified by IUCN.

Amazonia is the world's largest expanse of tropical forest, much of which is still pristine (photo by R.A. Mittermeier).

threatened or under imminent threat of large-scale devastation, includes a diverse range of habitat types, contains a significant proportion of species adapted to special edaphic conditions, or contains an important gene pool of plants of actual or potential value to people.

The centers of plant diversity approach will certainly help improve the effectiveness of protected areas by identifying a representative selection of those where species diversity and/or endemism is particularly high. It will also begin a process that will result in an interactive database that can respond to development needs of various countries.

Sites of Outstanding Diversity: Tropical Forests

Objective analysis of the patterns of distribution of total species diversity is always difficult, and in the case of highly diverse moist forest ecosystems it is an objective that cannot be attained quick enough to constitute a comprehensive tool for planning conservation action. At the same time, all data available on biologically important sites need to be applied systematically. It will be decades before conservation in the tropics can be based upon such a meticulous appraisal of species conservation needs as that which supports the National Nature Reserves network in the United Kingdom (Ratcliffe, 1977). A first priority must be to organize such information as does exist and, while implementing conservation action on the highest priorities thus identified, proceed in parallel with the gradual refinement of the information base. The IUCN protected area systems reviews (IUCN/UNEP, 1986a, b, c) constitute a first approximation that already identifies major areas where investments in conservation management are required.

A recent study undertaken by IUCN in the seven countries of the central African region (Congo, Zaire, Gabon, Cameroon, Central African Republic, São Tomé and Principe, and Equatorial Guinea) takes this one stage further. All sites recognized as being of concern for the conservation of any group of animals or plants, or sites having other critical ecological or biological factors, were inventoried, and data sheets were prepared summarizing the available information on the present status of the site, threats to its integrity, and measures required to ensure its conservation. These data were summarized into a regional action plan that is under consideration by the governments concerned (see Chapter VII).

The danger of excessive concentration of conservation resources in these so-called critical sites is that it could divert attention and resources away from the larger ecosystems. To the extent that the sites are often too small in isolation to constitute viable "conservation units," their protection as islands in a totally transformed landscape would not result in conservation objectives being met. Thus, while these sites can receive highest priority attention, and while the techniques applied to their conservation can provide a model applicable to the forest biome at large, it is essential that measures to protect them be an integral component of management strategies applied to the entire forest estate of the nations where they occur. In many moist tropical areas the optimum scenario would be a network of small, totally protected species-rich areas surrounded by much more extensive areas of managed natural forest.

A systematic approach to the determination of priority sites for forest conservation throughout the tropics, based on the most up-to-date vegetation maps and drawing upon a broad range of conservation expertise within the countries as well as the data collected for the centers of plant diversity effort, is now being prepared by IUCN, in cooperation with British Petroleum, FAO, and several others. It seeks to prepare a database, with digitized maps, for all remaining areas of tropical forests. The information will be published in the form of a series of Tropical Forest Resource Atlases, beginning with Asia in October, 1990. In the Amazon region, a workshop in January of 1990 brought together scientific expertise to determine key areas for conservation in this tropical forest biome. Brazil's environmental ministry (IBAMA) and CI, with coordination from the Royal Botanic Gardens, Kew, are producing a publication and maps generated by the meeting's participants.

One could reasonably imagine a network of 500 carefully selected and managed protected zones in the 57 countries that contain tropical moist forests, with an average size of 200,000 ha, of which 100,000 ha might be in a totally protected core zone. These 500 reserves would total 100 million ha, of which 500,000 would be totally protected. This would represent nearly 9 percent of the 1,200 million ha of moist tropical forest estimated to remain in 1980, a reasonable target to aim for. (It should be noted, however, that even this figure is far short of what conservation biologists would consider necessary to conserve viable populations; many experts feel that the principal protected areas should be in the 1 million ha range. Such large areas are likely to be feasible only through a mix of management regimes which control inappropriate land uses.) Of the 280 million ha of tropical forest that originally existed in the Guinean and Congolian forest blocks of Africa, for example, a total of just over 10 million ha has been given complete legal protection. This represents some 8.4 percent of the forest cover remaining in 1980. Proposals are now under consideration for a further 2.5 million ha of reserves, bringing the total to 10.5 percent of the remaining west-central African forests.

Many tropical developing countries already have over 5 percent of their land area under some form of legal protection, and the above targets seem reasonable in light of the present situation throughout the moist tropics of Asia and Africa; far larger areas may be possible in Latin America. The critical issue is to ensure that the resources are mobilized to guarantee the long-term survival of such a network of reserves (see Chapter VIII).

Adequate management programs for such reserves, in very general terms, require an initial capital investment estimated at $5 to $10 per ha and a recurrent budget of $1 to $3 per ha/year (though these figures will vary widely on the basis of distance from urban centers, objectives for tourism development, uses of adjacent lands, possible resettlement, and so forth). The hypothetical network of tropical forest protected areas could therefore be established for between $500 million and $1 billion — some of which has already been invested — and would then require about $100 to $300 million a year to maintain.

Under this ideal scheme, an "average" tropical country might require $5 million annually for management, with appropriate figures for establishing any new reserves required. Such figures are not so far out of line with what is already being spent on conservation-related programs in at least some of the countries concerned, though resources are seldom employed in the most effective way and many conservation agencies are prevented from carrying out their assigned tasks because of conflicting policies in other sectors (such as agriculture and foreign trade).

Conclusion: Guidelines for Determining Priorities for Conservation Action

Biological resources provide the basis for sustainable forms of development in all countries, so strategies need to be developed for all the countries of the world acting together in the interests of all humanity. Clearly, the best policy is to conserve areas of maximum biological diversity and at the same time ensure that within these areas endangered species of plants and animals, or those that are suffering from severe genetic erosion, are subject to special management procedures that will ensure their survival.

At the same time, every nation — and every local, national, or international institution — has only limited resources at its disposal for dealing with conservation priorities. The dilemma is how to use these resources in the most effective way. No single scheme for establishing conservation priorities can be acceptable to all individuals, organizations, or nations because different perspectives, values, and goals influence the importance given to various considerations. But a decision framework can allow conscious evaluation of the tradeoffs and value judgments that are made in reaching a set of priorities. The following elements will often be useful:

- *Distinctiveness.* To maintain the variety of the world's life forms and processes, higher priority would be given to more distinctive elements of that diversity. A community containing many widespread species makes a smaller contribution to the conservation of biodiversity than a community with many endemic species. A subspecies of a polymorphic and widespread species deserves less conservation concern than a monotypic species, or a species that is the only representative of its genus, family, or order. Similarly, habitats that are rare or contain numerous endemic species deserve higher priority than habitats that are widespread or contain species common elsewhere. And higher priority should go to biogeographic units that have no or few protected areas than to such units with numerous protected areas.

- *Threat.* In various regions of the world biodiversity is subject to threats of very different magnitudes; areas where threats are greater should receive higher priority than areas with lesser threats. Other things being equal, an endangered species should be given priority over a vulnerable one; a vulnerable over a rare one; and a rare species over one that even if it is declining is considered insufficiently threatened to qualify for one of the three IUCN categories. The major weakness of this approach, as discussed in Chapter II, is that imminence of threat is often a matter of the state of knowledge about the species, and that in turn generally becomes less adequate as the number of species found in the area increases.
- *Utility.* Different subsets of the world's current biodiversity may be equivalent in the "amount" of biodiversity that is maintained, but very different in their current or future utility. In assessing priorities, particularly in tropical countries, highest priority needs to go to the species whose loss will have the greatest negative impact on humanity. This, admittedly, is an anthropocentric perspective, but those species most likely to earn the necessary political support for their conservation will include wild plant species related to domestic food crops, wild relatives or forms of domestic animals, medicinal plants, species harvested by people, animal species that are useful research models, and fodder plants for domestic animals. Similarly, it is easier to justify the conservation of an ecosystem that protects many threatened species or plays a critical role as the watershed for a major irrigation project than one that provides only indirect ecosystem services to humanity.

Each of these three categories contains sub-categories addressing specific issues. For example, distinctiveness can be subdivided into categories of genetic, species, and ecological distinctiveness. Considerations of threat can examine the level of endemism (range) of the species, susceptibility to impacts, and development pressure. Under utility, a resource can be assessed to evaluate its current utility, possible future utility, local value, and global value.

A resource can be evaluated for each element of this set, with the various components weighted as to their perceived importance. The scores could then be combined to yield the relative value of the area's biodiversity for the purposes of conservation planning.

The utility of such a framework lies less in the final "score" than in the process of arriving at the score. When greater weight is given to one element of the set, a planner reveals the value judgments that are incorporated in the priorities. For example, if a planner is primarily concerned with the "global value" of biodiversity, this category would be given higher emphasis than "local value." The resulting priorities would not be more or less correct than a scheme that provided more weight to the local values of biological resources; it would simply be based on different value judgments. The following principles may be useful in helping to guide decisions about priorities for specific project activities, suitably adapted to the particular needs of the country and agency involved.

- **Ensure that decisions are based on the best available assessment of information.**
  The first requirement for making informed decisons about priorities is good information. While action should seldom be delayed by a lack of information, it is essential that such information as is available be utilized fully. Often, such information is widely dispersed and unpublished, but surveys of existing government institutions can lead to relatively complete information (e.g., forest surveys submitted by concessionaires, trade statistics, etc.). At the international level, the World Conservation Monitoring Centre and UNEP's Global Environment Monitoring Service provide a useful entry point. Information on a number of subjects — including species, habitats, local human communities, patterns of resource use, population trends, and local development projects — is necessary to provide a balanced picture. During the process of collecting information, gaps can be identified for future research.
- **Establish objectives for conservation.**
  Once the available information has been collected, objectives for conservation can be determined. This step, which seems relatively simple, is often ignored or left unstated; this can lead to misunderstandings about what is intended. Determining objectives is best done as part of a process of consultation involving those who will be affected by how a resource is to be managed, so both managers and consumers should be involved in the process (see Chapter VII for further discussion).
- **Design support activities that build the self-reliance of the recipient, rather than build dependence.**
  Earlier discussions have stressed the point that long-term success in conserving biological resources will depend on the cooperation of the people who are most directly concerned with those resources. In order to build sustainable relationships between rural people and their resources, local communities must be provided with the tools with which they can build their own conservation action. Building the capacity to manage resources is far better than providing a "turn-key" gift; for example, a training workshop on how to prepare management plans for species or protected areas is usually far better than sending in an

expert to prepare such a plan. Activities need to be designed to ensure the long-term economic viability of protected areas, including designing systems of sustainable utilization of biological resources (in areas that can support such harvests).

- **Ensure that the need for support has been clearly identified by the recipient.**
  The worst sort of international support is that which is foisted upon an unsuspecting tropical recipient; the best sort is that which is identified by the relevant authorities themselves as being essential to their ability to carry out their assigned duties more effectively. When the need is clearly stated, political support for the activity is far more likely to be forthcoming, as is follow-up action. Incentives may also be required to encourage countries to seek outside support for conservation action, and to afford biodiversity an appropriately high priority in development assistance programs.

Following page, overleaf: An ocelot *(Felis pardalis)*, one of 36 species of wild cats to be featured in a forthcoming IUCN/SSC Action Plan for Species Conservation (photo by A. Young).

CHAPTER VII

THE ROLE OF STRATEGIES AND ACTION PLANS IN PROMOTING CONSERVATION OF BIOLOGICAL DIVERSITY

Strategies and action plans can be very useful in presenting an agreed agenda for attention by various institutions and individuals. They are most successful when they are generated by those who are closest to the problems, and who are involved in implementing solutions.

Chapter VI suggested a number of approaches to determining priorities, emphasizing the importance of determining objectives for conservation of biological resources at the appropriate level. One of the best ways to ensure that the various institutions involved in conservation are in general agreement on priorities is to prepare a strategy that defines the basic problems and paints a broad picture of appropriate objectives. Strategies are turned into action through a more tactical process of planning specific activities to address the broad strategies; this often involves the preparation of an action plan.

Large numbers of strategies and action plans have been prepared, at local, national, regional, and global levels. Some have been quite useful (such as the *World Conservation Strategy*), while others (such as the Desertification Action Plan) have fallen far short of expectations. In fields that relate to biodiversity, action plans have tended to be global (such as the Bali Action Plan (Annex 4), Marine Mammal Action Plan, and the Tropical Forestry Action Plan), regional (such as the various protected area action plans prepared by IUCN/CNPPA, or the regional plans prepared by WWF), or taxonomic (such as the species action plans prepared by IUCN/SSC).

The biggest problem with strategies and action plans is lack of implementation, and this relates in turn to the process through which they are prepared. Experience has demonstrated conclusively that action plans — whether for an area, a species, a nation, or a region — need to be developed in the closest possible collaboration with those who are most directly affected by the action proposed.

This chapter examines a number of the current strategies and action plans for species and habitats, both to illustrate the current level of such planning and to provide background for the preparation of other such plans.

Strategies and Action Plans for Conserving Species or Species Groups

The Botanic Gardens Conservation Strategy

As discussed in Chapter IV, botanic gardens have great potential for contributing to the conservation of plants. In order to tap this potential, the Botanic Gardens Conservation Secretariat has prepared a Botanic Gardens Conservation Strategy. This document, published in 1989:

- recommends that each individual garden clarify its commitment to conservation in a Mission Statement and adopt more professional standards of management to achieve its Mission.
- provides the basis for a more coherent Accessions Policy that takes account of conservation needs and of what plants are held in other botanic gardens.
- outlines ways to improve the documentation of plant records and the verification of plant holdings, including computerization to improve management of the collection and to facilitate exchange of data between institutions.
- explores the relationships between wild and managed conservation for botanic gardens' efforts; for wild (*in situ*) conservation, outlines the role of the garden in habitat evaluation, rare species monitoring, "habitat gardening," and management of protected areas; for managed (*ex situ*) conservation, proposes strict rules and procedures for the establishment of reserve collections, gene banks, and other germplasm collections, and outlines methods of sampling populations to maintain adequate genetic variation.
- emphasizes the facilities that botanic gardens can offer for educating their estimated 150 million visitors each year.
- recommends that each garden provide a service to its local community as a resources and information service.

- Provides a framework for training of personnel, with emphasis on conservation.

Species Action Plans

For some of the most important plants and animals, IUCN's Species Survival Commission has established more than 100 "Specialist Groups" that are working to assess the status and priorities of these taxa (Box 27). Drawing on a worldwide network of specialists who are working on the taxa involved, SSC's Specialist Groups cover species groups such as Antelopes, Primates, Cats, Crocodiles, Cranes (with ICBP and the International Crane Foundation), and Palms. Each group is assigned the task of preparing an Action Plan, which includes conducting a thorough overview of the status of all the species within their brief, establishing a system of priorities, and defining the most relevant projects for addressing the priorities. These Action Plans fully involve experts in the countries where the species live, and are supported by a wide range of organizations, including WWF, UNEP, Wildlife Conservation International, the International Fur Trade Federation, the Chicago Zoological Society, the U.S. Fish and Wildlife Service, the Center for Marine Conservation, the American Association of Zoological Parks and Aquaria, and Conservation International.

The Species Action Plans serve several purposes:

- by establishing the priorities within a taxon, an Action Plan ensures that the right actions are addressed first by those with the ability to deal with the issues raised.
- the Action Plans organize masses of information, some of it unpublished or obscure, in ways that can be readily available to other organizations or groups.
- Action Plans are being produced in attractive formats, which can help raise funds for addressing the priority actions.
- Action Plans enable IUCN to respond quickly to opportunities for linking conservation of species to other major development projects, such as tourism, rural development, agroforestry, and resource management.
- a set of Action Plans from several groups will enable "species conservation hotspots" to be identified, where high-priority actions are required for a number of different taxa; identifying such hotspots can lead to very cost-effective conservation efforts.
- each Action Plan is used by the Captive Breeding Specialist Group to produce its own Captive Breeding Action Plan for the taxa concerned.
- an Action Plan can provide the glue that holds together specialists working in a wide range of countries and habitats, bringing the Specialist Group members together to assess where needs are being met and where serious obstacles still exist (Stuart, 1987).

It is too early to tell how the various Action Plans will come together to reinforce each other, and whether they will indeed combine to identify relatively few "hot spots" requiring urgent attention. The taxonomically oriented Action Plans are being supplemented by regional and national Action Plans on biodiversity; those on Venezuela and Madagascar are at an advanced stage, and work on an Afrotropical biodiversity strategy is a major emphasis of SSC's work during 1989 (see below).

Being species-specific, these Action Plans inevitably have a fairly narrow focus (Box 28). They address only part of the problem, typically the part that can be addressed by those who prepared the plan. Thus most plans give heavy emphasis to additional surveys and research, and to the establishment and strengthening of protected areas. While these actions are certainly necessary to conserve the species in question, they are not sufficient to do so. Few of the species Action Plans have addressed the indirect but nonetheless fundamental causes of species depletion, those that lie in such areas as international trade, agricultural policy, land tenure systems, and economic equity; instead, they are directed at clarifying

Box 27: IUCN/SSC Action Plans for Species Conservation.

The current status (as of February 1990) of SSC Action Plans for species is as follows:

Published

African Primates	Asian Primates
Duikers	East African Antelopes
Mustelids and Viverrids	River Dolpins
European Bats	Dolphins, Porpoises & Whales
Kouprey	Southern African Antelopes
African Forest Birds	Tortoises & Fresh Water Turtles
Rodents	River Dolphins
	Asian Rhinos

In final stages of preparation

African Elephants and Rhinos

Final draft being prepared

Asian Elephants	Molluscs
West & Central African Antelopes	
African Insectivores and Elephant Shrews	

First draft being prepared

Canids	Cats
Seals	Otters
Procyonids	African Rodents
Lagomorphs	Pigs & Peccaries
Sirenia	Equids
Caprinae	Cranes
European Reptiles & Amphibians	North American Plants
	Malagasy Lemurs
Neotropical Primates	

In concept stage

Australian Marsupials & Monotremes	Fruit Bats
Tapirs	Hippos
North African & Asian Antelopes	Deer
	Crocodiles
Parrots	Orchids
Marine Turtles	Bears
Palms	

ing species priorities as a contribution to more broad-based plans. The implementation of species Action Plans therefore needs to be closely integrated with broader economic and social aspects if the plans are to meet their conservation objectives.

Box 28: A Typical Species Action Plan: African Primates.

The following is the contents of the Action Plan for African Primate Conservation 1986-1990:

Introduction
Classification
Priority Ratings of Species and Subspecies for Conservation Action
Distinct Communities and Ecosystems
Recommended Conservation Action
Conclusion
References
Annex 1: Species Lists for Countries with 10 or More Primate Species
Annex 2: Distribution and Status of the Most Threatened Primate Species

Source: Oates, 1985.

Action Plans for Conserving Habitats

The Tropical Forestry Action Plan

The Tropical Forestry Action Plan was prepared through the combined efforts of governments, forestry agencies, UN agencies, and non-governmental organizations. The TFAP was conceived as a framework and an instrument for stimulating commitment and action within forestry and other sectors to address forest resource management challenges in developing countries. It was intended to act as a catalyst to maximize the potential contributions of sustainable use of forest resources to rural livelihoods, food and energy security, income and employment, and other national development priorities.

The Tropical Forestry Action Plan was launched in 1985 by FAO's Committee on Forest Development in the Tropics, an inter-governmental body dealing explicitly with questions of tropical forestry. It has contributed to raising political awareness of the seriousness of tropical deforestation and has stimulated a broad-based effort to develop national forest strategies. Over 50 countries across Africa, Asia, and Latin America are involved in developing and implementing national forest strategies. The purpose of the national strategy exercises is to foster a dialogue among governments, aid agencies, and NGOs on priority areas for policy reform and investment to control deforestation and improve forest resource management. The World Bank and other multilateral and bilateral aid agencies are increasing their support in this area.

The main issues that require action, identified within the framework of the Plan, fall within five related priority areas, of which one is the Conservation of Tropical Forest Ecosystems. The program on conservation covers the following objectives:

- to develop and adopt methods that enable the forest to be used for the production of wood, food, fodder, and other non-wood products in a sustainable manner.
- to select and adopt a series of protected areas covering the whole range of variation of tropical ecosystems.
- to expand the concept of conservation policy and management to include maintenance of intra-specific variation of species of actual and potential socio-economic importance, and adopt measures that conserve as much as possible of other species whose qualities are not yet known.
- to consider national parks and other protected areas within the context of the general pattern of land use of areas that surround them, and to design and operate them in ways that are acceptable to local people and bring benefits to them in the short as well as the long term.
- to develop closer links between policies for the conservation of ecosystems and genetic resources of priority species, and to promote measures that encourage the recovery of natural vegetation to provide protection for soil and water catchment areas.
- to assemble basic biological information for the conservation of germplasm.
- to raise awareness, at all levels, of the importance of ecosystem and genetic resource conservation.
- to train staff to implement the objectives listed above.

The TFAP has been criticized for being insufficiently critical in its perception of the role of development assistance in forest lands in the tropics. It failed to point out the serious shortcomings of past attempts by international agencies to "develop" tropical forest resources, instead focusing excessive attention on the objective of generating more finance for conventional approaches to forestry. Proponents of the TFAP now recognize these problems and are putting more emphasis on qualitative improvements in forest management and less in quantitative increases in aid. However, conservatism on the part of forestry professionals and inertia in both national and international institutions is making it difficult to realize the ecosystem conservation and social benefits to which the TFAP aspires.

Several opportunities exist for strengthening the TFAP planning process. First, although the TFAP called for grassroots participation in planning, actual participation in TFAP exercises has tended to be largely restricted to governments and external aid agencies. Second, the TFAP exercises have tended to focus too narrowly on the forestry sector. Many of the solutions to the problems of deforestation lie in agriculture, planning, finance, and other sectors and these must become involved in the process. Finally, greater emphasis must be placed on policy reforms, particularly policies outside the forestry sector. Often, government policies provide strong incentives that encourage deforestation (see, e.g., Repetto and Gillis, 1988). Policy reform to control deforestation and to promote sustainable forest resource use is essential if expanded investment and technical assistance programs are to be effective (Hazlewood, 1989; WRI, 1989b).

Action Plans for Protected Areas

Drawing on the WCMC database, and on some 400 experts in the respective biogeographical realms, IUCN has conducted detailed reviews of the protected area system of the Oceanian (IUCN/UNEP, 1986a), Afrotropical (IUCN/UNEP, 1986b), and Indomalayan (IUCN/UNEP, 1986c) realms. These reviews, prepared in collaboration with UNEP, were designed to:

- evaluate the representational coverage and conservation importance of the existing protected areas system of the realm;
- identify gaps and shortcomings in the system;
- evaluate the conservation importance of existing and proposed reserves and other areas of biological richness and recommend where additional protected areas are needed;
- identify priorities for strengthening protection;
- consider the suitability of the status, boundaries, design, and effectiveness of the existing reserve system; and
- identify conservation management needs in critical habitats outside protected areas.

The reviews evaluated protected area data in terms of: how much of each biogeographical sub-division is protected; coverage of the regional and altitudinal range of each sub-division and inclusion of other features (e.g., physical or ethnic interests) that need protection; coverage in relation to species richness, centers of high biological distinctiveness or endemism and in relation to threats to habitat; coverage in relation to commercial interest or value of content (e.g., genepools); the category status of the individual protected areas; evaluation of the designs of protected area systems on the basis of island biogeographical theory; the effectiveness of management in individual reserves; and consideration of adjacent land-use and critical habitat requirements. In essence, these reviews were exercises in applied conservation biology.

Since these reviews cover entire realms, they need to be supplemented by national reviews; indeed, one of their objectives is to help promote such detailed national-level reviews. Reviews of national systems of protected areas have been prepared, or are in preparation, in countries as diverse as Thailand, Saudi Arabia, Sri Lanka, Gabon, Madagascar, Oman, and Indonesia.

Priorities are determined on the basis of the information derived from the processes described above. The general perspective on protected areas was provided by the Bali Action Plan, developed at the World Congress on National Parks held in Bali, Indonesia, in 1982, to provide the guidance to enable protected areas to meet the needs of the 1980s (see Annex 4). This Action Plan has proved extremely effective in helping guide IUCN's activities in the international aspects of protected areas; CNPPA (1988) has presented a summary of progress in the first five years of implementing the Bali Action Plan.

Members of CNPPA have also prepared a set of action plans for each of the four tropical regions (Africa, tropical Asia, tropical America, the Pacific), each deriving from a meeting of protected area managers and scientists from within the region. Each of the action plans contains a list of objectives and high priority actions for the entire realm, country priorities, and recommended international actions.

The protected area action plans have been reasonably effective in addressing the major problems within the sector, and a number of the priority activities have been implemented. For example, on the basis of the South Pacific Action Plan, the government of New Zealand seconded a full-time park planner to the South Pacific Commission to help implement the plan. In Africa, the action plan helped generate considerable support for training efforts. And in tropical Asia, the action plan helped encourage a number of governments to join the World Heritage Convention.

Cross-Sectoral Strategies and Action Plans

SSC has recognized the limitations of its taxa-based action plans, so it has embarked on preparing national or regional biodiversity action plans that seek to bring together information on all taxa. Venezuela and Madagascar (in final draft) are the first two national biodiversity action plans being prepared; the latter is contributing to work of the World Bank on conserving biodiversity in Africa. In addition, IUCN is working with the International Council for Bird Preservation to prepare a biodiversity strategy for the Afrotropical Realm. The strategy is directed towards decision-makers in African governments. It seeks to distill the many recommendations made by SSC, CNPPA, ICBP, and other institutions into a short and readable document.

But even these more comprehensive plans still tend to treat conservation of biological diversity as if it were a sector. And indeed, sectoral agencies — such as national parks and wildlife management departments — do have an important, even dominant, role to play. But earlier chapters in this book have pointed out that significant parts of the real problems still are not being sufficiently addressed by taxa-oriented or protected area plans and strategies.

The World Commission on Environment and Development (WCED) has pointed out that environment and development are not separate challenges, but are inexorably linked. WCED (1987) stated that: "Development cannot subsist upon a deteriorating environmental resource base; the environment cannot be protected when growth leaves out of account the costs of environmental destruction. These problems cannot be treated separately by fragmented institutions and policies. They are linked in a complex system of cause and effect."

Many of the problems in conserving biological resources

are related to the fact that responsibilities are divided into sectoral units, leading to fragmentation, poor coordination, conflicting directives, and waste of human and financial resources. This can only be overcome by integration, by examining the impact of decisions in one sector on the ability of another sector to depend on the same resources. Integration is not easy, and in some respects it is not very practical. Still, an optimal balance point can be found where the benefit of considering secondary impacts (or externalities) is overtaken by the cost of doing so. In most cases, this balance point lies well beyond the current practice of taking decisions based on a very narrow range of sectoral considerations.

As one step in this direction, IUCN's members and collaborators in the Central Africa region have collaborated in the preparation of an Action Plan for the conservation of the continuous block of forest, the Guineo-Congolean, which extends into the six countries of the region. This action was based on the recognition that actions in each country influenced forest resources in those of its neighbors and that knowledge and experience of conservation management could usefully be shared among the countries. The Action Plan proposes a network of sites of critical importance that will be brought under conservation management as part of a $30 million regional program to be funded by the EEC. In each site, a critical forest area for biological diversity conservation will be safeguarded by developing sustainable agricultural and forestry practices on surrounding land. Each project will focus on a different approach to managing these lands and experience will be shared through annual workshops that will rotate among the sites. The Action Plan identifies numerous other critical sites where funding agencies might invest in replicating successful approaches, thus extending the network. Complementary policy measures needed to create conditions favorable to the success of the program are identified in the Action Plan (IUCN, 1989a).

A Global Strategy for the Conservation of Biological Diversity

It is apparent from the discussion above that action plans and strategies, when designed appropriately and implemented with vigor, can make important contributions to conservation. As noted earlier, a collaborative effort of the World Resources Institute, IUCN, and UNEP, working with other institutions, is leading to the preparation of A Global Strategy for the Conservation of Biodiversity, as a companion to the new version of the *World Conservation Strategy* now being prepared.

The aim of the Strategy is to provide a comprehensive framework to stimulate urgent, positive, innovative, and coordinated action to stem the loss and degradation of the world's biological resources and enhance the contribution of these resources to human well-being. The Strategy will be developed by and for national governments, NGOs, resource managers, scientists, international institutions, multilateral banks, and bilateral aid agencies. The development of the Global Strategy will be centered around a series of regional workshops in Asia, Africa, Europe, Latin America, and North America; several of these will also contribute to Regional Biodiversity Conservation Strategies. The Global Strategy will include considerations of a variety of factors influencing biological resource conservation, such as international financing, international cooperation, research, education, training, public awareness, and ecological restoration. However, the development of the strategy will place major emphasis on six pivotal issues:

a) *Root Causes*

Patterns of biological resource use are influenced by the incentives and disincentives that exist within the framework of national and international policies pertaining to agriculture, forestry, land tenure, foreign assistance, trade agreements, tariffs, defense, and so forth (McNeely, 1988). Nations are suffering serious economic losses, individual well-being is declining, and future generations are losing invaluable assets as a result of incentive structures that often favor unsustainable patterns of resource use (e.g., Repetto, 1988; Mahar, 1988) and that discourage local adaptation to environmental conditions. How can biological diversity be assigned appropriate prices, so that cost-benefit analyses can be carried out as a basis for designing incentive systems? What options for national and international policy reform are available to design and implement incentives and disincentives that will ensure sustainable patterns of resource use? What economic incentives are available to encourage conservation of biological resources by people living around protected areas? What is required to ensure that senior government officials become familiar with economic incentives that might be used to conserve biological diversity? How might national and international institutions be structured to achieve these ends?

b) *Sustainable Development*

In a period in earth's history that is characterized by very rapid changes in human land use, technology, climate, and a series of other factors, predictions about the future are problematic at best. Under such conditions, what are sustainable patterns of biological resource use? At the level of the local community, what are the obstacles to the development and persistence of sustainable local production systems and the opportunities for the development of policies permitting and fostering the development of such systems?

It appears that the most useful unit of analysis is the local rural community, because these are the units most directly dependent on the resources available within a fairly circumscribed area for most of their requirements (with many technological and energy inputs from afar). How can such communities manage biological resources to become more self-reliant, without making undue sacrifices in comparative standard of living? Clearly, the structure of the incentives that influence patterns of resource use must be tailored to the nature of local ecological and social systems. The policy framework cannot attempt to dictate specific patterns of sustainable resource use; the variety of ecological and social systems demands unique solutions for each setting. Instead, policies must permit and foster the development and persistence of sustainable local production systems, encourage the search for means of increasing their contribution to human needs, and encourage innovation and the development of alternative methods of use of biological resources.

Developing patterns of local resource use that are sustainable and that enhance the resource base will require that:

- appropriate systems of management responsibility are established within local communities;
- the benefits and costs of biological resource use that are normally external to the market be measured and incorporated into economic models and into the public consciousness;
- the substantial knowledge possessed by human cultures regarding the use of their local resources be maintained and enhanced as the basis for further development; and
- science and economics be applied to the identification of new values (products, foods, commodities) that might accrue to local people as a result of biological resource conservation (including factors such as marketing and trade that will ensure the sustainability of resource use).

In support of principles of community self-reliance, national and international policies are required that will enable governments, industry, and private enterprise to contribute to conserving biological diversity; the first step may be acceptance that local resource-using units are to be shielded from external interference (or, rather, the conditions under which such units are to be shielded). In addition, many of the problems facing biological diversity are global in nature — climate change, deforestation, environmental pollution, species extinctions — and require global solutions.

c) *Science and Its Application*

Knowledge of the status of biological resources and methods for their management represent the foundation on which policy decisions are made and the means by which their outcome is monitored (Reid and Miller, 1989). New approaches are required by which science can:

- greatly accelerate the identification and description of the millions of species still unknown (and therefore unavailable for scientific investigation);
- contribute to the identification of methods of increasing the capacity of systems to provide services to humanity;
- guide the protection, inventory, study, use, and monitoring of biodiversity and biological resources;
- provide the knowledge that is necessary for establishing priorities for action;
- develop biological indicators, or measurements, that can provide decision-makers with accurate and timely information on the effects of policy decisions;
- assess the effects of various forms of habitat management and utilization of biological resources on the composition and diversity of species communities;
- synthesize existing knowledge in forms that can be used by planners, managers, and local people; and
- develop a research agenda that will meet the needs identified in the Strategy.

Many local communities already contain an excellent basis of knowledge of how to manage the resources within their local ecosystems (Warren *et al.*, 1989; BOSTID, 1986). However, they are unlikely to be aware of innovations from outside their local systems that could enhance yields of useful products and increase biological diversity. Science therefore has very important contributions to make to local adaptations, through identifying the ecological functions of the various components of ecosystems and the way new and improved agro-ecosystems can be designed for specific localities.

Science can help mobilize traditional knowledge through:

- preparing guidelines on methods for obtaining, assessing, and presenting traditional conservation knowledge;
- preparing an inventory of traditional knowledge systems, highlighting those aspects that can contribute to conservation and development, and giving special attention to traditional knowledge systems in danger of being lost;
- documenting the role of women in resource utilization in traditional societies and ensuring that knowledge held by women is given appropriate recognition; and
- translating traditional means of conservation — sacred groves, community responsibility, taboos, etc. — into forms useful to development planners and managers of biological resources.

d) *Enhancing the Management of Biological Resources*

The responsibility for managing the world's biological resources falls on numerous institutions and individuals. Various activities pertaining to resource conservation have different goals and objectives, yet together the spectrum of land uses and conservation activities — ranging from national parks to agricultural and grazing land, from multiple-use protected areas to private forest land, and from zoos to botanical gardens to seed banks — must form a stable and integrated system meeting human needs through sustainable patterns of resource use. What patterns and types of land use and *ex situ* conservation activities will meet these needs? What institutional and policy framework will provide the necessary integration between sectors?

In many parts of the world, systems of land and water use need to be developed urgently, to meet long-term development goals before key resources and habitats are lost or degraded. While many natural habitats are being converted into uses — such as agriculture, aquaculture, and forestry — that yield greater productivity to humans, the natural value of some areas is so significant that they need to be converted with great care, or left in their natural state. Some of these areas may prosper through "benign neglect" while others will require intensive management to restore or maintain their natural value. Some areas will require legal designation as protected areas, while proper incentive structures may ensure the conservation of others in the status of private or common property.

The Strategy will seek means to identify the location and management requirements of areas important for genetic materials, the perpetuation of species, and the regulation and purification of water flows, as well as areas whose conversion would contribute little to humanity because of their inherently low productivity or their susceptibility to erosion.

e) *Information for Action*

A vast amount of information regarding biological resource use and status is available from many sources. While all recognize the potential value of that information to decision-makers if it were integrated and continually updated, attempts to do so have achieved little success. The technological capacity now exists to link databases together in a network that can be both integrated and continuously updated by those who are using the data actively. The demand for such information will grow as the importance of biological conservation issues is demonstrated (a and b, above) and as the way the information can be applied to solving real-life resource management problems is shown. Based on the discussion in Chapter V, the Strategy will address questions such as:

- What information is needed to support policy reform?
- What information is needed to help identify sites important for conservation?
- What information is needed to manage these sites?
- What information is needed to monitor the biological resource management policies?
- How should information be packaged so that it has the desired effect on decision-makers? On local communities? On the general public?

f) *Formulating an Effective Response to Problems Facing Conservation of Biological Diversity*

Detailed analyses of the cause of a problem and definition of steps toward its solution do not guarantee that the solution will be adopted, as many failed action plans can attest. Such action plans and strategies have failed not because of inappropriate recommendations but because of the failure to consider issues of process, constituency, and commitment. On the other hand, some very positive changes have been brought about in public and government behavior, often very quickly. Anti-litter and anti-smoking campaigns have worked reasonably well in some places, and government responses to the problems of human-induced changes in the atmosphere have been dramatic. Based on a critical review of existing action plans and strategies, and of social movements that have been reasonably successful, the Strategy will consider:

- How can a stronger constituency be developed for the conservation of biological diversity?
- Where are the real pressure points for conserving biological diversity: Industry? Commerce? Industrial governments? The defense establishment?
- What are the "images" that need to be packaged to gain more public support for conserving biological diversity?
- What are the regional mechanisms that will be most effective in promoting implementation of the Strategy?

Conclusions

Action plans and strategies can be influential mechanisms for stimulating and coordinating conservation activities. Among the factors that may contribute to the utility of biodiversity strategies and action plans, the following seem particularly important:

- the degree to which the action plans were prepared by those who will be responsible for implementing them (the "bottom-up" approach usually being more effective than the "top-down" approach);
- the degree of political and financial support for the plan, among both governments and other institutions that may be able to contribute;
- the extent to which the action plan addresses real needs, either of the areas or species concerned or the implementing institutions;
- the effectiveness of mechanisms to follow up on the recommended actions, and to generate the necessary funding; and
- the degree to which the strategy or action plan contains both necessary and sufficient activities to solve the problems being addressed.

Action plans are required to address the specific needs of geographic areas, such as nations or regional seas, and to address particular topics, such as the global network of protected areas and groups of species and varieties. National conservation strategies, environmental profiles, river basin and regional development plans and other existing approaches can be amended where necessary to incorporate biodiversity considerations. The regional protected areas strategies of IUCN, the Bali Action Plan, the Biosphere Reserve Action Plan of Unesco, and the various Regional Seas action plans of UNEP all need support in funding and implementation. Similarly, plans for selected wetlands, the Tropical Forestry Action Plan, and other ongoing initiatives should receive further reinforcement.

All of these strategies and action plans can contribute to conserving biological diversity, but none of them are likely to have very great impact by themselves because of the very complicated nature of linkages between the sectors that affect biological diversity. Even if all of their actions were to be implemented, most action plans can address only a part of the problem and often can provide only symptomatic relief.

The Global Strategy for the Conservation of Biodiversity is designed to take several significant steps farther, in attempting to identify the root causes of the problems of conserving biological diversity and suggest ways of mobilizing resources for their solution. A very considerable investment of time, energy, and resources will be required to stop the erosion of our planet's biological wealth. Careful planning at both strategic and tactical levels can help ensure that the investments made will yield the greatest possible return.

CHAPTER VIII

HOW TO PAY FOR CONSERVING BIOLOGICAL DIVERSITY

While many will agree to the desirability of conserving biodiversity, governments still have difficulty in finding sufficient financial resources for addressing the problems of conservation in a manner commensurate with the needs of society. Innovative funding mechanisms are required.

This book has attempted to demonstrate that biological diversity is a global resource as well as a national and local one, with conservation bringing benefits to all of humanity. Yet current threats to biodiversity are greatest in developing countries that have great biological diversity coupled with severely restricted financial means for supporting conservation efforts. At the same time, many governments are providing heavy subsidies to activities that have severe negative impacts on biological diversity; subsidized cattle ranching in Amazonia (Binswanger, 1987) and Botswana (Perrings *et al.*, 1988) are notorious examples. Action is therefore required at both national and international levels to identify ways to provide additional funding for conserving biological diversity.

Two programs are currently directly addressing the issue of international financing of biodiversity conservation. The International Conservation Financing Program of WRI (supported by CIDA, MacArthur Foundation, NORAD, Organization of American States, Pew Charitable Trusts, UNDP, UNEP, and USAID) recently released a report examining financial approaches to international conservation needs, with biodiversity an important component of those needs (WRI, 1989b). In addition, IUCN and UNEP are spearheading an effort to develop an international convention on the conservation of biodiversity, including a funding mechanism based on the use of biological materials, that can contribute to the financing of conservation activities (described in Chapter IV). The potential contribution of other international initiatives, such as "debt for nature swaps," must be examined as well.

Conservation has brought considerable and sustainable benefits to local communities. But conserving biological resources requires investments, in staff, in infrastructure, in benefits postponed, in education, and so forth. These investments are often very sound, showing high benefit-cost ratios; the more complete the economic analysis, the higher such ratios are likely to be (USAID, 1987).

Opposite page: A saltwater crocodile (*Crocodylus porosus*). Crocodiles are increasingly grown on ranches to avoid their overexploitation in the wild (photo by S.D. Nash).

Current conservation programs are usually implemented through resource management agencies whose budgets are generally insufficient to implement their mandates effectively, and are subject to considerable fluctuation from year to year. Such funding difficulties severely hamper the effectiveness of conservation agencies. To produce acceptable results and become fully operational, conservation agencies must have sufficient and reliable sources of support.

Unfortunately, in today's economic climate, the government agencies responsible for conservation are chronically under-funded, leading to abuses of natural resources. Significant new funding is clearly required, from both within the nation involved and from the international community. International support is particularly important. Some have contended that far greater benefits from conserving native gene pools, especially in the wilds of the tropics, will be gained by wealthy temperate countries than by the often poverty-stricken nation conserving them (Prescott-Allen, 1986); agriculture, medicine, and forestry in industrialized countries are able to afford the investments required to turn germplasm into profit. Further, those in industrial countries often care more about conserving elephants, tigers, and monkeys in the tropical countries than do the farmers who face daily conflicts with wildlife that is preying on their crops and domestic animals.

While protected areas provide significant local benefits in terms of watershed protection, tourism, harvest of renewable resources, and so on, it is a fact that the countries conserving living natural resources often receive much less benefit from them than those consuming their products at some distance. Further, within the tropical countries the people living on the edges of protected areas, and prohibited by law from harvesting the resources in the area, often earn virtually no benefits even where the protected areas are bringing in plentiful tourist revenue. Hence, a major incentive for conservation is often lacking where it is most vital. Clearly, the economic evaluation of conservation needs to incorporate an international perspective on costs and benefits, and systems of providing appropriate incentives to local communities need to be devised (see McNeely, 1988, for one effort in this direction).

Finally, some forms of financial support for conservation involve bilateral agreements or cooperation with international agencies, such as food for work programs. In many developing countries, large externally supported development projects can often include elements that support conservation of biological resources.

In seeking to promote more funds for conserving biological diversity, the following points need to be considered:

- in some cases, community development activities are already being planned or implemented in communities in or near areas important for conserving biological resources, in which case elements to promote changed behavior toward conservation can be incorporated in the development project with little additional cost (see Reid *et al.*, 1988, for examples of this).
- it is apparent that any funding mechanism will need to emanate from the competent government authority, either in terms of enabling legislation or administrative fiat; considerable coordination among various ministries — from Finance to Natural Resources — may be required.
- conservation needs to pervade all rural-based activities; it is not something that happens only in national parks and other protected areas. Therefore, economic incentives aimed at encouraging rural people to conserve biological resources outside of protected areas can be very cost-effective in terms of conservation achievement. While such incentives may not bring funding to the conservation agency, they may enable the agency to be more effective in managing protected areas (McNeely, 1988).
- finally, funding is seldom the only major constraint to conservation achievement. While conservation agencies never have sufficient funding, and additional funding is certainly called for, even generous budgets will not lead to conservation if government policies in other sectors are incompatible with conservation. Therefore, any new funding mechanisms need to be part of a package that includes necessary policy changes in land tenure, energy, frontier settlement, foreign trade, transportation, and so on.

The Issue of Property Rights to Biological Resources

Many biological resources can be conserved through actions taken to meet the immediate needs of the rural poor. But it is inevitable that a gap will exist between the conservation that can be achieved through its compatibility with rural development and the action that is desirable for the good of humanity.

For example, the benefits of the establishment of seed banks for crops of international importance are enormous. However, it would not be in the interest of any nation except major agricultural ones to be the sole financier of a seed bank for such a crop because the benefits for that country would be relatively small and the expense high. Thus, the financing of a substantial portion of agricultural germplasm conservation is best achieved through international mechanisms such as the Consultative Group on International Agricultural Research (CGIAR).

Several policy options exist between the extremes of conserving biological resources as a by-product of immediate economic considerations and international funding of conservation actions. These options center on the issue of property rights for certain biological resources.

The property rights issue clouds policy analysis in the areas of both genetic and species conservation. International seed banks, for example, have been subjected to criticism because new crop varieties produced from germplasm that is provided without charge by developing countries are then sold back to the contributing countries for a profit. Developing countries have argued that the varieties produced from seed bank material should be made freely available.

The problem could be solved either by increasing international financing of conservation efforts or by granting property rights to countries that implement effective programs to conserve their biological resources. This latter option has not been explored in sufficient detail, but the obstacles to such an approach are clear. In this era of biotechnology, it would be virtually impossible for a country to know or prove that genes from one of its species were in use in a given organism. However, it can also be argued that the issue would be no more complex than international copyright law (de Klemm, 1985). The quotation of a small passage from a book, like the use of a few genes, is not likely to be identified as a copyright infringement whereas the reproduction of an entire book in infringement of copyright would provide grounds for recompensation. Considerable work has been done on promoting the protection of plant varieties and parts as intellectual property. Williams (1984) concludes, "Because more and more private research funds are being poured into the development of plant varieties, stable and definitive protection for these varieties and parts thereof is very important. It remains to be seen whether adequate protection is available within the framework of the existing patent statutes or whether new legislation is required."

The essential point is to ensure that those who benefit from the use of wild plants pay some of the costs of ensuring that those species remain viable in the wild, where they can continue to evolve.

Mechanisms Useful Primarily at National and Local Levels

Although each country has its own legislation and its own ways of raising funds for conservation of biological diversity, the current period of budgetary restraint calls for innovative solutions to old problems. Each country will have its own history, traditions, and legislation, so funding mechanisms are likely to be highly variable and will require adaptation to local conditions; those suggested here have hundreds of permutations, and no doubt other mechanisms could be identified. Given these provisos, the following potential sources of funding can be identified (in addition to regular budgetary allocations from the central government).

Ecotourism in the Galápagos Islands, Ecuador (photo by R. Mast).

Charge Entry and Other Fees to National Parks

Most tourists appreciate the attractions of nature enough to pay for visiting outstanding natural areas. Galápagos National Park, for example, charges a fee of $40 per foreign visitor, which is still a tiny proportion of the total price the visitor is paying for the experience (Ecuadorians pay at a local scale). Fees charged to park visitors in Costa Rica were expected to generate $168,000 in 1988; foreign visitors are charged no more than local visitors, so some scope exists for increased fees. In Rwanda's Volcanoes National Park, famous for its mountain gorillas, a ticket for one gorilla visit (including three days in the park) currently costs about $180 per person; a second gorilla visit the next day is an additional $150. "Gorilla tourism" has now become the third largest foreign exchange earner for Rwanda, a major incentive for conservation. Poaching of gorillas and encroachment on the national park have been greatly reduced as a result (Vedder, 1989).

Strangely enough, many national parks do not charge entry fees, often because they do not want to discourage visitors who cannot pay and because they feel that they are providing a public service; parks are viewed as "merit goods" to which access is not denied on the basis of income. However, as costs of protected area management rise and budgets fall, most protected areas will need to consider charging fees.

Other user fees can also be charged, especially for those requiring maintenance or other management inputs. These can include campsites, bathing facilities, white-water rafting, lake cruises, spot-lighting wildlife at night, guided tours, car parks, and so on.

Cook (1988) quotes a number of arguments in support of applying user fees:

- the public tends to appreciate more fully those facilities and areas for which they are required to pay;
- fees and charges represent a means of having the user pay a proportionally greater share than the public at large;
- the willingness of the public to pay for certain activities or facilities is a useful guide for planning park programs;
- the collection of fees provides an opportunity for direct contact with the park visitor, increasing the possibilities for providing information and maintaining surveillance; and
- park programs may become increasingly limited and maintenance programs deferred unless additional funding is made available through user fees.

In determining the fee structure to charge for the various goods and services available within a protected area, the following points should be considered:

- What is the objective for charging fees? To supplement the regular government appropriation, or to enable the facility to be totally self-sufficient?
- How should the scale of fees compare with commercial institutions offering similar goods or services?
- How should the fee structure deal with special groups, such as children, school groups, senior citizens, low-income groups (especially local people), and foreign tourists?

The fee can be computed on the basis of actual cost of the good or service (when this can be determined); direct operating expenses, including staff time; interest and amortization of investment; support for the efficient management of the area, including necessary improvements; maintenance costs; or simply what the market will bear.

Funds thus earned should be returned to the protected area for management, including support for various economic incentives directed toward improving cooperation with surrounding communities. Unfortunately, the fees collected in most countries are deposited into the central treasury, and the funds appropriated for protected area operations or investments seldom correlate with the income generated by the protected area system.

Charge for Ecological Services

The ecological services provided by protected areas, natural forests, and wetlands are usually considered "public goods," but it is also possible to design systems to charge for these goods. The provision of high-quality water is probably the best example. For protected areas located in hilly or upland areas, watershed protection is an extremely valuable service. For example, Venezuela's Canaima National Park safeguards a catchment feeding hydroelectric developments that are so important that the government recently tripled the size of the park to 3 million ha to enhance its utility for watershed protection; replacing this hydroelectricity with petroleum would cost an estimated $3 billion per year (Garcia, 1984).

It would therefore seem appropriate for such areas to benefit from water-use charges from irrigation projects or hydroelectric installations whose water comes from the area. Such a mechanism can be both justifiable and useful, improving efficiency and equity of water use as well as generating funds for protecting the watershed. This may require studies to quantify the benefits the protected area is providing; for example, Hufschmidt and Srivardhana (1986) showed that annual expenditures of $1.5 million would be justified in terms of benefits to the Nam Pong reservoir in northeast Thailand. In Indonesia, the World Bank invested over $1 million to establish the Dumoga-Bone National Park to protect a major irrigation project (McNeely, 1987); water charges could be imposed to ensure that the running costs of the national park are met from the goods and services it is providing to the local community.

Additional examples abound. For instance coral reefs and mangroves support fisheries, so it would seem reasonable to return part of the profits from fishing to protecting the breeding grounds of the target fish. In some cases, it may be feasible to tax fisheries, perhaps in the form of an export tax (thereby avoiding taxing local consumers); in other cases, establishing such linkages could help convince fisheries departments of their need to invest in managing natural habitats important for fisheries.

Collect Special Taxes

In some countries, such as Costa Rica, special taxes on biological resources have proved useful. Taxes on timber extraction, wood trading, trade in wildlife and wildlife products, concession rights, or other activities connected with the sector can generate income that can then be invested within the sector. This can be made more flexible by allowing taxpayers to invest the amount in the kind of works that the tax is intended to promote. Special taxes can be used to set up development funds or national financing funds, e.g., for credit. An interesting example from the Côte d'Ivoire involves creating an Environment Fund using taxes imposed on ships, especially oil tankers, docking in the country; 50 percent of the tax goes to the Fund, which is then used to purchase equipment necessary for monitoring ecosystems, preventing pollution, or improving environmental management. Since its inception in 1986, the Fund has brought in about $300,000. In industrial countries, the dollar amounts involved can be far larger. For example, Florida's Recovery and Management Act establishes a Hazardous Waste Management Trust Fund to finance the correction of pollution problems should they occur. The Fund is financed by a 4 percent excise tax on disposal until the accrual reaches $30 million, and 2 percent thereafter.

In Costa Rica, legislation stipulates that all legal documents at the municipal level, newly issued passports, exit visas, first-time auto registrations, authenticated signatures registered at the Foreign Ministry, and operating licenses for all bars, nightclubs, dance halls, any other place that sells liquor, and all places of entertainment such as pool halls, cinemas, casinos, and public pools require fiscal stamps, with at least part of the revenue being returned to a Conservation Fund that supports protected area management. Additional fiscal stamps that contribute to conservation are required from annual vehicle registrations and from wildlife import and export permits (Barborak, 1988a).

In addition, Costa Rica collects excise taxes on arms and ammunition and income from fiscal stamps. These are potentially important, but have declined drastically in recent years. The stamp prices were set by law in 1977 and have not been increased since then, because a new law would have to be passed by the legislature to vary the amounts. The Costa Rican colon is now worth only 11.4 percent of its dollar value in 1981, and this devaluation has been accompanied by significant rampant inflation. The dollar value of fiscal stamp receipts in 1982 was over $86,000, nearly three times that expected for 1988. Much of the 1987 revenue had to be used to pay for a new issue of the stamps. Despite this difficulty, the funding mechanism of using fiscal stamps and excise taxes to support conservation would seem to hold promise for many tropical nations.

Industrialized nations also use revenue stamps to raise money for conservation. In the USA, for example, all duck hunters are required to purchase Federal Duck Stamps each year; these colorful stamps are extremely popular even with non-hunters, and have raised an average of $50 million per year over the past several years. The receipts are devoted to the Migratory Bird Conservation Fund, and are used to acquire habitat for the national system of refuges. The Duck Stamp program has proved so successful that some individual states have developed similar mechanisms.

Additional tax mechanisms, based on tourism involving natural areas, may include bed taxes for tourist hotels, departure taxes at airports, and many others.

Build Funding Linkages with Large Development Projects

Where major investments are made in rural development projects, linkages with conservation can often prove beneficial. In 1986, the World Bank promulgated a major new policy regarding wildlands, with elements specifically designed to build components into large projects — primarily for agriculture, livestock, transportation, water resources development, and industrial projects — for ensuring conservation of biological resources (see Annex 5). These components can include economic incentives for local communities affected by the project (Goodland, 1988).

Major hydroelectric projects, for example, can often build in a significant component to establish a protected area in an upland watershed. In Sri Lanka, USAID provided $5 million for a project to establish five new protected areas as part of a major effort to develop the agricultural resources of the Mahaweli river basin. Such support was not for altruistic motives; on the contrary, the protected areas were seen as essential to the success of the downstream development projects (McNeely, 1987).

One important linkage that might be established between conservation and major development projects might be an "environmental maintenance tax." Projects to build dams, irrigation networks, and roads might include explicit allocation of funds for thoroughly assessing the diversity of the area (thereby also supporting the development of local capacity to carry out such surveys), identifying and managing protected areas, and establishing a self-sufficient "endowment fund" for the continued management of the area.

A variant of such linkages is the obligatory investment of a percentage of the total costs in large-scale works that depend for their existence on environmental protection (water resources developments being the outstanding example). Sometimes an additional 10 percent allocated to reforestation and conservation works can lower the annual operating costs by increasing the useful life of the works and reducing requirements for maintenance.

Project support from development assistance agencies is often feasible when the living conditions of rural people are to be improved (recalling that many of these are the "poorest of the poor" and therefore of particular concern to many bilateral government agencies, and to various church, population, and hunger-related PVOs). A major point here is that effective incentives packages seldom require major funding, but rather effective funding aimed at very specific targets. Therefore, development assistance agencies may need to aggregate a significant number of community-level projects in order to attain the project magnitude that is administratively attractive. The major drawback to this approach is that it may breed dependence rather than self-reliance unless the support is provided with great sensitivity.

Return Profits from Exploitation of Biological Resources

Biological resources earn profits from tourism and harvesting, so creative ways and means need to be found to ensure that a fair share of these profits are returned to the local people who are paying the opportunity cost of not harvesting the resource themselves. Kenya, Zimbabwe, and Zambia (Box 29) have all developed appropriate funding mechanisms based on the principle that protected areas should earn a fair return on the money they bring into the economy. Many of these are already being tapped by governments to

Box 29: How Profits Reduced Poaching in Zambia's Luangwa Valley.

In Luangwa Valley of Zambia, a Wildlife Conservation Revolving Fund was established in 1983 to enable the National Parks and Wildlife Service to employ additional staff beyond the Government-approved civil servants. Income to the Fund comes from the harvest of hippos and from auctions among safari hunting companies for the rights to hunt in the Lower Lupande Game Management Area, with terms of the auction including quotas on animals that could be taken and minimum levels of employment from the local communities. Forty percent of the proceeds from the auction was handed over to the local Chiefs for community projects of their choosing and 60 percent was devoted to wildlife management costs.

Results have been remarkable. Personnel increased from 11 to 26 from 1985 to 1987, and the number of field-days by staff increased from 176 to 717. Annual mortality of elephant and black rhino, expressed as the number of poached carcasses found per year per hundred hectares, decreased by 90 percent in the same period. In 1987, the total earnings for the Revolving Fund were $48,620, of which $4,840 was devoted to wildlife management, including $4,410 for the village scout program. Overall recurrent costs of wildlife management for the year was $9,870, considerably less than was earned by the Revolving Fund. Villagers started supporting the National Parks and Wildlife Service management effort, and local tribal leaders established security committees to prevent poachers from entering their areas (McNeely, 1988).

Once economic benefits started to flow to the local villages, the reduced poaching of elephants led to an increase of their populations to the level where sustainable harvests could far exceed the total costs of effective management programs. In addition, about half the costs of supporting the village scouts was equivalent to the total derived from revenue from ivory collected by scouts from elephants that died naturally. While this source of revenue did not go back into the Revolving Fund, it does illustrate to the government the magnitude of funds that could be recovered by this form of local involvement in wildlife management.

In summary, the Wildlife Fund in Zambia acts as a legal mechanism for charging concession fees, selling wildlife products, and engaging in commercial ventures related to wildlife development. The Fund can then direct the income into appropriate channels to serve the interests of managing the biological resources of the area, as well as the interests of local communities co-existing with the wildlife. It therefore reduces the need to depend on Central Treasury for funds, which in recent years has been unable to meet the growing cost of conservation.

cover other expenditures; the point is that a more equitable return needs to go to conserving the biological resources that are bringing in the funds, even when the benefits of conservation are indirect.

Trophy hunting in many countries of Africa has brought in considerable funds, some of which have been returned to managing the resource. In response to increased poaching, some governments have banned trophy hunting, but this is primarily a political response, which often has a negative impact on the wildlife. In conservation terms, an absolute ban on sport hunting is often a misguided strategy because healthy populations of wildlife produce a harvestable surplus, because the number of animals taken by legal hunting is only a small fraction of those taken illegally, because the earnings from game hunting can compensate local populations for any sacrifices they make in the name of conservation, and because the presence of legal hunting parties can deter illegal hunting. The sale of between 100 and 200 licences to foreign hunters to shoot elephants in Zambia would have raised a sum equivalent to the external support provided to the country by donors, even ignoring the value of the ivory or meat for local people (Leader-Williams and Albon, 1988).

In other countries, mechanisms have been established for returning funds from logging activities to reforestation efforts. In Indonesia, for example, a levy of $4 per cubic meter of timber is collected from timber concession holders, to be repaid when they have reforested their concession area (unfortunately, this "deposit" is usually forfeited because it is cheaper to write off the loss than to reforest the concession).

Build Conditionality into Concession Agreements

This mechanism can be an effective instrument in countries that have such extensive timber or fisheries resources that concessions are sold to private investors. As part of such agreements, the concession holder could be required to provide support to various incentive programs aimed at maintaining the long-term productivity of the area being logged or fished. Where concessions are given for forest use, governments must ensure that they realize a significant proportion of forest rents and that, at a minimum, a proportion of such rent is returned to managing the forest to ensure its long-term productivity. In general, governments should design incentive systems that encourage sustainable use of the biological resources of the forest ecosystems.

Profits earned from non-extractive concessions as from hotels, tours, and restaurants, can often provide sufficient funds for running a protected area. Such concessions should be granted on the basis of conditions that do not detract from the natural values of the protected area, and the profits from such concessions should be returned to the resource management agency. Such concessions might also be required from tour companies bringing tourists into protected areas, even if they do not stay overnight; this could supplement admission fees.

Seek Support from the Private Sector

In many countries, the private sector earns considerable economic benefit from biological resources and may be able to provide voluntary support to conserving those resources. Contributions from enterprises involved in resource extraction or in non-consumptive uses of biological resources (such as tourism) can be effective, though such voluntary support is difficult to predict and incorporate in planning efforts. Such voluntary support might be particularly appropriate where a number of tourist enterprises rely on protected areas for their livelihood.

The private sector often provides significant incentives for conservation by providing grants to activities that lead to enhanced management of biological resources. One outstanding example is the International Trust for Nature Conservation, established by the Tiger Mountain Group (a nature tourism organization operating primarily in Nepal). This trust was designed to recycle a portion of the profits from nature tourism into activities that would promote the protection of wildlife and its habitat.

One of the principal activities has been a conservation education program aimed at the villages that surround Royal Chitwan National Park, where Tiger Tops Hotel is the flagship of the Tiger Mountain Group. More recently, the scope of the Trust has been expanded to include more general concern with sustainable development in the areas surrounding the Group's operations. The Trust is putting into practice its belief that wildlife must increasingly pay for itself if it is to survive in today's crowded world (Roberts and Johnson, 1985).

A variant of such support involves donations from multinational corporations investing in resource-based activities in developing countries. Such multinationals can contribute to conservation activities, both to protect their own investments and to contribute to host-country conservation goals. Such donations are often facilitated if the government conservation agency, or a private institution, has established a mechanism for receiving them; experience has shown that private industry is less eager to provide voluntary funds to regular government programs than to an independent foundation (especially if the donations are tax-deductible).

Establish Foundations for Conservation

In some cases, foundations established by or for a protected area or protected area system can be a useful stimulus for generating non-governmental sources of funding (many of which might come from sources discussed above). In Indonesia, for example, the Indonesian Wildlife Fund is supported by voluntary contributions from the timber trade. It was established by the Ministry of Forestry but operates independently under a board of directors that allocates the funds in support of various conservation projects. In Zambia, in contrast, an essential element in the success of its Wildlife Fund has been its establishment within the National Parks and Wildlife Service.

Some foundations have international linkages. For example, the Charles Darwin Foundation, established in 1960, collects funds from a variety of foreign donors and was responsible for managing Galápagos National Park until the early 1970s, when the National Park Service of Ecuador took over that responsibility. The Foundation continues to maintain the Charles Darwin Research Station and conduct research on a wide range of topics of great interest to the management of the Park (WWF, 1986). Similarly, the Seychelles Islands Foundation, under an international Board of Trustees and with funding from international sources, has responsibility for managing Aldabra Strict Nature Reserve; it also receives an annual grant from the government of Seychelles.

Collect Interest from Investments Made by a Protected Area

In many cases, a major protected area can establish an endowment fund, to be managed either by the management authority or by an appropriate NGO. Janzen (1988) suggests that tropical conserved wildlands can diversify their endowment portfolios through the ownership of agricultural lands adjacent to the protected area; the agricultural profits would support management of the area. This has the ancillary benefit of the protected area controlling the kinds of agriculture carried out on adjacent lands, thereby providing a public showcase on the relationship between protected areas and agriculture.

Mechanisms Useful Primarily at the International Level

As noted earlier, funds generated at the national level need to be supplemented by funds from international sources. A number of mechanisms are available for transferring funds from industrialized nations to the tropics.

Use International Conventions to Provide Financial Support

A number of international conventions provide some funding for conserving biodiversity, usually through the mechanism of projects. The World Heritage Convention, for example, gives over $1 million per year to projects in natural sites of great international importance for biological diversity. Project funding under the Convention on Wetlands of International Importance amounts to about $600,000 per year. Several of the Regional Seas Conventions established by UNEP involve Trust Fund agreements that provide significant funding to conservation activities.

The draft convention on the conservation of biological diversity developed by IUCN and now under consideration by UNEP, contains a major element on funding. Under the convention, an International Fund for the Conservation of Biological Diversity would be established. The Fund would be used to advance the objectives of the Convention (broadly, to enhance the conservation of biological diversity). It could draw on four main sources, based on the principle of enabling those who benefit from biological resources to pay the costs of ensuring that such resources are used sustainably:

- levies on activities that use a resource within the biosphere as a dispersion system (for example, for carbon dioxide from fossil fuel combustion);
- levies on general trade in natural living materials or products derived directly from them;
- levies on patentable new genetic material, or on synthetic products derived from wild sources; and
- voluntary contributions, gifts, or bequests made by any state, inter-governmental organizations (including development aid agencies), or public or private bodies or individuals.

While voluntary contributions could be earmarked for specific projects or areas, no political conditions would be attached to contributions to the Fund. Further, governments would agree that the Fund would be tax-exempt and freely transferrable from country to country. The Fund would be administered by a Conference of the Parties, with a Board established by the Conference. Payments from the fund will be made to the State from which the biomaterial (or species) originated, with those payments being applied to the conservation of biological diversity. In addition, payments would be made to States requiring financial support for the conservation of biological diversity, with priorities based on a long-term conservation program adopted by the Conference and on criteria established by the Conference.

A key element in the success of such a fund is persuading those paying the money that the charge is equitable and would be used in an effective fashion from which they would benefit, perhaps based on the principle that development is dependent on ensuring that biological resources are used sustainably, and that depletion of such resources is an externalized cost. The proposed charges for conservation would then be seen as an internalization of cost, and as a provision for future welfare and benefit. Establishing the administrative machinery within user countries would probably be most effective, with the monies being transferred via an international fund to conservation efforts in the countries of origin. Target industries might include timber (the European timber traders have already proposed a voluntary import duty, with the proceeds to be administered by ITTO), seeds, pharmaceuticals, fossil fuel burners, and tourism.

Any system of funding an international convention must be equitable and publicly acceptable, operate in a straightforward way, be compatible with the General Agreement on Tariffs and Trade, raise funds on an appropriate scale (in the hundreds of millions of dollars per year), provide benefits to the payer, and raise funds in industrialized countries but provide expenditure for genuinely relevant and properly monitored projects in developing countries.

Seek Direct Support from International Conservation Organizations

People living in industrialized countries earn considerable benefits from biological resources in tropical countries, and often have considerable interest in conservation that can be expressed through donations to conservation organizations. These donations can be allocated to conservation activities in the tropics, and have often involved significant contributions.

Such support has tended to focus on biological resources rather than on people, but this is beginning to change, and organizations such as WWF, Conservation International, New York Zoological Society, Frankfurt Zoological Society, The Nature Conservancy, and many others are now becoming more aware of the linkages between people and conservation, and indeed Conservation International's Ecosystem Conservation approach focuses mainly on these linkages. Such groups can often provide at least seed funding to get appropriate incentives projects started. IUCN, through its work in National Conservation Strategies, may be able to promote funding mechanisms being developed for incentives packages. Finally, private conservation agencies may have access to blocked funds owed to private companies operating in developing countries, and be able to apply such funds to conservation work.

A variant of this approach is a new initiative from IUCN and the International Union of Directors of Zoological Gardens known as the "Heritage Species Program." It is based on the premise that certain species are special, either because they are highly endangered or are of particular importance to people. A few of these will be designated "Heritage Species" and will be adopted by designated "Heritage Species Centers" that will assume special responsibility for raising funds to support conservation action, especially in the country or region where the species originates. The action proposed for each species will be developed with the best available technical advice from IUCN's Species Survival Commission and other relevant organizations.

Arrange Debt-for-Nature Swaps

The growing international debt held by developing countries is having serious consequences for economic development, political stability, and resource conservation. Particularly in Latin America, where debt burdens are highest, economies are stagnating and fiscal reform measures in several countries have stimulated public protests. The pressure of meeting debt payments has contributed to the rate of biological resource degradation in many countries. Forest lands are often managed for immediate export returns, depleting what should be a renewable resource. Similarly, the conversion of forest land to agricultural and ranching land is often subsidized in part to support export sales; in the long run, this depletes the resource and economic base of these countries.

The debt-swap mechanism involves a conservation organization (WWF, Conservation International, The Nature Conservancy, and others have been involved) buying a country's debt notes that are being discounted on the secondary market (Box 30). These notes are presented to the debtor country in exchange for local currency in the amount of the face value of the debt, with the local currency being invested in conservation under the management of local national institutions. While this mechanism is most useful in countries whose debts are heavily discounted (and therefore penalizes debtor countries that have sound financial management), it is still useful in a number of countries with significant biological resources. However, they are sometimes perceived as yet another type of conditionality imposed from abroad.

Box 30: Arranging a Debt-for-Nature Swap.

Arranging a debt-for-nature swap involves a number of stages and a broad spectrum of variables that, with persistence, can be fine-tuned to orchestrate an agreement satisfactory to everyone. The first step is to obtain approval in principle from the debtor country — specifically, from the government, the central bank, and a private conservation organization that will receive the funds and manage the conservation program. The host country must decide what exchange rate to apply in converting debt into local currency, what conditions of payment to use in exchange for the debt, and whom to designate or accept as a local agent to control the funds and dispense the proceeds. The conservation program is established based on local priorities; it may include site-specific projects or a list of general conservation activities (for example, training of park managers) to be undertaken when the local agent deems them appropriate.

Next, the debt to be acquired must be identified. Potential swappers must shop for debt notes that are of the right denomination, are acceptable to the debtor-country government, and have an acceptable maturation schedule. If the debt is not donated, it must be purchased — itself a technically complex transaction — at an acceptable discount.

Once obtained, the debt must be converted into a local currency instrument by the host government's central bank, in the manner specified in the agreement. Finally, the actual conservation program can begin.

Source: WWF-US, 1988, and Conservation International, 1989.

Ecuador is a small South American country with extraordinary levels of biological diversity, containing nearly twice as many species of plants and animals as all of North America. As with many Latin American countries, Ecuador is suffering from significant external debt; its debt balance has increased eightfold in the past decade. Ecuador is having such difficulties repaying the debt that the lending banks slashed the price in half in the last six months of 1988. After examining the situation, a small group of Ecuadorian professionals mobilized a private foundation, Fundación Natura, to use the debt crisis as an opportunity to attract financial resources to be invested in conservation of biological diversity (Sevilla, 1988). Fundación Natura will be in charge of obtaining funds abroad through donations in hard currency; WWF-US has played a major role in supporting the effort. With these funds, a fraction of the Ecuadorean external debt will be purchased at discount value on the secondary financial market (fluctuating between 30 and 38 percent of the face value). The debt notes thus obtained will be exchanged by Fundación Natura for stabilization bonds; the interest from these bonds will be invested in conservation projects. The first year's proceeds from the interest-bearing fund created by the agreement are targeted to implement a National Conservation Strategy, with special emphasis on Sangay National Park, Yasuni National Park, and Cotacachi-Cayapas Ecological Reserve.

The first ever debt-for-nature swap was negotiated with Bolivia by Conservation International in 1987. Since then, similar debt swaps have been arranged for Costa Rica, the Philippines, Madagascar, Zambia, and elsewhere, often with support from U.S.-based NGOs such as WWF, Conservation International, The Nature Conservancy, and the National Wildlife Federation. The mechanism could also be adapted to debts contracted by Third World governments with multilateral financial institutions such as the World Bank, the International Monetary Fund, the Interamerican Development Bank; with bilateral aid agencies such as USAID, CIDA, SIDA; and with other governments. Debt swaps enable the lender to write off debts if the debtor guarantees to invest the same amount of funds in projects aimed at conserving biological resources.

Use Restricted Currency Holdings

In many countries, excess profits or local currency held by multinational corporations, or even by foreign governments, must be spent within the country. Given proper mechanisms, such profits can be allocated to conservation. For example, funds derived from PL 480 (a U.S. Public Law that enables certain nations to pay in local currency for food imports from the U.S.A., with the local currency to be spent in the importing nation) and other public sector international assistance operations can often be used to support conservation efforts, including incentives packages. Kux (1986) has pointed out that for USAID, at least, it should be relatively painless to increase investments in conservation considerably through greater use of local currencies generated from sales of agricultural commodities provided by the U.S.A. to some developing countries. These funds could be used for activities such as the purchase of land for protected areas, inventories of tropical forests, education and training, and support for alternatives to destructive land-use practices.

Rent "Conservation Concessions"

Such concessions, parallel to those for forestry or mining, might be provided to international conservation organizations for areas of outstanding international importance, in exchange for a rent that would be provided to the resource management agency for funding other areas. The concession agreement would specify standards of management, access to the public, permissible developments (usually non-extractive), etc., and the international agency would assume full responsibility for living up to the concession agreement.

A major problem with this approach is possible charges of "imperialism" or outside influence (which ignores the fact that outside influence is the major factor involved in the overexploitation of most local ecosystems). One way to overcome this concern would be for development agencies to consider providing support to local NGOs or other local agencies for purchasing concessions on a few outstanding areas of local interest, and using them as a demonstration of how an area can be developed so that its biological resources can be managed in an economically sustainable manner. Obviously, this would require the local NGO to have demonstrated its competence in managing its own affairs, and to have the capacity to manage a small natural area; universities managing demonstration natural areas for research might be one appropriate model.

As a variant, property rights for species or protected areas of outstanding importance might be issued to conservation organizations or relevant UN agencies, with payments being made to the government and the concession holder being required to manage the species or area to a high international standard (and subject to a contractual agreement with the government).

Conclusions

In general, conservation should be supported to the maximum extent possible through the marketplace, but the marketplace needs to be established through appropriate policies from the central government. One problem faced by all the funding mechanisms described above is that they face opportunity costs; any funds earned might be used by the government in other ways that the government considers of higher priority. The attraction of the methods suggested is that the income is being earned by the biological resources, and some of the funding is being provided by the public in expression of their support for non-consumptive uses of biological resources.

The major requirement from government policy makers is that they recognize the many values of biological resources, and take advantage of opportunities to invest in the continued productivity that such resources require. They also need to be persuaded to create conditions whereby the private or

NGO sector can assume total management control of important biological resources or areas and seek their own funding in an attractive tax climate. Through the use of innovative funding mechanisms, backed by compatible government policies, one of the major obstacles to progress in conservation can be overcome.

The funding of the conservation of biological diversity needs urgent and realistic discussion by experts able to negotiate proposals that governments will accept as part of an overall convention or other binding international agreement. Such experts should pay particular attention to:

- the need to place an appropriate economic value on biological resources, properly reflected in inventories of national capital wealth and properly accounted for in national revenues when the resources are used;
- the need to provide economic incentives for the conservation of biological diversity, at the international level (by transfer of resources), nationally, and locally by ensuring that local communities benefit from the biological diversity of their regions;
- the prospects for charging and taxing systems linked in various ways to the use of biological diversity;
- machinery for ensuring that the poorest countries, or those with very limited commerce based on biological diversity, are exempted from charging systems;
- the case for an International Fund (supported by such charges), administered under a Convention; and
- the case for voluntary contributions to such a Fund, possibly assessed in proportion to the economic benefits the contributing countries derive from the exploitation of biological diversity.

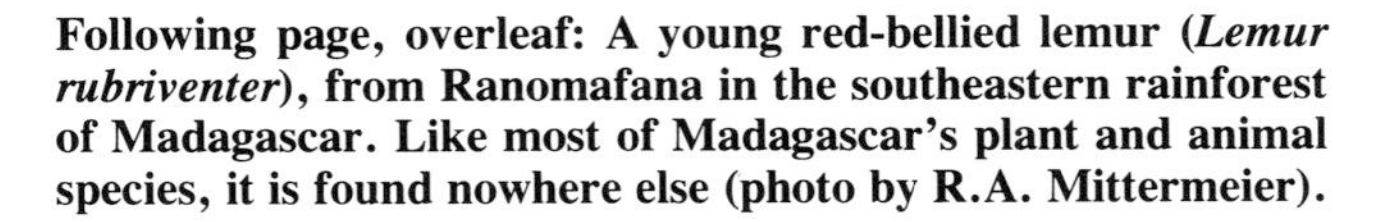

Following page, overleaf: A young red-bellied lemur (*Lemur rubriventer*), from Ranomafana in the southeastern rainforest of Madagascar. Like most of Madagascar's plant and animal species, it is found nowhere else (photo by R.A. Mittermeier).

CHAPTER IX

ENLISTING NEW PARTNERS FOR CONSERVATION OF BIOLOGICAL DIVERSITY

Conservation makes fundamental contributions to sustaining society, but while the benefits are widely shared, only a few institutions are given responsibility for conservation. A far wider range of collaborators is required, involving all ministries and departments that depend directly on biological resources.

Earlier chapters have presented evidence that an essential foundation of development is improved resource management aimed at ensuring a sustainable flow of goods and services from natural ecosystems. This will often involve maintaining areas under relatively natural vegetation, but such areas need to be supplemented by improved resource use in the fields of agriculture, timber production, fisheries, coastal zone management, and so forth.

It is apparent from the preceding chapters that the more strictly protected areas — nature reserves, national parks, and monuments (IUCN categories I, II and III) — need to be managed far more effectively, and to be brought into the mainstream of overall rural development without destroying the values for which they were established. In addition, the categories of protected areas that include extraction of biological resources as a management objective need to be implemented widely to provide goods and services to the local communities — and the world at large — on a sustainable basis. This will require increased resources, and Chapter VIII suggested some sources of additional funding.

However, in addition to these, new partners in conservation need to be sought among the line agencies. This chapter suggests the benefits that could flow to such partners if they became more active in conserving the natural resource base upon which their prosperity depends.

The Contributions of Biological Resources to "Non-Conservation" Sectors

Why should a ministry of agriculture, defense, or health worry about conserving biological resources? Some arguments were presented in Chapter II. The following examples further illustrate the point, though of course specific applications will vary from country to country and community to community.

Watershed management. MacKinnon (1983) examined the condition of the water catchments of 11 irrigation projects in Indonesia for which development loans were being requested from the World Bank. The condition of the catchments varied from an almost pristine state to areas of heavy disturbance due to deforestation, logging, or casual settlements. By using standard costing for the development of the protected areas, reforestation where necessary, and any resettlement of families required, the costs of providing adequate protection for the catchments were estimated. These ranged from less than 1 per cent of the development costs of the individual irrigation project in cases where the catchment was more or less intact, to 5 percent where extensive reforestation was needed, and to a maximum of about 10 percent in cases where resettlement and reforestation were required. Overall these costs were trivial compared with the estimated 30 to 40 per cent drop in efficiency of the irrigation systems expected if catchments were not properly safeguarded.

Tourism development. Natural areas — mountains, rivers, wetlands, forests, savannas, coral reefs, deserts, beaches — are major attractions for tourists. Tourism can bring numerous socio-economic benefits to a country, in terms of creating local employment, stimulating local economies, generating foreign exchange, stimulating improvements to local transportation infrastructure, and creating recreational facilities. Positive effects on the environment often derive from these socio-economic benefits (Goldsmith, 1975; McNeely and Thorsell, 1987).

Agricultural development. While many, even most, agricultural development projects deal primarily with farmsteads or arable lands, the success of agricultural development will often involve linkages with natural areas important for biological diversity. Each agricultural village is part of an ecosystem. This ecosystem varies widely — from the broad expanses of river deltas where year-round irrigation is possible, to areas where seasonally irrigated fields are interspersed among forests, to areas where rain-fed crops dominate. Legumes, medicinal plants, other cereals, tubers, tree crops, livestock, wild animals (such as pigs, monkeys, and rats), and fish all play important roles in most agricultural villages, so agricultural development projects need to consider all these factors. Further, each agricultural community has ecological relationships far beyond the village. For example, Sattaur (1987) points out that in the hills of Nepal, each hectare of farmland needs 3.48 hectares of forest to support it. Many Nepalese forests are ecologically sensitive, requiring expert management if they are to continue providing benefits in terms of fodder, firewood, construction materials, fruit, and medicinal plants. Agricultural development projects that incorporate means of protecting the larger ecosystem within which agricultural communities survive and flourish are far more likely to succeed than those that are too narrowly based. Such considerations will often involve ensuring that the relevant communities are given management responsibility for the natural areas upon which their continued prosperity depends.

Conserving crop relatives. Responsibility for protecting areas that harbor extremely important populations of wild relatives of domestic plants often needs to be assigned to the appropriate arm of the Ministry of Agriculture. These areas can be extremely important, such as the location in India that supported the sole known population of the wild rice *Oryza nivara,* the only source of resistance to grassy stunt virus. Wild populations of rice that are salt-tolerant could help adapt the crop to saline soils or brackish irrigation water, and long-stemmed populations of floating rice may help adaptation to the deeper waters that may come with rising sea levels. Natural areas important for wild relatives of domestic plants or animals, or for protecting wild populations of insects useful in integrated pest control, should be established and managed by agriculture ministries to ensure that all wild relatives of domestic plants are conserved as a basis for adapting to future changes.

Fisheries. The establishment and management of protected areas in coastal and marine habitats is still in its infancy, with most such areas being merely an extension seaward of existing terrestrial protected areas, such as Ujung Kulon in Java. However, many critical habitats in the coastal zone need protection so that they can provide services to humanity on a continuous basis; in addition to shoreline protection (as in the Sundarbans of Bangladesh) and sustainable harvesting of construction materials, such areas can be especially important as fish breeding grounds (particularly when the surrounding waters are over-harvested)(Hamilton and Snedaker, 1984; Ketchum, 1972). Virtually all wetland habitats are important for fisheries, but of particular relevance are inland floodplains that are often affected by development projects (Goulding, 1980). Dams, irrigation systems, and other measures affect both inland and coastal wetlands important for fisheries, and alternative means of managing these systems need to be developed. Fisheries departments need to take a far more active role in managing such areas, including allocating some habitats for strict protection.

Energy. While energy is typically seen as a highly technological field, much of the energy needs of rural households in the tropics are still met by traditional sources, and many of these come from natural habitats important for conservation of biological resources. In addition to the hydroelectric implications of natural areas mentioned above, many forests provide firewood to local people; for example, over 90 percent of the energy needs of Nepal, Tanzania, and Malawi are still met by firewood (Pearce, 1987a). In 1983, over 1.6 billion cubic meters of fuelwood were consumed in the world, amounting to some 54 percent of total roundwood production from forests (FAO, 1985). While the traditional energy sector is not commercialized, it still can form an important part of national energy policies, and improved regimes for managing natural forests to provide firewood may bring major benefits to rural people; the management of natural forests is therefore of considerable interest to ministries involved in energy.

Public health. Many tropical countries remain highly dependent on medicinal plants (some 5,000 medicinal plants have been catalogued in China, 2,500 in India, and 6,500 in Southeast Asia), and many of these are found in natural forests. The World Health Organization estimates that 80 percent of the people in the Third World depend for their primary health care on medicinal plants, either grown locally or collected from nature (Farnsworth, 1988). Since most rural people still depend on traditional medicines to some extent, protecting the sources of medicinal plants could be a productive part of rural health projects. In Sri Lanka, the Ministry of Traditional Medicine has established a series of special small reserves to protect areas important for local medicinal plants.

Industry. In addition to the hydroelectric energy benefits mentioned above, natural habitats can also contribute a wide range of raw materials for industrial processes. Tropical forests produce gums, fats, oils, starches, resins, rattans, fibers, dyes, tannins, and many others. Coral reefs and other marine habitats produce hundreds of products useful to industry. Ensuring the sustainable production of such products should be of considerable concern to those industries dependent on products available from nature, and this may require investments to be made by industry in protecting certain areas of particular value. These areas may help sustain an industrial base that can be largely self-sufficient in terms of raw materials, and since such industries are often highly profitable, this provides a mechanism for the costs of production to be internalized.

Pollution control. Some natural areas, notably wetlands near urban centers, are effective natural sewage treatment centers. For example, Calcutta's sewage has been naturally purified in the 4,000-ha ''Salt Lakes'' marshland east of the city for over 50 years. The wetlands serve as highly efficient oxidation ponds and support a thriving fishery that provides employment for 20,000 fishermen and produces an annual catch of 6,000 tons. Coliform bacteria from feces are reduced by 99.9 percent in the well-stocked ponds (Maltby, 1986). The Salt Lakes therefore make an extremely important economic contribution to the people of Calcutta, and similar functions are served by numerous wetlands throughout the world. Investing in the maintenance of such systems can often make far more economic sense for a ministry of public health or sanitation than developing expensive new sewage treatment plants.

Disaster prevention. Natural areas important for conserving biological resources often help prevent disasters such as landslides and avalanches (areas in mountain forests) or dampen the impacts of typhoons (coastal mangroves). Since prevention is often far less expensive than disaster relief, especially in terms of human costs, appropriate investments in protecting such areas can often be included as part of disaster prevention programs.

Land titling. Land tenure governs the use and disposal of land and its products so that the use of the land can be stabilized. When villagers do not have secure title to their land, they have little incentive to make investments that would ensure sustainable use, and insecure tenure may bias the choice of crops against perennials, tree crops, and forest plantations. Villagers lacking secure tenure are therefore forced to clear new land continually, often destroying natural areas and leaving little but wasteland behind. Land titling projects therefore have an important contribution to make to conserving natural areas (Kennedy, 1980).

The Special Case of the Military

In most parts of the world the defense services are a dominant force politically, socially, and economically. While their primary task is to defend the nation's political viability, the defense services are increasingly coming to recognize that political, economic, and ecological viability are closely interrelated. Yet they have seldom been systematically approached to provide their support for positive action in conservation of biological resources. Indications that such an approach would be both useful and productive include the following (realizing that considerable variations exist from country to country).

- The officer corps of the military is the source of many government leaders who make fundamental decisions that affect conservation and sustainable development (this is most obviously the case during periods of direct military rule, but also holds in most developing countries generally).
- The military controls large areas of land, as training facilities, military reservations, border ''buffer zones,'' etc., and such areas are often of considerable biological and ecological value.
- In many countries, the military is an active participant in rural development programs, providing logistics, labor, and stability.
- The military includes large numbers of impressionable post-adolescent males as recruits, who are put through intensive training programs that could easily incorporate social and ecological considerations; military training programs continue throughout an individual's military career, with increasing sophistication through staff colleges.
- The defense services have access to excellent information on landforms, vegetation, and other geographically based information useful for conservation purposes.
- Some international legislation relevant to conservation, such as the Law of the Sea, can only be enforced with military support.
- Many individuals in the defense services are from rural areas and have particular affinity for nature and the outdoors, making them well predisposed for conservation; properly motivated, such individuals working in remote areas can make significant contributions to both *in situ* and *ex situ* conservation of biological diversity.
- The military is concerned primarily with national security, and it is increasingly apparent that many threats to national security have their roots in inappropriate ways and means of managing natural resources; the military might therefore reasonably be expected to have a serious interest in resource management issues.
- As conflicts between people and resources increase in the coming years, the military will require detailed understanding of the biological, ecological, social, and economic issues involved if they are to deal effectively with these conflicts.

In short, the various national military establishments operate for the benefit of their respective nations. Since conservation of biological resources is essential to the well-being of a nation, the military should also support conservation and sustainable development in the name of national security. That they have seldom done so, at least explicitly, could well be due to a lack of the right approach being made to them.

One approach might be to develop a series of case studies in which the military are having a positive influence on conservation of biological resources (possibilities include Burma, China, India, Madagascar, Pakistan, Peru, Sri Lanka, Venezuela, and Zimbabwe). A plan might be prepared for influencing the military, including actions such as developing conservation-oriented curriculum materials for recruit training programs, providing top military leaders with material demonstrating how conservation affects national security, and developing guidelines and manuals on how to manage areas under military control for conservation objectives. A group of military leaders who have demonstrated a sensitivity to environmental issues might be brought together with conservation professionals, to recommend how the defense services can be approached most effectively to promote conservation interests.

New Approaches to Managing Areas for Sustainable Production of Biological Resources

A number of examples linking natural areas important for conserving biological resources to various types of development projects have been presented in this chapter. Many more could be provided, but the sample is sufficient to indicate that the long-term success of development projects — in other words, their sustainability — very often depends on ensuring that natural areas are sufficiently well managed to provide a continuous flow of benefits to society. New approaches to management may be required to ensure that these benefits are actually delivered to local communities and to the global community at large.

The various sectors that depend directly on the biological resources of natural areas need to become more responsible for ensuring that these areas are managed to deliver sustainable benefits. While national parks and wildlife departments should be strengthened, they need to concentrate their efforts on the areas most important for conserving biological diversity. In addition, numerous sectors need to be involved in managing natural habitats. Thus, national parks departments should to be joined in habitat management by a wide range of other institutions to represent all interests.

Furthermore, other line agencies need to develop the capacity to manage biodiversity of particular relevance to their respective missions. Forestry departments need to ensure that annual felling plans incorporate conservation activities; fisheries departments need to be concerned with natural nurseries in mangroves; tourism departments need to be concerned about the quality of coral reefs; departments of industry and commerce need to be concerned about their sources of raw materials; departments of health need to be concerned about the wild sources of medicinal plants; irrigation departments need to be concerned about the source of water; and the list goes on.

A major effort is therefore required to develop sufficient technology and expertise in the line agencies so that they can manage the areas for which they are responsible, and thereby ensure the sustainability of their own development efforts. In many cases, a high-level coordinating and oversight body may be required to ensure that the management plans of the various line agencies are prepared in accordance with national objectives for conserving biological diversity.

Conclusions

The governments of many nations have recognized the contributions of natural areas to their development programs. Wetlands, coral reefs, mangroves, mountains, and tropical forests are important for social, economic, political, and ethical reasons, and with proper management they can support sustainable resource use in sectors ranging from forestry to tourism to rural development.

However, many natural areas are being abused rather than nurtured, and a number of general policy changes are required to enable the most important areas to be identified, and for the most appropriate management regimes to be designed and implemented. Each nation will have its own particular opportunities and constraints. No recipe book will automatically provide the right answers. But the basic principle should be that the distribution of costs and benefits of both conservation and exploitation should be equitable and should lead to long-term sustainable use. Local support for protecting natural areas must be increased through such measures as education, revenue sharing, participation in decisions, complementary development schemes adjacent to protected areas, and, where compatible with conservation, access to resources.

New approaches to linking protected areas to surrounding lands are required if the appropriate benefits are to flow to society, involving a wide range of government and private institutions in managing natural areas of various management categories. Concrete steps can be taken to ensure that such areas are managed in ways that will bring sustainable benefits to people, thereby contributing to forms of development that will be durable in the long run.

The elements now exist that will reverse the trend toward the biotic impoverishment of the world. Novel approaches, new financial mechanisms, and new policies need to be applied at the appropriate level of responsibility to translate the good intentions into a reality of improved human well-being and a secure biotic heritage.

The 1990s may be the last decade during which decisions, activities, and investments can be made to ensure that many of the world's species and ecosystems are maintained, examined for their material and ecological value, and promoted for sustainable use to support new and innovative approaches to development. The combination of maintaining the maximum possible biological diversity, the maximum possible cultural diversity, and the greatest possible scientific endeavor would seem the most sensible approach toward dealing with the dynamic future facing humanity.

We are at a crossroads in the history of human civilization. Our actions in the next few years will determine whether we take a road toward a chaotic future characterized by overexploitation and abuse of our biological resources, or take the opposite road — toward maintaining great biological diversity and using biological resources on a sustainable basis. The future well-being of human civilization hangs in the balance.

ANNEX 1: CLASSIFICATION OF LIFE ON EARTH BY PHYLUM

KINGDOM: PROKARYOTAE

Phylum: Methanocreatrices
Halophilic and Thermoacidophilic Bacteria
Aphragmabacteria
Spirochaetae
Thiopneutes
Anaerobic Phototrophic Bacteria
Cyanobacteria
Chloroxybacteria
Nitrogen-fixing Aerobic Bacteria
Pseudomonads
Omnibacteria
Chemoautotrophic Bacteria
Myxobacteria
Fermenting Bacteria
Aeroendospora
Micrococci
Actinobacteria

KINGDOM: PROTOCTISTA

Phylum: Caryoblastea
Dinoflagellata
Rhizopoda
Chrysophyta
Haptophyta
Euglenophyta
Cryptophyta
Zoomastigina
Xanthophyta
Eustigmatophyta
Bacillariophyta
Phaeophyta
Rhodophyta
Gamophyta
Chlorophyta
Actinopoda
Foraminifera
Ciliophora
Apicomplexa
Cnidosporidia
Labyrinthulomycota
Acrasiomycota
Myxomycota
Plasmodiophoromycota
Hyphochytridiomycota
Chytridiomycota
Oomycota

KINGDOM: FUNGI

Phylum: Zygomycota
Ascomycota
Basidiomycota
Deuteromycota
Mycophycophyta

KINGDOM: ANIMALIA

Phylum: Placozoa
Porifera
Cnidaria
Ctenophora
Mesozoa
Platyhelminthes
Nemertina
Gnathostomulida
Gastrotricha
Rotifera
Kinorhynca
Loricifera
Acanthocephala
Entoprocta
Nematoda
Nematomorpha
Ectoprocta
Phoronida
Brachiopoda
Mollusca
Priapulida
Sipuncula
Echiura
Annelida
Tardigrada
Pentastoma
Onychophora
Arthropoda
Pogonophora
Echinodermata
Chaetognatha
Hemichordata
Chordata

KINGDOM: PLANTAE

Phylum: Bryophyta
Psilophyta
Lycopodophyta
Sphenophyta
Filicinophyta
Cycadophyta
Ginkgophyta
Coniferophyta
Gnetophyta
Angiospermophyta

ANNEX 2: THE WORLD CHARTER FOR NATURE

The General Assembly of the United Nations

Reaffirming the fundamental purposes of the United Nations, in particular the maintenance of international peace and security, the development of friendly relations among nations and the achievement of international cooperation in solving international problems of an economic, social, cultural, technical, intellectual or humanitarian character.

Aware that:

a) Mankind is a part of nature and life depends on the uninterrupted functioning of natural systems which ensure the supply of energy and nutrients.

b) Civilization is rooted in nature, which has shaped human culture and influenced all artistic and scientific achievement, and living in harmony with nature gives man the best opportunities for the development of his creativity, and for rest and recreation.

Convinced that:

a) Every form of life is unique, warranting respect regardless of its worth to man, and to accord other organisms such recognition man must be guided by a moral code of action.

b) Man can alter nature and exhaust natural resources by his action or its consequences and therefore, must fully recognize the urgency of maintaining the stability and quality of nature and of conserving natural resources.

Persuaded that:

a) Lasting benefits from nature depend upon the maintenance of essential ecological processes and life support systems, and upon the diversity of life forms, which are jeopardized through excessive exploitation and habitat destruction by man.

b) The degradation of natural systems owing to excessive consumption and misuse of natural resources, as well as to failure to establish an appropriate economic order among peoples and among States, leads to the breakdown of the economic, social and political framework of civilization.

c) Competition for scarce resources creates conflicts, whereas the conservation of nature and natural resources contributes to justice and the maintenance of peace.

Reaffirming that man must acquire the knowledge to maintain and enhance his ability to use natural resources in a manner which ensures the preservation of the species and ecosystems for the benefit of present and future generations.

Firmly convinced of the need for appropriate measures, at the national and international, individual and collective, and private and public levels, to protect nature and promote international cooperation in this field.

Adopts, to these ends, the present World Charter for Nature, which proclaims the following principles of conservation by which all human conduct affecting nature is to be guided and judged.

I. GENERAL PRINCIPLES

1. Nature shall be respected and its essential processes shall not be impaired.
2. The genetic viability on the earth shall not be compromised; the population levels of all life forms, wild and domesticated, must be at least sufficient for their survival, and to this end necessary habitats shall be safeguarded.
3. All areas of the earth, both land and sea, shall be subject to these principles of conservation; special protection shall be given to unique areas, to representative samples of all the different types of ecosystems and to the habitats of rare or endangered species.
4. Ecosystems and organisms, as well as the land, marine and atmospheric resources that are utilized by man, shall be managed to achieve and maintain optimum sustainable productivity but not in such a way as to endanger the integrity of those other ecosystems or species with which they coexist.
5. Nature shall be secured against degradation caused by warfare or other hostile activities.

II. FUNCTIONS

6. In the decision-making process it shall be recognized that man's needs can be met only by ensuring the proper functioning of natural systems and by respecting the principles set forth in the present Charter.
7. In the planning and implementation of social and economic development activities, due account shall be taken of the fact that the conservation of nature is an integral part of those activities.
8. In formulating long-term plans for economic development, population growth and the improvement of standards of living, due account shall be taken of the long-term capacity of natural systems to ensure the subsistence and settlement of the population concerned, recognizing that this capacity may be enhanced through science and technology.
9. The allocation of areas of the earth to various uses shall be planned, and due account shall be taken of the physical constraints, the biological productivity and diversity and the natural beauty of the areas concerned.

10. Natural resources shall not be wasted, but used with a restraint appropriate to the principles set forth in the present Charter, in accordance with the following rules:
    a) Living resources shall not be utilized in excess of their natural capacity for regeneration;
    b) The productivity of soils shall be maintained or enhanced through measures which safeguard their long-term fertility and the process of organic decomposition, and prevent erosion and all other forms of degradation;
    c) Resources, including water, which are not consumed as they are used, shall be reused or recycled;
    d) Non-renewable resources which are consumed as they are used shall be exploited with restraint, taking into account their abundance, the rational possibilities of converting them for consumption, and the compatibility of their exploitation with the functioning of natural systems.
11. Activities which might have an impact on nature shall be controlled, and the best available technologies that minimize significant risks to nature or other adverse effects shall be used. In particular:
    a) Activities which are likely to cause irreversible damage to nature shall be avoided;
    b) Activities which are likely to pose a significant risk to nature shall be preceded by an exhaustive examination; their proponents shall demonstrate that expected benefits outweigh potential damage to nature, and where potential adverse effects are not fully understood, the activities should not proceed;
    c) Activities which may disturb nature shall be preceded by assessment of their consequences, and environmental impact studies of development projects shall be conducted sufficiently in advance, and if they are to be undertaken, such activities shall be planned and carried out so as to minimize potential adverse effects;
    d) Agriculture, grazing, forestry and fisheries practices shall be adapted to the natural characteristics and constraints of given areas;
    e) Areas degraded by human activities shall be rehabilitated for purposes in accord with their natural potential and compatible with the well-being of affected populations.
12. Discharge of pollutants into natural systems shall be avoided and:
    a) Where this in not feasible, such pollutants shall be treated at the source, using the best practicable means available;
    b) Special precautions shall be taken to prevent discharge of radioactive or toxic wastes.
13. Measures intended to prevent, control or limit natural disasters, infestations and diseases shall be specifically directed to the causes of these scourges and shall avoid adverse side-effects on nature.

III. IMPLEMENTATION

14. The principles set forth in the present Charter shall be reflected in the law and practice of each State, as well as at the international level.
15. Knowledge of nature shall be broadly disseminated by all possible means, particularly by ecological education as an integral part of general education.
16. All planning shall include, among its essential elements, the formulation of strategies for the conservation of nature, the establishment of inventories of ecosystems and assessments of the effects on nature of proposed policies and activities; all of these elements shall be disclosed to the public by appropriate means in time to permit effective consultation and participation.
17. Funds, programs and administrative structures necessary to achieve the objective of the conservation of nature shall be provided.
18. Constant efforts shall be made to increase knowledge of nature by scientific research and to disseminate such knowledge unimpeded by restriction of any kind.
19. The status of natural processes, ecosystems and species shall be closely monitored to enable early detection of degradation or threat, ensure timely intervention and facilitate the evaluation of conservation policies and methods.
20. Military activities damaging to nature shall be avoided.
21. States and, to the extent they are able, other public authorities, international organizations, individuals, groups and corporations shall:
    a) Cooperate in the task of conserving nature through common activities and other relevant actions, including information exchange and consultations;
    b) Establish standards for products and manufacturing processes that may have adverse effects on nature, as well as agreed methodologies for assessing these effects;
    c) Implement the applicable international legal provision for the conservation of nature and the protection of the environment;
    d) Ensure that activities within their jurisdictions or control do not cause damage to the natural systems located within other States or in the areas beyond the limits of national jurisdiction;
    e) Safeguard and conserve nature in areas beyond national jurisdiction.
22. Taking fully into account the sovereignty of States over their natural resources, each State shall give effect to the provisions of the present Charter through its competent organs and in cooperation with other States.
23. All persons, in accordance with their national legislation, shall have the opportunity to participate, individually or with others, in the formulation of decisions of direct concern to their environment, and shall have access to means of redress when their environment has suffered damage or degradation.

24. Each person has a duty to act in accordance with the provisions of the present Charter; acting individually, in association with others or through participation in the political process, each person shall strive to ensure that the objectives and requirements of the present Charter are met.

ANNEX 3: INTERNATIONAL LEGISLATION SUPPORTING CONSERVATION OF BIOLOGICAL DIVERSITY

Since international relations often work best within a framework of agreed legal instruments, considerable effort has been devoted to developing a series of conventions and other international instruments that promote the conservation of biological diversity. This annex briefly describes the main components of the existing international legal system.

THE SCOPE OF EXISTING CONVENTIONS

At global level, the Ramsar Convention on Wetlands of International Importance especially as Waterfowl Habitat and the Paris Convention on the Protection of the World Cultural and Natural Heritage deal with aspects of habitat conservation. A number of regional measures also touch on or cover this field, notably:

a) the Convention on Nature Protection and Wildlife Preservation in the Western Hemisphere (Washington, 1940);
b) the Antarctic Treaty, with its subordinate Agreed Measures on Conservation of Antarctic Flora and Fauna;
c) the Convention on the Conservation of Nature in the South Pacific (Apia, 1976);
d) the African Convention on the Conservation of Nature and Natural Resources (Algiers, 1968);
e) the Convention on the Conservation of European Wildlife and Natural Habitat (Berne, 1976); and
f) certain European Community Directives, notably on the conservation of bird habitats.

Of these, the World Heritage Convention is of special value in giving added status and some additional supporting resources to outstanding sites that are already protected, but it has so far placed more emphasis on cultural than on natural sites and it is not designed as an instrument for the protection of the world's biological diversity per se. The Ramsar Convention has been the means of designation of over 400 sites covering some 30 million hectares, although these too are invariably listed after they have gained protection under national legislation. The Convention covers fresh water, estuarine, and coastal marine habitats that are important for both the diversity of wild species they support and as the location of relatives of key cultivated plants (notably rice). As it has developed, the application of the Ramsar Convention has been broadened and it has become the most important global measure concerned with habitat protection, but it is clearly only able to cover a small part of the world's total biological diversity. In a similar way, while the regional conventions listed above and the designation of Biosphere Reserves under a Unesco program provide valuable protection or public recognition of a range of sites, taken collectively they meet only a fraction of the needs we identify in the text.

Various other international measures conserve particular species or groups of species or protect the living resources of designated marine areas. At global level these include the Convention on International Trade in Endangered Species of Wild Flora and Fauna (Washington, 1973) and the Convention on Conservation of Migratory Species of Wild Animals (Bonn, 1979), and on a narrower geographical scale:

a) the Convention on Conservation of Antarctic Marine Living Resources (Canberra, 1982); and
b) various agreements preserving species or classes, etc., e.g.:

- International Convention for the Regulation of Whaling (Washington, 1946), establishing the International Whaling Commission;
- agreements protecting birds (International Convention for the Protection of Birds (Paris, 1950); Benelux Convention on the Hunting and Protection of Birds (Brussels, 1970));
- agreements concerning measures for protection of marine and polar region species (Prawns, Lobsters and Crabs (Oslo, 1952); Fur Seals (Washington, 1957); Antarctic Seals (London, 1972); Convention on Fishing and Conservation of Living Resources in the Baltic Sea and the Belts (Gdansk, 1973); Polar Bears (Oslo, 1973); Salmon (Reykjavik, 1982)); and
- agreements protecting vicuña (Lima, 1979).

Since the biological diversity of the earth is also at risk from pollution, notice also needs to be taken of the substantial number of agreements in this field. At global level these include:

a) the Convention on the Prevention of Marine Pollution by Dumping of Wastes and Other Matters (London, 1972); and
b) the Convention on the Protection of the Ozone Layers (Vienna, 1985), with its Protocol on the regulation of the manufacture and use of chlorofluorocarbons (Montreal, 1987);

and at regional level a substantial number of measures including the Convention on Long-Range Trans-Boundary Air Pollution in the European region (Geneva, 1979, with protocols), Conventions on the prevention of marine pollution in the waters of the north-east Atlantic (Oslo, 1972), and in the Baltic (Helsinki, 1983), the Regional Seas Conventions of UNEP, and a host of specific measures undertaken especially in Europe and North America.

The Regional Seas Conventions prepared under UNEP auspices embrace both the protection of particular areas of the marine environment, the safeguarding of marine and coastal species there, and the coordination and strengthening of action against marine pollution in these areas. As such they extend across all three of the categories noted above.

DETAILS OF THE MAJOR INTERNATIONAL INSTRUMENTS FOR CONSERVING BIODIVERSITY

The four major international instruments that make significant contributions to conserving biodiversity are so important that they deserve some expansion. The following highlights the main elements of these conventions.

1. *The Convention on Wetlands of International Importance especially as Waterfowl Habitat (Ramsar, 1971).* Depositary: Director-General, Unesco. Secretariat: Provided by IUCN, with a branch at International Waterfowl Research Bureau headquarters in the United Kingdom (currently five permanent staff).
   - The only global nature conservation convention designed to cover a particular broad habitat type (inland, coastal, and marine wetlands).
   - Broad in scope, as wetlands are defined to encompass a wide variety of areas including rivers, lakes, swamps, coastal areas, tundra, floodplains, and areas of the sea that are less than 6 m deep at low tide. This breadth in scope is both a strength and a weakness of the Convention, as management approaches vary so widely with the habitat type that it is difficult for all relevant agencies to be represented at meetings.
   - Contracting Parties undertake to use wisely all wetland resources under their jurisdiction (the wise use requirement has been subject to much analysis by a Conference of Parties with guidelines developed for national implementation policies); Contracting Parties also agree to designate for conservation at least one wetland of international importance, under criteria provided by the convention for identifying such wetlands.
   - By January 1989, the 54 Contracting Parties had designated 421 sites (covering almost 30 million hectares) onto the List of Wetlands of International Importance; no site has been removed from the List despite the possibility to do so in "urgent national interest."
   - Requires the establishment of reserve areas for wetlands whether or not included on the List.
   - Monitoring procedure adopted by Standing Committee for secretariat to review status of listed sites and assist Contracting Parties in maintaining ecological character of sites.
   - Requires cross-border cooperation for shared wetland resources and international cooperation along flyways for migratory waterfowl.
   - Financial regime established as from 1 January 1988 based upon mandatory and voluntary contributions from Contracting Parties. Annual budget of SFr 600,000 plus project funding gives annual turnover in the magnitude of SFr 1 million.
2. *The Convention Concerning the Protection of the World Cultural and Natural Heritage (Paris, 1972).* Depositary: Director-General, Unesco. Secretariat: Provided by Unesco, with specialist assistance provided by IUCN and the International Council on Monuments and Sites.
   - Currently has 109 State Parties, the most of any conservation convention.
   - Unique in its use of international NGOs (ICOMOS and IUCN) as technical "arbitrators."
   - Recognizes the obligation of all states to protect unique natural and cultural areas, and recognizes the obligation of the international community to help pay for them. The combination of both cultural and natural sites makes the Convention more comprehensive, but also weakens its focus; participation in meetings has tended to be far stronger from the cultural side than the natural side.
   - Establishes exceptional World Heritage Sites, of which the natural properties (as opposed to cultural properties, which are far more numerous) protect wild animals and plants and their gene pools in those sites; convention is reinforced by national legislation in some countries, especially those with Federal systems.
   - Includes on the list several of the most biologically diverse sites in the world, including Manu National Park (Peru), Queensland Rainforests (Australia), Dja National Park (Cameroon), Serengeti National Park (Tanzania), Great Smokies National Park (U.S.A.), Iguaçu (Brazil), and Sinharaja Forest Reserve (Sri Lanka).
   - Establishes the World Heritage Fund to ensure that areas are not lost because of a local lack of money

or skills (the Fund disperses nearly US$ 2 million per year). Each Party must contribute to the Fund, currently calculated at 1 percent of their contribution to the annual budget of Unesco.

3. *The Convention on International Trade in Endangered Species of Wild Fauna and Flora (CITES) (Washington, 1973).* Depositary: Government of Switzerland. Secretariat: Provided by UNEP; currently located in Lausanne, with eight full-time staff.
   - As of March 1989, some 99 States Parties to the convention.
   - Establishes lists of endangered species for which international trade is to be controlled via permit systems, as a means of combating illegal trade and over-exploitation; revised appendices (listing protected species) now include 406 animals and 146 plants on Appendix I (which prohibits trade), and about 2,500 animals and 25,000 plants on Appendix II (which monitors trade).
   - Encourages international cooperation between governments and organizations to control such trade, particularly through informal consultations at the regular meetings of the Parties.
   - Establishes a network of national Management Authorities (to deal with mechanics of trade), and Scientific Authorities (to deal with biological aspects of trade) which operate in direct communication with each other and the secretariat.
   - Recommends that multilateral and bilateral development agencies assist development countries on request and facilitate exchange of administrative and scientific experience among trading countries.
   - Provides for a Trust Fund (established in 1979 under UN procedures) to finance the Secretariat and meetings of the Conference of the Parties.
   - Directed at species rather than habitats, so has neither the goal nor the effect of protecting large areas of the environment from degradation.
   - Has sometimes been criticized for being counterproductive in preventing tropical countries from marketing their wildlife products in industrialized countries, even though the species in question are under effective management regimes.
   - Supported by technical advice from IUCN's Wildlife Trade Specialist Group, from WWF's TRAFFIC network, and from trade data managed by WCMC's Wildlife Trade Monitoring Unit.
4. *The Convention on Conservation of Migratory Species of Wild Animals (Bonn, 1979).* Depositary: Government of the Federal Republic of Germany. Secretariat: Provided by UNEP; currently located in Bonn, FRG, with a full-time staff of one.
   - As of August 1988, 24 State Parties had joined.
   - Convention addresses a wider range of threats to migratory species than is to be found in any other global convention. Obligates parties to protect endangered migratory species and to endeavor to conclude international conservation agreements for the conservation of migratory species.
   - Provides a framework for (1) international cooperation between range states for the conservation of certain species of wild animals that regularly migrate across or outside national boundaries, and (2) coordinated research, management, and conservation measures such as habitat protection and hunting regulation under regional and/or species-specific agreements.
   - Provides an important adjunct to wetlands and waterfowl conservation because of the high number of species of waterfowl that are migratory.
   - Provides an important species-by-species complement to a comprehensive scheme for the conservation of biological diversity, as international conventions are the only effective means of protecting animals that cross national boundaries.
   - Limitations include: insufficient Parties, lack of financial support, does not deal with fisheries.

ANNEX 4: THE BALI ACTION PLAN

How to Enable Protected Areas to Meet the Needs of the 1980s

By

IUCN's Commission on National Parks and Protected Areas

INTRODUCTION

The World National Parks Congress met in Bali, Indonesia, from 11 to 22 October, 1982, with the primary objective to define the role of national parks and other protected areas in the process of social and economic development. The Bali Action Plan builds on the reports of working groups from the world's eight biogeographic realms on priorities for each realm. The Plan recognizes that most countries already have competent government agencies whose responsibility is the management of national parks and other categories of protected areas, and that each of these agencies is already carrying out a program of work relevant to the needs and priorities of the country involved; the total budgets of these agencies exceeds $2 billion. However, the 450 professionals attending the Congress also recognized that there was a serious lack of understanding of management tools (biogeography, zoning, monitoring, training procedures, protected area economics, etc.), that budgets are not always allocated to the most important priorities, that clear objectives exist for relatively few protected areas, that management plans are the exception rather than the rule, that relevant information is not flowing as well as it should, that training is lagging far behind needs, and that government officials and the public generally undervalue the role of protected areas in environmentally sound development.

The Bali Action Plan aims to provide guidance and assistance to those national agencies which are seeking to improve their own management effectiveness in meeting the objectives for which their protected areas were established. Clearly, this is not the work of the IUCN Secretariat alone; it must involve all parts of the Union — State Parties, Government Agencies, and Non-Governmental Organizations — as well as IUCN's major international partners in conservation: UNEP, Unesco, FAO, and the World Wildlife Fund.

The Bali Action Plan has ten *Objectives*. Under each objective is a series of *Activities*, and under each activity is a series of *Priority Projects;* the lists of projects are far from exhaustive, but they do indicate the sorts of projects that will be necessary for the activity to be carried out.

THE BALI ACTION PLAN

***Objective 1.* TO ESTABLISH BY 1992 A WORLDWIDE NETWORK OF NATIONAL PARKS AND PROTECTED AREAS, TO COVER ALL TERRESTRIAL ECOLOGICAL REGIONS.**

Activity 1.1 Develop and make available to all responsible for protected areas, tools and guidelines for the identification and selection of natural areas critical for meeting the objectives of conservation and for supporting development.

Priority Projects

1.1.1. Preparation and publication of *Managing Protected Areas,* an IUCN publication on the concepts of protected area management, based on a workshop held at the World National Parks Congress.

1.1.2. Preparatory workshop and publication of an IUCN document on identifying and selecting natural areas for conservation, (a practical, field-level manual for direct application on the ground).

1.1.3. Research to develop more detailed criteria for identification and selection of each of the eight IUCN categories for protected area management.

Activity 1.2 Promote necessary technical, scientific and financial support for the identification, selection, planning and management of protected areas which fit strategically into the world network.

Priority Projects

1.2.1. Identification of priority areas which require additional support.

1.2.2. Focusing support on priority countries and regions (tropical forests, areas threatened by desertification, wetlands, and tundra environments).

Activity 1.3 Further develop and distribute a biogeographical classification system for use in the global analysis of protected area coverage.

Priority Projects
1.3.1. Complete final development of the IUCN global system of biogeographic provinces.
1.3.2. Publish an Atlas of Biogeography, based on the work under 1.3.1 and 2.3.1.

Activity 1.4 Develop and distribute a more detailed biogeographical classification system with a flexibility of scale which can be used in the analysis of protected area coverage at a variety of regional and national levels.

Priority Projects
1.4.1. Consultancy to develop a biogeographical classification system which can be widely applied at the country level for conservation purposes.
1.4.2. Application of the IUCN biogeographical classification system to selected priority countries.
1.4.3. Application of the IUCN biogeographical classification system to selected priority countries.
1.4.4. Review of all resource maps in Latin America and preparation of a comprehensive review of the biogeography of the Neotropics for resource planning purposes.

Activity 1.5 Promote the detailed evaluation at the regional and country level of protected area coverage.

Priority Projects
1.5.1. Examination of protected area coverage in selected priority countries.

Objective 2. TO INCORPORATE MARINE, COASTAL AND FRESHWATER PROTECTED AREAS INTO THE WORLDWIDE NETWORK.

Activity 2.1 Develop and distribute concepts and tools for the establishment of protected areas in marine, coastal and freshwater environments.

Priority Projects
2.1.1. Preparation and publication of *Managing Coastal and Marine Protected Areas,* an IUCN publication on the concepts of protected area management in marine habitats.
2.1.2. Preparation and publication of an IUCN document on establishment and management of protected areas in freshwater environments.
2.1.3. Organization and holding of Marine Sanctuaries Symposium, to develop further the scientific tools for the establishment of marine protected areas.
2.1.4. Design and stimulation of research programs directed at functioning of marine ecosystems, the paths and effects of pollutants, and how to utilize such knowledge in management.
2.1.5. Preparation and publication of marine resources and conservation atlases for the North Sea and the Baltic.
2.1.6. Preparation of a guide for protected area managers and planners, showing in clear and concise terms the importance of protected areas for conservation of coastal and marine genetic resources.
2.1.7. Preparation of case studies to provide information on establishing and developing coastal and marine resources to planners and managers, particularly in the Pacific, Central America, and the Caribbean.

Activity 2.2 Develop a classification system for categories of marine, coastal and freshwater protected areas.

Priority Projects
2.2.1. Further development and publication of IUCN system of categories of coastal and marine protected areas.

Activity 2.3 Further develop and distribute biogeographical classification systems for marine, coastal and freshwater protected areas, at both the global level and at the regional/national level.

Priority Projects
2.3.1. Complete final development of the IUCN global system of coastal and marine biogeographic provinces.
2.3.2. Consultancy to develop a coastal and marine biogeographical classification system which can be widely applied at the country and region level for conservation purposes.
2.3.3. Application of the IUCN coastal and marine biogeographical classification system to selected priority countries and regions (Indonesia, Melanesia, Antarctica, Caribbean, UNEP Regional Seas).

Activity 2.4 Incorporate scientists, managers, administrators and supporters of marine, coastal and freshwater conservation into the protected areas community.

Priority Projects
2.4.1. Identification of those professionally involved in aquatic conservation, for inclusion into IUCN Commissions, CDC Consultant Roster, IUCN project screening and programme development procedures, and other work of the Union.
2.4.2. Promote the inclusion of advisers from non-governmental organizations interested in the Antarctic environment on national delegations of the Antarctic Treaty Powers, and establish close working relationships between IUCN and SCAR.

Activity 2.5 Promote the establishment of marine, coastal and freshwater protected areas by all states, including the extension of all currently protected littoral areas into the aquatic environment.

Priority Projects
2.5.1. Survey of legal and administrative procedures required to extend currently protected littoral areas into the aquatic environment.

2.5.2. Identification of all currently protected littoral areas which could be extended into the aquatic environment.
2.5.3. Preparation and publication of document for decision-makers encouraging the extension of currently protected littoral areas into the aquatic environment.
2.5.4. Promotion of establishment, under the Law of the Sea, of large sanctuaries in the open sea.
2.5.5 Development of model education programme focused on the significance of marine protected areas, the need for the wise use of marine resources, and an increased awareness of human relationships with such areas.
2.5.6. Planning and establishment of marine reserves in the Mediterranean.

Activity 2.6 Promote cooperation between neighboring nations sharing resident and migratory species to establish networks of protected areas and other regulations to meet the critical needs of those species, with special priority for threatened and endangered species.

Priority Projects
2.6.1. Development of political, administrative, and legal means for promoting cooperation between neighboring nations, in the context of the Law of the Sea, UNEP's Regional Seas Programme, and FAO's Marine Mammal Plan.

***Objective 3.* TO IMPROVE THE ECOLOGICAL AND MANAGERIAL QUALITY OF EXISTING PROTECTED AREAS.**

Activity 3.1 Develop and make available tools and guidelines for the evaluation of the ecological capacity of protected areas to maintain living resources, and the evaluation of area management to ensure that appropriate measures are being applied.

Priority Projects
3.1.1. Further development of ecological theory of protected area design.
3.1.2. Development and publication of methodology for determining carrying capacity of key wildlife species.
3.1.3. Design a system for evaluating protected area management, for application by managers.

Activity 3.2 Promote the development of concepts and methods which will lead to scientific principles for management and support the continuous analysis of conservation requirements for each area.

Priority Projects
3.2.1. Establishment and operation of task force to develop ways and means of applying scientific principles to protected area management.
3.2.2. Promote appropriate measures to ensure that effective eradication or control measures are undertaken in regard to introduced species in protected areas.

Activity 3.3 Document the living resources contained in protected areas, including preparing and disseminating inventories of wild species and populations of known or likely value as genetic resources.

Priority Projects
3.3.1. In East Africa, national inventory of ecosystems, in particular mangroves and other coastal wetlands, lagoons and coral reefs; national inventory of threatened or endangered coastal and marine species and descriptions of related critical habitats with proposals for preserving them.
3.3.2 Documentation of living resources contained in protected areas in priority tropical forest countries.
3.3.3. Documentation of plant genetic resources in priority countries identified under the Plants Programme.
3.3.4. Based on 3.3.2 and 3.3.3, review the extent to which wild genetic resources are adequately maintained by existing protected area systems.

Activity 3.4 Develop and implement a system of reporting on protected areas under particular threat.

Priority Projects
3.4.1. Development of system of reporting on protected areas under particular threat.
3.4.2. Preparation and publication of annual IUCN List of Threatened Protected Areas.

Activity 3.5 Support a systematic approach to the preparation of area and system management plans which provide for management and development to be in accordance with an appropriate range of conservation objectives.

Priority Projects
3.5.1. Preparation of guidelines for designing systems plans and area management plans.
3.5.2. Support to preparation of management plans in priority countries.

Activity 3.6 Reinforce measures to reduce the external threats to protected areas.

Priority Projects
3.6.1. Design of model legislative and administrative measures regarding environmental impact assessments prior to project finalization.
3.6.2. Design of model measures to safeguard the integrity of the environment when a development project which affects a protected area is deemed acceptable in principle.
3.6.3. Promotion of measures of sustainable social and economic development which will relieve the pressures of local populations around protected areas.

3.6.4. Promotion of rehabilitation of degraded lands and the regeneration and recovery of damaged natural areas through reforestation and other programs.
3.6.5. Preparation of manuals for guidance of planners, managers, and decision-makers outside the protected area system describing integrated environmental approaches and other techniques for enhancing the security of protected areas.

Objective 4. TO DEVELOP THE FULL RANGE OF WILDLAND MANAGEMENT CATEGORIES.

Activity 4.1 Develop and make available the concepts and tools necessary for the design and implementation of each category, in both terrestrial and aquatic habitats.

Priority Projects
4.1.1. Elaboration of design of each of the IUCN Categories for both terrestrial and aquatic habitats.
4.1.2. Preparation and publication of guidelines for implementation of each of the IUCN categories.
4.1.3. Promotion of the World Heritage Convention, including technical evaluation of nominations and promotion of additional nominations.
4.1.4. Preparation of management guidelines for natural World Heritage Sites.
4.1.5. Cooperation with the Council of Europe on the Palaearctic protected areas.

Activity 4.2 Establish pilot protected areas for each category, within each realm, to demonstrate to political leaders and local peoples the importance of these alternatives for supporting social and economic development through sustainable approaches to resource management.

Priority Projects
4.2.1. Identify and select appropriate areas for establishing pilot protected areas in each category (both terrestrial and marine) in each realm (total of 160 areas).
4.2.2. Establishment of pilot protected areas in each category in each realm.

Activity 4.3 Include all 10 wildland management categories on the *United Nations List of National Parks and Protected Areas.*

Priority Projects
4.3.1. Establishment of network to report on categories not currently included on the UN List (categories V, VI, VII, and VIII).
4.3.2. Preparation and publication of *UN List of Protected Areas.*

Activity 4.4 Provide for the establishment of *in situ* gene banks.

Priority Projects
4.4.1. Prepare a model plan of an *in situ* gene bank in one of the biogeographic realms.
4.4.2. Prepare plans for the establishment of *in situ* gene banks, including as appropriate the zoning of existing protected areas and the designation of new ones.

Objective 5. TO PROMOTE THE LINKAGE BETWEEN PROTECTED AREA MANAGEMENT AND SUSTAINABLE DEVELOPMENT.

Activity 5.1 Develop and make available the tools for the survey of ecological processes, habitat requirements, and other components of protected area integrity to enable managers to critically examine the context for area conservation, and be able to associate conservation with development in adjacent lands.

Priority Projects
5.1.1 Development of easily applied methodology for determining population status and trends and habitat requirements of wildlife species.
5.1.2 Development of easily applied methodology for assessing ecological processes within protected areas.

Activity 5.2 Work with governments and development assistance agencies to achieve the incorporation of protected area considerations and support within development projects.

Priority Projects
5.2.1. Preparation of guidelines on incorporating protected area considerations within development projects.
5.2.2. Cooperation with development assistance agencies to incorporate protected area concerns within development projects, including in Sri Lanka (Mahaweli), Ivory Coast (Tai), and others to be identified.
5.2.3. Review of all major national and international projects in the Afrotropical Realm in order to select a limited number which could most profitably benefit from IUCN input.
5.2.4. Preparation of a Consultant's Roster which would facilitate the provision of appropriate expertise when required for protected area matters.

Activity 5.3 Develop policy guidelines and legal instruments regarding the use of protected areas for research, environmental monitoring and the collection of scientific materials.

Priority Projects
5.3.1. Development of model guidelines for promoting and managing research in protected areas.
5.3.2. Development of model legislation regarding research, environmental monitoring and the collection of scientific materials.

Activity 5.4 Develop tools and guidelines for the practical incorporation of new objectives for protected area management of particular relevance to sustainable development, including environmental monitoring and genetic resources conservation.

Priority Projects
5.4.1. Determination of technical feasibility of making genetic stocks available from protected areas.
5.4.2. Determination of legal and policy guidelines based on technical feasibility identified in 5.4.1.
5.4.3. Determination of administrative guidelines based on 5.4.1 and 5.4.2.
5.4.4. Develop guidelines for controlled wildlife cropping in Category IV areas.

Activity 5.5 Investigate and utilize the traditional wisdom of communities affected by conservation measures, including implementation of joint management arrangements between protected area authorities and societies which have traditionally managed resources.

Priority Projects
5.5.1 Further investigation of the role of traditional societies in the management of living resources.
5.5.2. Preparation and publication of case studies on the role of traditional societies in protected areas.
5.5.3. Pilot projects to implement involvement of traditional societies in conservation management.

Activity 5.6 Carry out research to determine ways to foster appropriate recreation and tourism in protected areas for which tourism has been deemed an objective, and to minimize the adverse impacts of such activities.

Priority Projects
5.6.1. Preparation of guidelines for management of tourist activities in the Antarctic.
5.6.2. Assessment of effects of tourism on East African national parks.

Activity 5.7 Develop ways and means of promoting greater public support for protected areas.

Priority Projects
5.7.1. Provision of assistance — financial, technical, and information — to voluntary conservation organizations for enlisting public support for protected areas
5.7.2. Promotion of youth activities in support of protected areas (including tree-planting campaigns, work-study camps, field studies, and curricular elements).
5.7.3. Develop model interpretive programs for wide dissemination and adaptation to local conditions which emphasize the social and scientific values of protected areas, giving specific attention to issues of public concern.

***Objective 6.* TO DEVELOP THE FULL CAPACITY TO MANAGE PROTECTED AREAS.**

Activity 6.1 Promote the establishment and recognition of protected area management as a professional career of vital relevance to society.

Priority Projects
6.1.1. Publication of Training Protected Area Personnel, based on workshops held at World National Parks Congress.

Activity 6.2 Develop and promote training seminars, courses and workshops at the regional and local levels for protected area managers.

Priority Projects
6.2.1. Continued support for International Seminar on National Parks.
6.2.2. Design model training seminars/workshops on new area establishment/evaluation of area management/working with people in adjacent lands.
6.2.3. Holding of training seminars and workshops in the tropics.

Activity 6.3 Strengthen support to regional and national training schools.

Priority Projects
6.3.1. Promote the implementation of proposed training programs in the Neotropics.
6.3.2. Provision of support to regional training schools in Africa and Asia.

Activity 6.4 Promote the establishment of local, in-service training efforts for all personnel.

Priority Projects
6.4.1. Development of model curricula and training methods for all levels of protected area personnel.

***Objective 7.* TO DEVELOP ECONOMIC TOOLS FOR SUPPORTING PROTECTED AREAS.**

Activity 7.1 Develop and distribute tools for the analysis of values, tangible and non-tangible, monetary and non-monetary, associated with protected natural areas.

Activity 7.2 Promote the quantification of values which relate conservation to development, specifically watershed protection but also including genetic resources, pollution control, soil formation, amelioration of climate, provision of recreation and tourism, and others of nature's services.

Activity 7.3 Explore and publish concepts and tools which relate ecology and economics to promote a more consistent perspective for analyzing and explaining the role of protected areas in sustaining development.

Objective 8. TO IMPLEMENT AN EFFECTIVE INVENTORY AND MONITORING SERVICE.

Activity 8.1 Expand and develop the Protected Areas Data Unit (PADU) and related components of the World Conservation Monitoring Center (WCMC), to provide information on protected areas, guide the determination of priorities, and support development agencies (both national and international) in relating the design of development projects to critical protected areas.

Priority Projects
8.1.1. Provide support to the Protected Areas Data Unit at WCMC.

Activity 8.2 Publish and distribute realm-based directories and periodic reports to inform and support national and international organizations in their planning activities.

Priority Projects
8.2.1. Preparation of data sheets and publication of one realm-based directory per year.
8.2.2. Preparation of special reports to inform and support national and international organizations in their planning efforts.

Activity 8.3 Promote arrangements by international organizations, governments, and regional associations of nations for the long-term development and use of data collection systems, such as satellite remote sensing, covering all protected areas.

Priority Projects
8.3.1. Consultancy to apply satellite remote sensing techniques to protected area needs.

Activity 8.4. Promote and implement methodology for implementing monitoring systems.

Objective 9. TO IMPLEMENT INTERNATIONAL COOPERATION MECHANISMS.

Activity 9.1 Integrate and strengthen ties between protected areas management and the Man and the Biosphere Programme, the Global Environmental Monitoring System, and the World Heritage Convention, to realize the full potential of these instruments for the common objectives of conservation and sustainable development.

Priority Projects
9.1.1. Participation in World Heritage Committee meetings.

Activity 9.2 Encourage and advise all States on the preparation, use, and, where required, updating of international legal instruments which support protected areas.

Priority Projects
9.2.1. Preparation of concise descriptive material on the international legal instruments which support protected areas.
9.2.2. Preparation of guidelines on the implications of the Law of the Sea for coastal and marine protected areas.

Activity 9.3 Explore and promote the development of tools and mechanisms for the fair sharing of costs and benefits associated with protected areas management, both among nations and between protected areas and adjacent communities.

Priority Projects
9.3.1. Consultancy to advise on ways and means to develop mechanisms for the fair sharing of costs and benefits associated with protected area management.

Activity 9.4 Explore the potential for new agreements and instruments needed to further strengthen international cooperation, particularly in relationship to genetic resources.

Objective 10. TO DEVELOP AND IMPLEMENT A GLOBAL PROGRAMME TO SUPPORT PROTECTED AREA MANAGEMENT.

Activity 10.1 Design and implement regional action programs to ensure practical accomplishments close to the ground, taking into account relevant cultural and institutional diversity, and the necessary responsiveness to local needs.

Priority Projects
10.1.1. Provision of support to IUCN Regional Councilors and CNPPA Regional Vice-Chairman to prepare and coordinate regional action plans.

Activity 10.2 Provide technical and scientific guidance through the publication of a series of documents on practical subjects of global concern to protected area management such as those noted in the preceding items.

Priority Projects
10.2.1. Publication of *Proceedings of the World National Parks Congress*
10.2.2. Publication of reports and documents resulting from the Bali Action Plan.

Activity 10.3 Establish a communications network with the global community responsible for or supporting protected areas to ensure the flow of information and support the identity of the protected area profession.

Priority Projects
10.3.1. Investigate the ways and means available to establish and operate an international protected areas communication network.
10.3.2. Maintain the publication of *Parks* magazine (in English, Spanish, and French) as one of the best means for communication among those involved or interested in protected areas.

Activity 10.4 Build the institutional support necessary to carry out these activities as follow up from the Congress.

Activity 10.5 Initiate steps for the celebration of the next World Congress on National Parks in 1992, with intermediate international, regional and national events designed to further the Bali Action Plan.

Activity 10.6 Charge the IUCN to monitor the implementation of this Plan and to report on progress at the next Congress.

ANNEX 5: THE WORLD BANK WILDLANDS POLICY — *WILDLANDS: THEIR PROTECTION AND MANAGEMENT IN ECONOMIC DEVELOPMENT*

(Effective June 1986)

INTRODUCTION

The maintenance of specific natural land and water areas in a state virtually unmodified by human activity, hereafter termed wildland management, is an important subset of the broad environmental concerns addressed in "Environmental Aspects of Bank Work" (1984). The conversion of wildlands to more intensive land and water uses (through land clearing, inundation, plantations, or other means) continues to meet important development objectives, and is an element of certain World Bank-supported projects. At the same time, wildlands are rapidly diminishing in many Bank member countries. The remaining wildlands can often contribute significantly to economic development, particularly in the longer term, when maintained in their natural state. The Bank's policy therefore is to seek a balance between preserving the environmental values of the world's more important remaining wildlands, and converting some of them to more intensive, shorter term human uses.

The World Bank already has considerable experience of wildland management in Bank-supported projects. This policy codifies existing practices and provides operational guidance concerning conservation of wildlands.

JUSTIFICATION

There are two principal justifications for wildland management. First, wildlands serve to maintain biological diversity (i.e., the full range of the world's biota). Second, wildlands provide environmental services important to society. In addition, certain wildlands are essential for maintaining the livelihood of tribal peoples.

Biological Diversity

Wildland management is necessary to prevent the untimely and often irreversible loss of a large proportion of the world's remaining biota, including the more visible plant and animal species. Because their wildland habitats are today rapidly disappearing, a large and growing number of plants and animals face extinction. Appropriate, low-cost wildland management measures can greatly reduce current extinction rates to much lower (perhaps almost "natural") levels, without slowing the pace of economic progress. By preserving the integrity of the biotic community and its plant and animal species, wildlands are important for the replenishment of surrounding degraded or abandoned areas.

Preserving biological diversity is important to development because of the economic potential of species that are currently undiscovered, undervalued, or underutilized. Many previously unknown or obscure, and often threatened, species have turned out to have major economic benefits. But less than 20 percent of the world's plant and animal (largely invertebrate) species have ever been inventoried, and even fewer screened for possible human uses. They therefore present valuable development opportunities if they are not irreversibly destroyed. In addition, there are important scientific, aesthetic, ethical, and practical reasons to avoid or minimize the extinction of the remaining biotic stock. While some species can be conserved *ex situ* (such as in zoos or seed banks), wildland management is the only technically and economically feasible means of preserving most of the world's existing biological diversity.

Environmental Services

In addition to maintaining biological diversity, many wildlands also perform important "environmental services," such as improving water availability for irrigated agriculture, industry, or human consumption; reducing sedimentation of reservoirs, harbors, and irrigation works; minimizing floods, landslides, and coastal erosion (and possibly droughts in some regions); improving water quality; and providing essential habitat for economically important fishery species. Despite their economic value and importance in meeting human needs, such environmental services are not always accorded adequate attention because they are usually public goods that tend to be poorly understood, undervalued, and even overlooked. When environmental services are lost due to wildland elimination, remedial measures are almost always far more expensive than prior maintenance. While many environmental services can also be maintained by establishing more intensive water and/or land use systems (e.g., bio-oxidation sewage treatment, tree plantations), wildland management is frequently more cost-effective.

Wildlands of Special Concern

Wildlands of special concern are areas that are recognized to be exceptionally important in conserving biological diversity or perpetuating environmental services. They can be classified into two types. First are wildlands officially designated as protected areas by governments, sometimes in collaboration with the United Nations or the international scientific community. These are National Parks, Biosphere Reserves, World Heritage Natural Sites, Wetlands of International Importance, areas designated for protected status in national conservation strategies or master plans, and similar "wildland management areas" (WMAs), i.e., areas where wildlands are protected and managed to retain a relatively unmodified state.

Second are wildlands as yet unprotected by legislation, but recognized by the national and/or international scientific and conservation communities, often in collaboration with the United Nations, as exceptionally endangered ecosystems, known sites of rare or endangered species, or important wildlife breeding, feeding, or staging areas. These include certain types of wildlands that are threatened throughout much of the world, yet are biologically unique, ecologically fragile, or of special importance for local people and environmental services. Wildlands of special concern often occur in tropical forest, Mediterranean-type brushlands, mangrove swamps, coastal marshes, estuaries, sea grass beds, coral reefs, small oceanic islands, and certain tropical freshwater lakes and riverine areas. Within the spectrum of tropical forest, lowland moist or wet forest are the most species-rich and often the most vulnerable. Wildlands of special concern also occur in certain geographical regions that have been reduced to comparatively small patches and continue to undergo rapid attrition. As a result, these regions harbor some of the most threatened species in the world.

THE BANK'S INVOLVEMENT TO DATE

During the last 15 years, the World Bank has assisted with financing of upwards of 40 projects with significant wildland management components. Most of them have involved establishment or strengthening of WMAs. Bank-supported WMAs include national parks, nature reserves, wildlife sanctuaries, and those forest reserves managed primarily for their watershed or biological values, rather than for wood harvest. Other wildland management components of Bank projects have involved management of wildlife and the humans that utilize it, including anti-poaching measures, management of water flows from reservoirs to maintain wildlife habitat, and relocation of certain species.

Wildland management components have two principal objectives: first, to prevent, minimize, or partially compensate for wildland elimination, thereby conserving biological diversity; second, to preserve or improve the environmental services provided by wildlands, thereby enhancing the project's economic or social benefits. Most Bank-supported projects emphasize one or the other objective; however some Bank-assisted projects have wildland components seeking both objectives.

Cost of wildland management components in Bank-assisted project have typically been low. They have normally accounted for less than three percent of total project costs, and in half of the cases for less than one percent. In many instances, it is difficult to separate out the cost of the wildland component because of its integration with other components.

In one case, wildland management was the sole objective, thus accounting for 100 percent of project costs. At the other extreme, a large number of Bank projects have achieved significant wildland management objectives at zero additional cost. For example, manipulation of a hydroelectric project's water release schedule costs little or nothing, even though it provides major downstream benefits for wildlife, as well as for people and cattle.

Wildland management components require additional Bank staff time and can increase project complexity, but they have rarely caused significant delays at any stage of the project cycle. Moreover, the failure to incorporate adequate wildland components can result in much greater delays and complexity later on. Furthermore, the failure to incorporate adequate wildland components can substantially reduce project benefits and might result in project failure. As wildland management components within Bank-supported projects become more routine, the additional staff effort required to manage them successfully is expected to decrease further.

The Bank's track record in implementing wildland management components is encouraging. According to project completion reports or environmental post-audits, implementation of only three out of 43 wildland components has been markedly slower than for most other projects components. In at least four cases, the wildland component has been implemented with less difficulty than other project components.

Lessons Learned

A number of important lessons have emerged from the Bank's experience with wildland management to date. First, wildland management components should be routinely and systematically incorporated into certain types of Bank projects. Up to now, this has not always been done and some projects which would have benefitted from wildland components have not included them.

Second, wildland components should be incorporated as early as possible within the project cycle to minimize costs and facilitate implementation. While inclusion of wildland components in later stages of the project cycle may at times be necessary because of unforeseen circumstances, it is more effective and less costly to incorporate them as early as possible in the project cycle.

Third, meeting wildland management goals requires effective management "on the ground," not simply on paper. Colonists and resource extractive companies have rapidly moved into such "paper parks" (parks existing only on a legal document or map, rather than on the ground) unless they were inaccessible for other reasons. The wildland management objectives have to be translated into specific measures with a budget for their implementation. These measures include hiring and training of personnel, provision of necessary infrastructure and equipment, development of a scientifically sound management plan for each particular wildland, and a policy environment — legal, economic and institutional — which supports the wildland preservation objective. The mere declaration of intent to protect wildlands or wildlife, or even the designation of WMAs on a map, does not ensure effective management unless specific supporting measures are implemented.

Fourth, the multiple objectives of wildland management are most successfully attained if the WMA is carefully designed. For example, a WMA cannot preserve biological and genetic diversity, evolutionary processes, and environmental services if it is too small. While some Bank-supported WMAs clearly appear sufficiently large to accomplish most or all of their objectives, others are so small that their ability to conserve biological diversity or provide environmental services or other benefits is questionable. Besides size, the specific location and shape of a WMA can be important factors in determining its success. Appropriate WMA design features are best determined for each case by a conservation specialist.

Finally, the success of a WMA, as of other project components, is contingent upon government commitment. This in turn, often depends upon the degree of financial support provided by some direct support for establishing or strengthening WMAs by the bank. Most of the Bank-supported wildland components have provided some direct support to establishing or strengthing WMAs. However, in some cases, the costs of the WMA establishment were assumed entirely by the Government, and the Bank took no specific measures to ensure the continued availability of such financing. By taking measures to ensure counterpart financing, or by providing the financing itself, the Bank can help ensure the availability of the relatively modest sums necessary for WMA establishment and continuation.

Financial support is usually not sufficient, however. It is often also necessary to maintain dialogue with governments, affected local people, and environmental advocates about the importance of conservation and the benefits of WMAs (tourism, watershed protection, etc.) and to include local people in the planning and benefits. Government commitment to the WMA is fostered by such dialogue, by supervision, by monitoring of national legal provisions, and by loan conditionality. In addition, two complementary and parallel activities contribute to WMA success: (1) rural development investments that provide farmers and villagers in the vicinity of the WMA an alternative to further encroachment, and (2) coherent national and sectoral planning and policies that promote wildland conservation.

POLICY GUIDANCE

The World Bank's general policy regarding wildlands is to seek to avoid their elimination and rather to assist in their preservation. Specifically, (1) the Bank normally declines to finance projects involving conversion of wildlands of special concern even if this conversion occurred prior to the Bank being invited to consider financing. (2) When wildlands other than those of special concern may become involved, the Bank prefers site projects on lands already converted (e.g., logged over, abandoned, degraded, or already cultivated areas) sometime in the past, rather than in anticipation of a Bank-financed project. Deviations from this policy must be explicitly justified. (3) Where development of wildland is justified, then less valuable wildlands should be converted rather than more valuable ones. (4) When significant conversion (e.g., 100 sq km, or a significant proportion of the remaining wildland area of a specific ecosystem, if smaller) of wildlands is justified, the loss should be compensated by inclusion of wildland management components in the project concerned, rather than in some future project. This component should directly support preservation of an ecologically similar area. This policy pertains to any project in which the Bank is involved, irrespective of whether the Bank is financing the project component that affects wildlands.

The success of projects that do not eliminate any wildland often depends on the environmental services provided by wildlands. In such cases, the Bank's policy is to require borrowers to include a project component to conserve the relevant wildland in a WMA, rather than leaving its preservation to chance. In areas without remaining wildlands, alternative conservation measures may be needed to provide similar project benefits. In other cases, where the wildlands do not directly benefit or serve the objectives of the project, the project may be improved by supporting management of wildlands to provide socio-economic benefits in the general project area. Projects with wildland management as the sole objective should also be encouraged.

Types of Projects Needing Wildland Management Components

Based upon these criteria, projects with the following aspects should normally contain wildland components:

a) *Agriculture and livestock projects involving:* wildland clearing, wetland elimination, wildland inundation for irrigation storage reservoirs; watershed protection for irrigation; displacement of wildlife by fences or domestic livestock; *fishery projects* involving elimination of important fish nursery, breeding, or feeding sites; over-fishing or introduction of ecologically risky exotic species within aquatic wildlands; *forestry projects* involving access roads, clear-felling or other intensive logging of wildlands, wildland elimination.
b) *Transportation projects involving:* construction of highways, rural roads, railways, or canals which penetrate wildlands, thus easing access and facilitating spontaneous settlement; channelization of rivers for

fluvial navigation; dredging and filling of coastal wetlands for ports projects.

c) *Hydro projects involving:* large-scale water development, including reservoir, power, and water diversion schemes; inundation or other major transformation of aquatic or terrestrial wildlands; watershed protection for enhanced power output; construction of power transmission corridors.
d) *Industry projects involving:* chemical and thermal pollution which may damage wildlands; wildland loss from large-scale mining; wildland conversion for industrial fuels or feedstocks.

Types of Wildland Management Components

The most effective type of wildland management component is support for the conservation of ecologically similar wildlands in one or more WMAs. In cases where a WMA already exists in the same type of ecosystem that is to be converted by a Bank-supported project, it may be preferable, for administrative or biological conservation reasons, to enlarge the existing WMA, rather than to establish a new one. The government's wildland agencies, local university wildlife departments, and various international organizations can often advise in such judgments.

A wildland management component could also involve the creation of wildlife habitat, in addition to or rather than preservation of already existing habitat. For example, marginal land on the fringes of irrigation projects could be converted to wildlife reserves by taking advantage of the water supply created by the projects. Natural depressions or seasonal swamps could be exploited by diverting water from the canal systems (probably a very small part of the total supply). Such reserves attract significant numbers of migratory and residential waterfowl with minimal additional project costs and land.

A useful option is to improve the quality of management of existing WMAs. Many WMAs in Bank member countries receive insufficient on-the-ground management, due to lack of adequately paid staff, training, staff housing, other infrastructure, equipment, spare parts, fuel, or a well-developed management plan through which efficient resource allocation decisions can be made. Small components can often help correct these deficiencies. In countries where effective management is clearly lacking, it is generally preferable to improve the management of existing WMAs than to create new units "on paper," thereby further overextending the limited capabilities of the responsible agencies. Whenever a new WMA is established as a project component, provisions are needed to ensure effective management. Since many wildland agencies (e.g., departments of national parks or wildlife) are not as operationally effective as necessary, institutional strengthening (particularly support for training) should be an important element of Bank-supported wildland management components.

The establishment or strengthening of WMAs is particularly effective when the Government includes these wildland areas in a national conservation or land use plan. A growing number of Bank member governments have undertaken some type of systematic land use planning for wildland management. Such planning can take various forms, ranging from "master plans" for a system of national parks and other WMAs, to "National Conservation Strategies" which address wildland management as only one component of a broad range of natural resource planning concerns, and in which policy intervention such as economic incentives are used to influence resource utilization. Bank assistance with such planning efforts greatly strengthens wildland management at the national level. When member governments agree to develop appropriate land use plans, it is important for the Bank to refrain from supporting projects which involve eliminating wildland and run counter to these plans.

In those relatively few Borrower countries in which wildland elimination pressures are still minor, the requirement of a compensatory wildland component can be interpreted more flexibly to involve measures other than the establishment or strengthening of one or more WMAs. Such alternative options include careful project siting to avoid converting the more environmentally sensitive wildlands, support for research on and management of particularly sensitive species, support for land use planning efforts, or institutional strengthening of the government's wildland management agency, and training in ecology, biological conservation, and wildland management.

DESIGN OF WILDLAND MANAGEMENT AREAS

Design Considerations

WMA design features include size, shape, and siting. Because an optimal design may vary greatly in different ecosystems, it is best determined in each case by a conservation specialist.

The size of a compensatory WMA should be sufficient to maintain the biological diversity or other important values present in the area to be converted. A WMA which is large enough to encompass a viable population of the largest local predator (e.g., eagle, tiger), or the seasonal territories and migration routes of the largest local herbivore, will most likely preserve all other pertinent ecological values. These objectives would most likely be achieved in a WMA larger than 1,000 sq km. Many values are conserved in moist forest WMAs of 500 sq km although possibly not all in perpetuity. Interim WMAs of less than 100 sq km can be useful short-term expedients for subsequent expansion into surrounding degraded areas. In general, the larger the WMA, the greater the number of ecological interdependencies and gene pools that will be preserved. Both are necessary to a healthy and self-perpetuating ecosystem. It is recognized that conflicting pressures for more intensive land use often make the establishment of large WMAs difficult. In any case, compensatory WMAs should be no smaller than the wildland area converted by the project.

The optimal shape of a WMA will depend upon its objectives. A more circular shape may preserve more biological diversity than other shapes of the same area. Shape is also determined by the location of centers of endemism and other wildlife resources. Boundaries are more effective when they coincide with natural surficial features, such as a river or watershed.

To ensure that the compensatory WMA is ecologically similar to the area to be converted, it is obviously necessary to site the WMA in the same ecosystem as the area to be converted. Moreover, siting the WMA some distance away from the converted area (separated by a managed buffer zone for example) helps reduce pressures for encroachment upon the WMA from people living in the converted area.

Management Categories

A variety of different use related categories can be used in establishing WMAs. The choice of category depends upon the particular objectives being accorded priority for management. The system of categories devised by IUCN indicate the variety of WMAs appropriate under different circumstances.

Personnel and Training Needs

The need for well-trained personnel in the proper management of WMAs cannot be overemphasized. Without adequate numbers of such trained people, WMAs cannot effectively serve their intended national or societal functions. Bank-supported wildland project components should therefore provide for staffing levels and training activities that ensure competent management of WMAs. The appropriate number and types of WMA personnel depend upon the category of WMA, its size and its intensity of management. The minimum adequate permanent staff size for a "modest to average" WMA is usually about eight.

Equipment, Infrastructure, and Budgetary Needs

Designation of WMAs on a map in no way ensures that they will be managed to provide their greatest possible benefits to society. Effective on-the-ground management requires a variety of physical inputs. In Bank-supported WMAs, efforts should be made to ensure that these inputs are provided as a project component in adequate supply and on a timely basis. Some types of WMAs will require a variety of additional inputs, according to specific management objectives.

The budgetary requirements for establishing and operating WMAs will vary according to size and the amounts of needed infrastructure, equipment, and personnel. The comparatively large (3,200 sq km) Dumoga-Bone National Park, financed by the Indonesia Irrigation XV Project, cost roughly US$1 million for establishment and initial operating costs; most smaller WMAs can be expected to cost considerably less.

In some instances, establishment or enlargement of WMAs may require additional funds for purchasing land from private or tribal owners. It may at times also be necessary to resettle and compensate people living within the boundaries of a newly-established WMA. Usually, however, WMAs are established on wholly government-owned properties on which people have not settled.

The largest recurrent cost of WMAs is usually staff salaries. It is important to maintain salaries at levels that encourage high productivity and a degree of permanence, and discourage corruption. Spare parts for machinery, while usually a relatively small budget item, are also a vital recurrent expenditure. Without a reliable supply of spare parts for often remote WMA areas, necessary equipment will often lie idle or may become cannibalized to provide spare parts. In some cases, salaries, spare parts, fuel, and other recurrent costs can be fully or partly met by fees collected from tourists, persons engaged in some form of harvesting, or scientific researchers. Otherwise, small annual outlays from the national or other government budget will be needed.

Management Plans

Wildland management areas typically need well-developed management plans to ensure efficient allocation of the scarce financial and skilled human resources devoted to their management. A management plan is a written document which guides and controls the use of the resources of a WMA and directs the design of subsequent programs of management and development. A thorough management plan will:

a) Describe the physical, biological, social, and cultural features of the WMA within a national, regional, and local context;
b) Identify those items of particular concern from which the objectives for managing specific areas of the WMA are derived;
c) Describe appropriate uses of the entire WMA through zoning; and
d) List in chronological order the activities to be carried out to realize the proposed management programs.

Preparation and implementation of management plans are carried out by the government wildland agency. Project staff should ensure that Bank-supported WMAs either have adequate management plans or will develop them early in the project. Some parts of a management plan can be completed in a few days, while others may take years to refine. While a longer-term management plan is being developed as soon as possible after loan signing, an "interim management plan" or "operational plan" may be used. PPDES can be of assistance in these matters.

Legal Considerations

The success of a WMA may depend upon how its design fits into an overall national legal framework concerning natural resources management in general and wildland management in particular. To maintain their legitimacy in the eyes of policy-makers and local populations, WMAs must have a firm legal foundation. National legislation, sometimes accompanied by regulations from the Head of State, is often needed to establish a WMA. Depending upon the particular situation, such legislation needs to establish precise WMA boundaries; specific management zones within the WMA, including buffer zones; a central management authority (at the national or sub-national level) with unambiguous responsibilities; and mechanisms to channel local participation in WMA management decisions. Bank staff should ensure that Bank-supported WMAs are established and managed within a compatible legal and policy context.

ANNEX 6: GLOSSARY

Alien: Belonging to another place; a foreign organism.

Allele: Any of all possible forms of the genetic code for a specific trait in organisms.

Allopatric: Having separate or mutually exclusive areas of geographical distribution (compare sympatric).

Autochthonous: From within; independent of external sources; self-produced, pertaining to the ecosystem (see Ecosystem).

Biological Diversity: The variety and variability among living organisms and the ecological complexes in which they occur (OTA, 1987); often shortened to "biodiversity." "Species diversity" refers to the number of species found within a given area, while "genetic diversity" refers to the variety of genes within a particular species, variety, or breed.

Biological Resources: Living natural resources, including plants, animals, and microorganisms, plus the environmental resources to which species contribute. Biological resources are the practical target of activities aimed at the principle of conserving biological diversity; they have two important properties, the combination of which distinguishes them from non-living resources: they are renewable if conserved, and they are destructible if not conserved (IUCN, 1980).

Buffer Zone: An area on the edge of a protected area that has land-use controls that allow only activities compatible with the objectives of the protected area; appropriate activities might include tourism, forestry, agroforestry, etc. The objective of such zones is to give added protection to the reserve, and to compensate local people for the loss of access to the biodiversity resources of the reserve (Oldfield, 1988).

Carrying Capacity: The maximum number of organisms that can use a given area of habitat without degrading the habitat and without causing social stresses that result in the population being reduced. When applied to humans, the maximum number of users that can be sustained by a given set of land resources at a particular level of technology.

Conservation: The management of human use of the biosphere so that it may yield the greatest sustainable benefit to present generations, while maintaining its potential to meet the needs and aspirations of future generations. Thus conservation is positive, embracing preservation, maintenance, sustainable utilization, restoration, and enhancement of the natural environment (IUCN, 1980).

Consumer Surplus: The difference between the total amount of money a consumer would be prepared to pay for some quantity of a good, and the amount the consumer actually has to pay. In economic analysis, consumer surplus is a consideration when the output of the project causes the market price of the product to fall. Those consumers previously paying the higher, old price (what they are willing to pay) will reap a benefit (consumer surplus) from the lower, new price, which must be added to the benefits accruing to the new consumers (USAID, 1987).

Cost-Benefit Analysis: The analytical technique used to appraise projects with quantifiable benefits and costs over a finite planning horizon. In project analysis, costs are goods or services used in a project that reduce the benefits of the project; benefits are any goods or services produced by a project that advance the project's objective. In economic analysis, benefits increase the national income of the society while costs reduce the national income of the society. A benefit forgone is a cost just as much as a cost avoided is a benefit. Costs and benefits may be either tangible (land, labor, materials, equipment are tangible costs and increased production of a good or service is a tangible benefit) or intangible (which by definition cannot be directly valued, though they may be quantified in some form).

Debt Swap: Mechanisms by which part of the external debt of a nation is purchased at a discount and is then sold back to the government in local currency, with the proceeds being used for conservation purposes.

Discount Rate: The interest rate used to determine the present value of a future value by discounting. The opportunity cost of capital is often taken as the discount rate. The "social discount rate," which expresses the preference of a society as a whole for present returns rather than future returns, is used in economic analysis to discount the incremental net benefit stream.

Disincentive (for conserving biological diversity): Any inducement or mechanism that discourages governments, local people, and international organizations from conserving biological diversity.

Ecology: A branch of science concerned with the interrelationships of living organisms with each other and their environment.

Economic Rent: A value in excess of the costs of production, including a return on the necessary investment. Highly relevant in forestry, where rents collected by concession-holders can be a powerful incentive for increasing production.

Ecosystem: The totality of factors of all kinds that make up a particular environment; the complex of biotic community and its abiotic, physical environment, functioning as an ecological unit in nature.

Ecotone: A transition (area) between two adjacent ecological communities.

Endemic: Native, restricted or peculiar to a locality or region.

Environment: All the physical, chemical, and biological factors impinging on a living organism.

Environmental Resource: Resources such as clean air, clean water, and scenic values that are not considered assets;

as a result, most interest is on activities involved in using these resources and to the ways in which the actions of some users affect the well-being of others.

Externality: A cost that is generated by that person, but not paid for by him; for example, extracting logs from a hillside may cause increased sedimentation of streams, the cost of which is borne by the downstream farmers instead of the logger. In project analysis, an effect of a project felt outside the project and not included in the valuation of the project. In general, economists consider an externality to exist when production or consumption of a good or service by one economic unit has a direct effect on the welfare of producers or consumers in another unit, without compensation being paid. Detrimental externalities arise if the action is harmful and the agent who carried it out is not charged for the damage done; beneficial externalities arise if the action is beneficial but the agent who carried it out receives no (or insufficient) payment for the benefit. When an externality is quantified in money terms and added to the project accounts, it is said to have been "internalized."

Extirpation: Local extinction; a species or subspecies disappearing from a locality or region without becoming extinct throughout its range.

Extinction: The evolutionary termination of a species caused by failure to reproduce and death of all remaining members of the species; the natural failure to adapt to environmental change.

Fauna: The total animal life of an area; usually the total number of animal species in a specified period, geological stratum, geographical region, ecosystem, habitat, or community.

Flora: The total plant life of an area; usually the total number of plant species in a specified period, geological stratum, geographical region, ecosystem, habitat, or community.

Food Web (Chain): Arrangement of the living organisms on ecological communities according to the order of predatory activity, in which each group of organisms uses the next members as a food source; e.g., carnivores eat herbivores, which eat plants.

Gene: A section of a chromosome containing enough DNA to control the formation of one protein; a gene controls the transmission of a hereditary character.

Gene Pool: The total of the alleles in a population of organisms.

Genetic Drift: Changes in the genetic composition of a population of an organism due to chance preservation or extinction of particular genes especially pronounced in small populations.

Genetic Resource: A genetic resource is the heritable characteristics of a plant or animal of real or potential benefit to people. The term includes modern cultivars and breeds; traditional cultivars and breeds; special genetic stocks (breeding lines, mutants, etc.); wild relatives of domesticated species; and genetic variants of wild resource species. A "wild genetic resource" is the wild relative of a plant or animal that is already known to be of economic importance. The reasons for conserving such a resource are evident, providing direct and immediate economic benefits; but the genetic material conserved by such a resource must be made available to the people who require it to improve the productivity, quality, or pest resistance of utilized plants or animals.

Genetics: That part of biology that deals with variation and heredity.

GIS: Geographic Information System, an information technology that stores, analyzes, and displays both spatial and non-spatial data. A GIS can transform data held in a database to produce new information internally; for example, fish population sizes, productivity levels, access, and other factors can be combined in a model to establish catch limits (Parker, 1988).

Habitat: The place or type of site where a plant or animal naturally and normally lives and grows; its home.

Herbivore: An animal that eats plants or parts of them.

Heredity: The organic relationships between successive generations of a population or species.

Hydrology: That branch of knowledge that deals with the properties, distribution, and circulation of water on the surface of the land, in the soil and underlying rocks, and in the atmosphere.

Incentive (for conserving biological diversity): An incentive is that which incites or motivates desired behavior; in this context, an incentive is that which incites or motivates governments, local people, and international organizations to conserve biological diversity. More broadly, an incentive is any inducement on the part of government that attempts to temporarily divert resources such as land, capital, and labor toward conserving biodiversity, and facilitates the participation of certain groups or agents in work that will benefit biodiversity.

Indigenous: Having originated in and being produced, growing, or living naturally in a particular region or environment; native.

Limnology: That branch of knowledge that deals with the physics, chemistry, biology, and ecology of inland waters (rivers, lakes, artificial lakes, swamps and so on).

Management: The efforts of humans to select, plan, organize and implement programs designed to achieve specified goals; activities can range from protective measures to ensure that nature remains uninterrupted by human influence, on into ever-more manipulative (active) tasks required to maintain diversity, install facilities, control populations, or eradicate aliens.

Monotypic: A taxon that has only one unit in the immediately subordinate category, e.g., a genus comprising only one species or a species not divisible into subspecies.

Mutualism: A relationship between two unrelated organisms (different species) in which both of them benefit.

Natural Resource: Includes renewable (forests, water, wildlife, soils, etc.) and non-renewable (oil, coal, iron ore, etc.) resources that are natural assets.

Natural Selection: The differential reproduction of individuals; the tendency for some individuals (plants or animals) to produce more successful offspring than others. Natural selection is generally acknowledged to be the primary force responsible for evolution.

Niche: The sum total of all physical and biological requirements for a species; different species occupying different niches; the ecological role of an organism in a community, especially in regard to food consumption. Literally, the "profession" of the species.

Opportunity Cost: The benefit forgone by using a scarce resource for one purpose instead of for its best alternative use.

Organism: Any living thing, animal or plant, that is capable of carrying out life processes.

Parasite: An organism living in, or on, another unrelated organism from which it obtains benefits and that it usually injures, sometimes fatally.

Perverse Incentive (regarding biological diversity): Any incentive that induces behavior leading to the reduction in biological diversity; obviously, "perverse" depends on the perspective, and most perverse incentives are designed to achieve positive policy objectives and the perversity is usually an external factor.

Photosynthesis: The process by which simple carbohydrates (sugars, starches) are formed from carbon dioxide, water, and essential nutrients in special plant cells, using sunlight as the energy source.

Pollination: The process involving the transfer of pollen from a stamen (male organ) to an ovule (female cell), promoting the production of fertile seeds in plants.

Pollinator: the agent (usually animal) responsible for pollinating flowers.

Population: A somewhat arbitrary grouping of individuals of a species, which is circumscribed according to a set of specific criteria; usually taken as all the individuals of a species in a given time and place.

Predator: An organism that preys upon and eats another organism (the prey).

Primary Productivity: The rate at which energy from light is absorbed and utilized together with carbon dioxide, water, and other nutrients in the production of organic matter in photosynthesis. Net production is given by the amount of organic matter formed in excess of that used in respiration. It represents food potentially available to the consumers of an ecosystem; it can be measured approximately by sampling vegetation at intervals and measuring the dry mass produced per unit area per unit time. (As opposed to secondary production, the amount of consumer (animal) tissue produced per unit area per unit time in any ecosystem).

Protected Area: Any area of land that has legal measures limiting human use of the plants and animals within that area; includes national parks, game reserves, multiple-use areas, biosphere reserves, etc.

Resource: A feature of the environment that contributes to an organism's fitness. Also, often used to describe a source of natural wealth or revenue which can be biotic and renewable (e.g., fish stocks in the ocean) or abiotic and non-renewable (e.g., gold).

Species: A group of actually or potentially interbreeding living organisms more or less isolated from other such groups; in simple terms, a "kind" of plant or animal.

Subsidy: Government economic assistance granted directly or indirectly to individuals or administrative bodies to encourage activities designed to satisfy the needs of the public. It is discretionary and revocable, and is conditional upon certain rules being observed. In contrast to grants, subsidies are usually much more institutionalized and are primarily aimed less at a particular, specific activity than at encouraging works in the public interest.

Sustainable Development: A pattern of social and structural economic transformations (i.e., "development") that optimizes the economic and other societal benefits available in the present without jeopardizing the likely potential for similar benefits in the future.

Symbiosis: The living together in more or less close association of two dissimilar organisms, in which one or both derive benefit from the relationship.

Sympatric: Having the same or overlapping areas of geographical distribution (compare *allopatric*).

Taxon: (plural: taxa) A term for any category used in classification. Taxonomy is the science of the classification of plants and animals. The fundamental taxon in biology is the species, which represents a real biological entity; this category can be defined generally in objective terms, whereas all other taxa are either subdivisions of the species or grouping of species, which cannot be defined except in terms involving subjective judgments.

Terrestrial: Of, or relating to, the land.

Vegetation: The total plant cover of an area.

Vertebrate: Any of a major group (Vertebrata) of animals (fish, amphibians, reptiles, birds and mammals) with a segmented spinal column (backbone).

Watershed: Area bounded peripherally by waters parting and draining to one or more watercourses; a dividing point or line.

Wetland: Temporarily or permanently inundated terrestrial systems bordering on aquatic systems and including shallow systems such as estuaries, salt marshes, bogs, sponges, mires, swamps, floodplains, and many coastal lakes and lagoons; systems that essentially are driven by littoral processes.

BIBLIOGRAPHY

The following contains both citations from the text and additional references that are relevant to how conserving biological diversity can contribute to development action. The most important references are marked with an asterisk (*).

Abbott, A.T., E.A. Kay, and C.H. Lamoureux. 1982. *Natural Landmarks Survey of the Islands of the Pacific*. National Park Service, Washington, D.C. 194 pp.

Abbott, I. 1977. Species richness, turnover, and equilibrium in insular floras near Perth, Western Australia. *Aust. J. Botany* 25:193-208.

Abbott, I. 1978. Factors determining the number of land bird species on islands around South-western Australia. *Oecologia* 33:221-233.

Abbott, I. 1980. Theories dealing with the ecology of landbirds on islands. *Adv. Ecol. Res.* 11:329-371.

Adamus, P.R. and G.C. Clough. 1978. Evaluating species for protection in natural areas. *Biological Conservation* 13(2):165-178.

Adsersen, Henning. 1989. The rare plants of Galápagos and their conservation. *Biological Conservation* 47:49-77.

Ahmad, Yusuf and G. K. Sammy. 1985. *Guidelines to Environmental Impact Assessment in Developing Countries.* Hodder and Stoughton, London. 52 pp.

Aiken, S.R. and C.H. Leigh. 1985. On the declining fauna of Peninsular Malaysia in the post-colonial period. *Ambio* 14:15-22.

Aiken, S.R. and C.H. Leigh. 1986. Land use conflicts and rain forest conservation in Malaysia and Australia. *Land Use Policy* 3:161-179.

Allard, R.W. 1960. *Principles of Plant Breeding.* Wiley, New York.

Allegretti, Mary Helena, and S. Schwartzman. 1986. *Extractive Reserves: A Sustainable Development Alternative for Amazonia.* Report to WWF-US, Washington, D.C.

Allen, William H. 1988. Biocultural restoration of a tropical forest. *BioScience* 38(3):156-161.

Altieri, M.A. and C.L. Merrick. 1987. In situ conservation of crop genetic resources through maintenance of traditional farming systems. *Econ. Bot.* 41:86-96.

Altieri, M.A., M.K. Anderson, and L.C. Merrick. 1987. Peasant agriculture and the conservation of crop and wild plant resources. *Conservation Biology* 1:49-58

Altman, I. and J.F. Wohlwill (eds.). 1983. *Behavior and the Natural Environment.* Plenum Press, New York. 346 pp.

Amacher, R.C., R.D. Tollison and T.D. Willett. 1974. The economics of fatal mistakes: fiscal mechanisms for preserving endangered predators. *Public Policy* 7:411-441.

American Institute of Biological Sciences. 1968. The role of the biologist in preservation of the biotic environment. *BioScience* 18(5):383-424.

Anders, G., W.P. Graham, and S.C. Maurice. 1978. Does resource conservation pay? *International Institute Econ. Research Paper* 14:1-42.

Anderson, D. 1983. Research goals for natural areas. *Natural Areas Journal* 3(1):27-32.

Anderson, I. 1987. Epidemic of bird deformities sweeps US. *New Scientist*: 3 September.

Arrow, K.J. and A.C. Fisher. 1974. Environmental preservation, uncertainty, and irreversibility. *Quarterly J. Economics* 88:312-319.

ASEAN. 1985. Agreement on the conservation of nature and natural resources. *Environmental Policy and Law* 15(2):64-69.

Ashton, P.S. 1981. Techniques for the identification and conservation of threatened species in tropical forests. Pp. 155-164 *in:* H. Synge (ed.), *The Biological Aspects of Rare Plant Conservation.* John Wiley and Sons, Chichester, England. 558 pp.

Ashton, P.S. 1984. Botanic gardens and experimental grounds. Pp. 39-48 *in*: V.H. Heywood and S.M. Moore (eds.), *Current Concepts in Plant Taxonomy.* Academic Press, London.

Ashton, P.S. 1988. Conservation of biological diversity in botanical garden. Pp. 269-278 *in:* E.O. Wilson and Francis M. Peter (eds.), *Biodiversity.* National Academy Press, Washington, D.C. 521 pp.

Astanin, L.P. and K.N. Biagosklinow. 1983. *Conservation of Nature.* Progress Publishers, Moscow. 148 pp.

Atkinson, S.F. 1985. Habitat-based methods for biological assessment. *The Environmental Professional* 7(3):265-282.

Aubreville, A. 1971. The destruction of forests and soils in the tropics. *Adansonia* 2(11):5-39.

Austin, M.P. and C.R. Margules. 1986. Assessing representativeness. Pp. 45-67 *in:* M.B. Usher (ed.), *Wildlife Conservation Evaluation.* Chapman and Hall, London. 394 pp.

Australian Conservation Foundation. 1980. *The Value of National Parks to the Community.* A.C.F. Hawthorn, Victoria. 223 pp.

Ayensu, E.S. and R.A. DeFilipps. 1981. Smithsonian Institution endangered flora computerized information. Pp.

111-122 *in:* L.E. Morse and M.S. Henifin (eds.), *Rare Plant Conservation: Geographical Data Organization.* New York Botanical Garden, Bronx. 377 pp.

Bachmura, F.T. 1971. The economics of vanishing species. *Natural Resources J.* 11:675-692.

Backus, E.H. 1985. *Computer Geographic Information Systems to Manage Biogeographic Data for the Design of Habitat Reserves: A Study of the Puerto Rico Conservation Data Center.* Tropical Research Institute, Yale School of Forestry and Environmental Studies, New Haven, Connecticut.

Backus, E.H., R.M. Alfaro, L.F. Corrales, Q. Jimenez. L.H. Elizondo, and W.H. Soto. 1988. *Costa Rica: Assessment of the Conservation of Biological Resources.* Fundación Neotropica and Conservation International, San José, Costa Rica, and Washington, D.C.

Bacow, L.S. and M.W. Wheeler. 1984. *Environmental Dispute Resolution.* Plenum Press, New York. 372 pp.

Bailey, R.G. 1983. Delineation of ecosystem regions. *Environmental Management* 7(4):365-373.

Bailey, R.G. and H.C. Hogg. 1986. A world ecoregions map for resource reporting. *Environmental Conservation* 13(3):195-202.

Baker, A.N., F.W.E. Rowe, and H.E.S. Clark. 1986. A new class of Echinodermata from New Zealand. *Nature* 321:862-864.

Baker, H.G., H.A. Mooney, and J.A. Drake. 1986. *Ecology of Biological Invasions of North America and Hawaii.* Springer-Verlag, New York, 321 pp.

Baker, M.L.B. and A. Ellington. 1985. *The World Environment Handbook: A Directory of Natural Resource Management Agencies and Non-Governmental Environment Organization in 145 Countries.* World Environment Center, New York, 280 pp.

Balser, D., A. Bielak, G. de Boer, T. Tobias, G. Adindu, and R.S. Dorney. 1981. Nature reserve designation in a cultural landscape, incorporating island biogeography theory. *Landscape Planning* 8(2):329-347.

Bannon, J.J. 1981. *Problem Solving in Recreation and Parks* (2nd ed.). Prentice-Hall, Englewood Cliffs, New Jersey. 396 pp.

Barborak, James R. 1988a. Innovative funding mechanisms used by Costa Rican conservation agencies. Paper presented at Workshop on Economics, IUCN General Assembly, 4-5 February 1988, Costa Rica.

Barborak, James R. 1988b. The role of economics in conserving biological diversity: A non-economist's view. Paper presented at Workshop on Economics, IUCN General Assembly, 4-5 February 1988, Costa Rica.

Barbour, Clyde D. and James H. Brown. 1974. Fish species diversity in lakes. *American Naturalist* 108:473-89

Barclay, A.S. and R.E. Perdue. 1976. Distribution of anticancer activity in higher plants. *Cancer Treatment Reports* 60(8):1081-1113.

Barrett, Scott. 1988. Economic guidelines for the conservation of biological diversity. Paper presented at Workshop on Economics, IUCN General Assembly, 4-5 February 1988, Costa Rica.

Baskin, J.M. and C.C. Baskin. 1978. The seed bank in a population of an endemic plant species and its ecological significance. *Biological Conservation* 14:125-130.

Baskin, J.M. and C.C. Baskin. 1986. Some considerations in evaluating and monitoring populations of rare plants in successional environments. *Natural Areas Journal* 6(3):26-30.

Bastedo, J.D., J.G. Nelson, and J.B. Theberge. 1984. Ecological approach to resource survey and planning for environmentally significant areas: The ABC method. *Environmental Management* 8(2):125-134.

Batisse, M. 1982. The biosphere reserve: a tool for environmental conservation and management. *Environmental Conservation* 9(2):101-11.

BCAS. 1989. *The Botanic Gardens Conservation Strategy.* IUCN/WWF, Gland, Switzerland.

Beanlands, G.E. and P.N. Duinker. 1984. An ecological framework for environmental impact assessment. *J. Environmental Management* 18(2):267-277.

Beddington, J.R. and R.M. May. 1980. Maximum sustainable yield in system subject to harvesting at more than one trophic level. *Math. Biosc.* 51:261-281.

Bell, B.C. 1981. Humanity in nature: Toward a fresh approach. *Environmental Ethics* 3:245-257.

Bella, Leslie. 1987. *Parks for Profit.* Harvest House Ltd., Montreal.

Bennett, J.W. 1982. Valuing the existence of a natural ecosystem. *Search* 13(9/10):232-235.

Bentkover, J., V. Covello, and J. Mumpower. 1986. *Benefits Assessments: The State of the Art.* D. Reidel, Boston.

Berck, R. 1979. Open access and extinction. *Econometrica* 47:877-882.

Bernstein, B.B. 1981. Ecology and economics: Complex systems in changing environments. *Annual Review of Ecology and Systematics* 12:309-330.

Beven, S.F., E.F. Connor and K. Beven. 1984. Avian biogeography in the Amazon Basin and the biological model of diversification. *J. Biogeog.* 11:383-399.

BGCS (Botanic Gardens Conservation Secretariat). 1987. *The International Transfer Format for Botanic Gardens Plant Records.* Hunt Institute for Botanical Documentation, Carnegie Mellon University, Pittsburgh, PA, 70 pp.

Binswanger, Hans P. 1987. Fiscal and legal incentives with

environmental effects on the Brazilian Amazon. *World Bank Report* ARU 69: 1-48.

Bishop, R.C. 1978. Endangered species and uncertainty: the economics of a safe minimum standard. *Am. J. Agricultural Economics* 60:10-18.

Bishop, R.C. 1980. Endangered species: an economic perspective. *Trans. North Am. Wildl. Nat. Resourc. Conf.* 45:208-218.

Blouin, M.S. and E.F. Connor. 1985. Is there a best shape for nature reserves? *Biological Conservation* 32:277-88.

Boardman, R. 1981. *International Organizations and the Conservation of Nature.* Indiana University Press, Bloomington. 215 pp.

Boecklen, W.J. 1986. Optimal design of nature reserves: Consequences of genetic drift. *Biological Conservation* 38(4):323-338.

Boecklen, W. J. and Nicholas J. Gotelli. 1984. Island biogeographic theory and conservation practice: species-area or specious-area relationships? *Biological Conservation* 29(1):63-79.

Bolin, B., Bo.R. Döös, J. Jäger and R.A. Warrick. 1986. *The Greenhouse Effect, Climatic Change and Ecosystems.* John Wiley, Chichester, UK.

Bonnicksen, T.M. and E.C. Stone. 1985. Restoring naturalness to national parks. *Environmental Management* 9(6):479-486.

Boom, B.M. 1985. Amazonian Indians and the forest environment. *Nature* 314:324.

Booth, William. 1989. Monitoring the fate of the forests from space. *Science* 243:1428-1429.

Bordwin, H.J. 1985. The legal and political implication of the international undertaking on plant genetic resources. *Ecology Law Quarterly* 12(4):1053-1070.

Borman, F.H. 1976. An inseparable linkage: Conservation of natural ecosystems and the conservation of fossil energy. *BioScience* 26:754-760.

BOSTID. 1983. *Little-known Asian Animals with a Promising Economic Future.* National Academy of Sciences Press, Washington D.C. 131 pp.

BOSTID. 1985. *Conservation of biological diversity in developing countries.* Report to USAID from the Board on Science and Technology for International Development, National Research Council, Washington D.C.

BOSTID. 1986. *Proceedings of the Conference on Common Property Resource Management.* National Academy Press, Washington D.C. 631 pp.

Boudet, G. 1972. Desertification de l'Afrique tropicale seche. *Adansonia* 2(12):505-24.

Bowonder, B. 1983. Environmental management conflicts in developing countries: An analysis. *Environmental Management* 7(3):211-22.

Bradshaw, A.D. 1983. The reconstruction of ecosystems. *J. Applied Ecology* 20(1):1-17.

Bradshaw, M.E. and J.P. Doody. 1978. Plant population studies and their relevance to nature conservation. *Biological Conservation* 14(2):223-242.

Bratton, S.P. 1982. The effects of exotic plant and animal species on nature preserves. *Natural Areas Journal* 2(3):3-12.

Bratton, S.P. and P.S. White. 1980. Rare plant management: After preservation what? *Rhodora* 82:49-75.

Bresard, P. F. 1982. The role of field stations in the preservation of biological diversity. *BioScience* 32(5): 327-330.

Breslin, O. and M. Chapin. 1984. Conservation Kuna-style. *Grassroots Development* 8(2):26-35

Briggs, D.J. and J. France. 1983. Classifying landscapes and habitats for regional environmental planning. *J. Environment Management* 17(3):249-261.

Briggs, J.C. 1974. *Marine Zoogeography.* McGraw-Hill, New York, NY.

Brockman, C.F. and L.C. Merriam, Jr. 1973. *Recreational Use of Wildlands* (2nd ed.). McGraw-Hill, New York. 329 pp.

Brookshire, D.S., A. Randall, and J. Stoll. 1980. Valuing increments and decrements in natural resources service flows. *Am. J. Agric. Econ.* 62:478-488.

Brookshire, D.S., L.S. Eubanks and A. Randall. 1983. Estimating option price and existence values for wildlife resources. *Land Economics* 59(1):1-15.

Brown, G.M. 1979. Notes on the economic value of genetic capital: What can the blue whale do for you? Department of Economics, University of Washington, Seattle (MS).

Brown, G.M. 1985. Valuation of genetic resources. Paper prepared for Workshop on Conservation of Genetic Resources, Lake Wilderness, WA.

Brown, G.M. and Jon H. Goldstein. 1984. A model for valuing endangered species. *J. Environmental Economics and Management* 11:303-309.

Brown, J.H. 1971. Mammals on mountaintops: nonequilibrium insular biogeography. *American Naturalist* 105:467-478.

Brown, J.H. 1984. On the relationship between abundance and distribution of species. *American Naturalist* 124(2):255-279.

Brown, J.H. and A. Kodric-Brown. 1977. Turnover rates in insular biogeography: Effect of immigration on extinction. *Ecology* 58:445-449.

Brown S. and A.E. Lugo. 1982. The storage and production of organic matter in tropical forests and their role in the global carbon cycle. *Biotropica* 14:161-187.

Brownrigg, L.A. 1985. Native cultures and protected areas: management options. Pp. 33-44 *in:* J.A. McNeely and D. Pitt (eds.), *Culture and Conservation: The Human Dimension in Environmental Planning.* Croom Helm, London.

Buckley, R. 1982. The habitat unit model of island biogeography. *J.Biogeog.* 9:339-44.

Budowski, G. 1976. Tourism and environmental conservation: Conflict, coexistence, or symbiosis? *Environmental Conservation* 3(1):27-31.

Bunyard, Peter. 1987. The significance of the Amazon basin for global climatic equilibrium. *The Ecologist* 17(4-5):139-141.

Burgman, M.A., H. R. Akcakaya, and S.S. Loew. 1988. The use of extinction models for species conservation. *Biological Conservation* 43:9-25.

Burton, J.A. 1984. A bibliography of Red Data Books. *Oryx* 18:61-64.

Caldecott, Julian. 1988. *Hunting and Wildlife Management in Sarawak.* IUCN, Gland. 172 pp.

Caldwell, L.K. 1984. *International Environmental Policy: Emergence and Dimensions.* Duke University Press, Durham, NC. 367 pp.

Caldwell, L.K. 1984. Political aspects of ecologically sustainable development. *Environmental Conservation* 11(4):299-308.

Campbell, F.T. 1980. Conserving our wild plant heritage. *Environment* 22(9):14-20.

Carlquist, S. 1974. *Island Biology.* Columbia University Press, New York. 660 pp.

Carlson, A.A. 1977. On the possibility of quantifying scenic beauty. *Landscape Planning* 4(2):131-172.

*Carpenter, R. (ed.). 1983. *Natural Systems for Development.* Macmillan, New York. 485 pp.

Carr, A.F. III. 1982. Tropical forest conservation and estuarine ecology. *Biological Conservation* 23(3):247-259.

Cartwright, J. 1985. The politics of preserving natural areas in Third World States. *Environmentalist* 5(3):179-186.

CEMP (Centre for Environmental Management and Planning). 1988. Preliminary paper on compensation for the establishment of protected areas within tropical forest ecosystems. UNEP, Nairobi.

Chalk, D.E., S.A. Miller, and T.W. Hoekstra. 1984. Multiresource inventories: Integrating information on wildlife resources. *Wildlife Society Bulletin* 12(2):357-364.

Charbonneau, J.J. and M.J. Hay. 1978. Determinants and economic values of hunting and fishing. *Trans. N. Am. Wildl. and Natl. Res. Conf.* 43.

Child, B. and G. Child. 1986. Wildlife, economic systems and sustainable human welfare in semi-arid rangelands in southern Africa. Pp. 81-91 *in: Report on the FAO/Finland Workshop on Watershed Management in Arid and Semi-arid Zones of SADCC Countries,* Maseru, Lesotho.

Child, G. 1984. Managing wildlife for people in Zimbabwe. Pp. 118-123 *in:* J.A. McNeely and K.R. Miller (eds.), *National Parks, Conservation and Development: The Role of Protected Areas in Sustaining Society.* Smithsonian Institution Press, Washington, D.C.

Child, G. 1988. Economic incentives and improved wildlife conservation in Zimbabwe. Paper presented at Workshop on Economics, IUCN General Assembly, 4-5 February 1988, Costa Rica.

Childress, J.J., C.R. Fisher, J.M. Brooks, M.C. Kennicutt II, R. Bidigare, and A.E. Anderson. 1986. A methanotropic marine molluscan (Bivalvia, Mytilidae) symbiosis: Mussels fueled by gas. *Science* 233:1306-1308.

Christiansen, M.I. 1977. *Park Planning Handbook.* John Wiley and Sons, New York.

Christiansen, Sofus and Gunnar Poulson (eds.). 1985. *Environmental Aspects of Agricultural Development Assistance.* Nordic Council of Ministers, University of Trondheim, Norway.

*Ciriacy-Wantrup, S.V. 1968. *Resource Conservation: Economics and Policies.* University of California Press, Berkeley. 395 pp.

Ciriacy-Wantrup, S.V. and W.E. Phillips. 1970. Conservation of the California Tule Elk: A socioeconomic study of a survival problem. *Biological Conservation* 3(1): 23-32.

Clark, Colin W. 1973a. The economics of overexploitation. *Science* 181: 630-634.

Clark, Colin W. 1973b. Profit maximization and the extinction of animal species. *J. Pol. Econ.* 81: 950-961.

Clark, Colin W. 1976. *Mathematical Bioeconomics: The Optimal Management of Renewable Resources.* John Wiley, New York.

Clark, J.R. 1974. *Coastal Ecosystems: Ecological Considerations for the Management of the Coastal Zone.* The Conservation Foundation, Washington, D.C. 178 pp.

*Clark, J.R. 1977. *Coastal Ecosystem Management: A Technical Manual for the Conservation of Coastal Zone Resources.* John Wiley and Sons, New York. 928 pp.

Clark, W.C. and R.E. Munn (eds.). 1986. *Sustainable Development of the Biosphere.* Cambridge University Press, London. 491 pp.

Clarke, B. 1979. The evolution of genetic diversity. *Proc. Roy. Soc. Lond. B. Biol. Sci.* 205:453-474.

Clarke, J.E. and R.H.V. Bell. 1986. Representation of biotic communities in protected areas: A Malawian case study. *Biological Conservation* 35(4):293-312.

Clawson, D.L. 1985. Harvest security and intraspecific diversity in traditional tropical agriculture. *Economic Botany* 39:56-67.

Clawson, M. 1980. Wilderness as one of many land uses. *Idaho Law Review* 16(3):449-468.

Clements, James. 1981. *Birds of the World: A Checklist.* Facts on File, New York. 562 pp.

CNPPA. 1988. The Bali Action Plan: The first five years. Pp. 10-28 *in:* J. Thorsell (ed.), *New Challenges for the World's Protected Area System.* IUCN, Gland. 189 pp.

Cocheba, Donald J., and W. A. Langford. 1978. Wildlife valuation: the collective good aspect of hunting. *J. Land Econ.* 54(4):490-504.

Cock, James H. 1982. Cassava: A basic energy source in the tropics. *Science* 218:755-762.

Cody, M.L. and J.M. Diamond. 1975. *Ecology and Evolution of Communities.* Harvard University Press., Cambridge. Mass.

Cohn, Jeffrey P. 1989. Gauging the biological impacts of the greenhouse effect. *BioScience* 39(3):142-146.

Collar, N.J. and S.N. Stuart. 1985. *Threatened Birds of Africa and Related Islands.* (3rd edn.) ICBP/IUCN, Cambridge. 761 pp.

Collar, N.J. and S.N. Stuart. 1988. *Key Forests for Threatened Birds in Africa.* ICBP/IUCN, Cambridge. 102 pp.

Collar, N.J. and J. Andrew. 1988. *Birds to Watch: The ICBP World Check-list of Threatened Birds.* ICBP Technical Publication No. 8. ICBP, Cambridge UK. 303 pp.

Collins, N.M. and M.G. Morris. 1985. *Threatened Swallowtail Butterflies of the World: The IUCN Red Data Book.* IUCN, Gland, Switzerland. 401 pp.

Connell, J.H. 1978. Diversity in tropical rain forests and coral reefs: High diversity of trees and corals is maintained only in a nonequilibrium state. *Science* 199:1302-1310.

Connell, J.H. 1980. Diversity and the coevolution of competitors or, the ghost of competition past. *Oikos* 35(2):131-138.

Connell, J.H. and E. Orians. 1964. The ecological regulation of species diversity. *American Naturalist* 98:399-414.

Connell, J.H. and R.O. Slatyer. 1977. Mechanisms of succession in natural communities and their role in community stability and organization. *American Naturalist* 111(4):1114-1119.

Connor, E.F. and E.D. McCoy. 1979. The statistics and biology of the species-area relationship. *American Naturalist* 133:791-833.

Connor, E.F., E.D. McCoy, and B.J. Cosby, 1983. Model discrimination and expected slope values in species-area studies. *American Naturalist* 122:789-96.

Conrad, Jon M. and Colin W. Clark. 1987. *Natural Resource Economics.* Cambridge University Press, Cambridge, UK.

Conservation International. 1989. The Debt-for-Nature Exchange: A Tool for International Conservation. Conservation International, Washington, D.C., 43 pp.

Conservation International. 1990. *The Rain Forest Imperative.* Conservation International, Washington, D.C., 15 pp.

Conway, William. 1986. The practical difficulties and financial implications of endangered species breeding programmes. *Int. Zoo Yearbook* 24/25:210-219.

Conway, William. 1988. Can technology aid species preservation? Pp. 263-268 *in:* E.O. Wilson and Francis M. Peter (eds.), *Biodiversity.* National Academy Press, Washington, D.C. 521 pp.

Cook, Clive. 1988. Applying the user pays principle to national parks. *Australian Ranger Bulletin* 5(1):5-6

Cooper, C. 1981. *Economic Evaluation and the Environment.* Hodder and Stoughton, London.

Corbet, G.B. and J.E. Hill. 1980. *A World List of Mammalian Species.* British Museum, London. 226 pp.

Corfield, T. 1985. *The Wilderness Guardian.* David Sheldrick Foundation, Nairobi. 621 pp.

Corner, E.J.H. 1978. Plant life. Pp. 112-178 *in:* D.M. Luping, C. Wen, and E.R. Dingley (eds), *Kinabalu: Summit of Borneo.* Sabah Society, Kuching.

Council on Environmental Quality (U.S.A.). 1980. Ecology and living resources: Biological diversity. Pp. 31-80 *in:* CEQ, *Annual Report.* Washington, D.C.

Crandall, Lee S. 1964. *Management of Wild Mammals in Captivity.* University of Chicago Press, Chicago. 769 pp.

Croisat, Leon. 1962. *Space, Time, Form: The Biological Synthesis.* Caracas, Venezuela. 881 pp.

Cropper, M.L., M.C. Veinstein and R.J. Zeckhauser. 1978. The optimal consumption of depletable natural resources: An elaboration, correction, and extension. *Quarterly J. Economics* 92: 337-353.

Crossen, T.I. 1979. A new concept in park design and management. *Biological Conservation* 15(2):105-125.

Crowe, T.M. and A.A. Crowe. 1982. Patterns of distribution, diversity and endemism in Afrotropical birds. *J. Zool.* London. 198:417-442.

Crowley, Thomas J. and Gerald R. North. 1988. Abrupt climate change and extinction events in earth history. *Science* 240:996-1002.

Crutchfield, J.A. 1962. Valuation of fishery resources. *Land Econ.* 38(5):145-154.

Cuddington, J.T., F.R. Johnson, and J.L. Knetsch, 1981. Valuing amenity resources in the presence of substitutes. *Land Economics* 57:526-535.

Cumming, D.H.M. 1985. Environmental limits and sustainable harvests. Paper presented to the Plenary Session of the Zimbabwe National Conservation Strategy Conference, Harare.

Cummings, R., D. Brookshire, and W. Schulze. 1986. *Valuing Environmental Goods: A State of The Art Assessment of the Contingent Valuation Method.* Rowman and Allenheld, Totowa, NJ. 270 pp.

Cummins, J.E. 1988. Extinction: The PCB threat to marine mammals. *Ecologist* 18(6):193-195.

Currie, D.J. and V. Paquin. 1987. Large scale biogeographical patterns of species richness of trees. *Nature* 329:326-327.

Dahl, A.L. 1980. *Regional Ecosystems Survey of the South Pacific Area.* SPC Technical Paper 179, Noumea. 99 pp.

Dahlberg, K.A. 1983. Plant germplasm conservation: Emerging problems and issues. *Mazingira* 7(1):14-25.

Dahlberg, K.A. 1987. Redefining development priorities: Genetic diversity and agroecodevelopment. *Conservation Biology* 1:311-323.

Daly, H.E. 1980. *Economics, Ecology and Ethics.* Freeman, San Francisco, CA.

Daniel, J.G. and A. Kulasingam. 1974. Problems arising from large-scale forest clearing for agricultural use. *Malaysian Forester* 37:152-160.

Darling, F.F. and N.D. Eichorn. 1967. *Man and Nature in the National Parks: Reflection on Policy.* The Conservation Foundation, Washington, D.C. 80 pp.

Darlington, P.J. 1957. *Zoogeography: The Geographical Distribution of Animals.* J. Wiley and Sons, New York.

Dasgupta, P. 1982. *The Control of Resources.* Harvard University Press, Cambridge, MA.

Dasmann, R.F. 1980. The relationship between protected areas and indigenous peoples. Pp. 667-671 *in:* J.A. McNeely and K.R. Miller (eds.), *National Parks, Conservation and Development: The Role of Protected Areas in Sustaining Society.* IUCN/Smithsonian Institution Press, Washington, D.C.

*Dasmann, R.F. 1984. *Environmental Conservation* (5th edn.). Wiley, New York. 486 pp.

Dasmann, R.F. 1985. Achieving the sustainable use of species and ecosystems. *Landscape Planning* 12(3):211-220.

*Dasmann, R.F., J.P. Milton, and P. Freeman. 1973. *Ecological Principles for Economic Development.* John Wiley, London. 252 pp.

Dasmann, R.F. and D. Poore. 1979. *Ecological Guidelines for Balanced Land Use, Conservation and Development in High Mountains.* IUCN, Gland, Switzerland. 40 pp.

Davidson, R.J. and T.J. Beechey. 1982. An agenda for a natural heritage conservation strategy. *Environments* 14(1):77-84.

Davies, A.G. 1987. *The Gola Forest Reserves, Sierra Leone: Wildlife Conservation and Forest Management.* IUCN, Gland. 126 pp.

Davis, R.K. 1964. The value of big game hunting in a private forest. *Transactions Twenty-Ninth North American Wildlife and Natural Resources Conference.*

*Davis, Stephen D., Stephen J.M. Droop, Patrick Gregerson, Louise Henson, Christine J. Leon, Jane L. Villa-Lobos, Hugh Synge, and Jana Zantovska. 1986. *Plants in Danger: What do we Know?* IUCN, Gland, Switzerland.

Davis, S.H. 1988. Indigenous peoples, environmental protection, and sustainable development. *IUCN Sustainable Development Occasional Paper* 1:1-26.

Day, D. 1981. *The Doomsday Book of Animals.* Viking Press, New York, N.Y.

de Camino Velozo, Ronnie. 1987. Incentives for community involvement in conservation programmes. *FAO Conservation Guide* 12:1-59.

de Groot, Rudolf. 1986. *A Functional Ecosystem Evaluation Method as a Tool in Environmental Planning and Decision Making.* Agricultural University, Wageningen, the Netherlands. 38 pp.

de Klemm, C. 1985. Preserving genetic diversity: A legal review. *Landscape Planning* 12(3):221-238.

Dearden, P. 1978. The ecological component in land use planning: A conceptual framework. *Biological Conservation* 14:167-179.

Dearden, P. 1981. Landscape evaluation: The case for a multi-dimensional approach. *J. Environmental Management* 13:95-105.

Decker, D. and G. Goff. 1987. *Valuing Wildlife: Economic and Social Perspectives.* Westview Press, Boulder, CO.

DeLong, R., W.G. Gilmartin, and J.G. Simpson. 1973. Premature births in California sea lions: Association with high organochlorine pollutant residue levels. *Science* 181:1168-1170.

Denniston, C. 1978. Small population size and genetic diversity: Implications for endangered species. Pp. 281-289 *in:* S.A. Temple (ed.), *Endangered Birds: Management Techniques for Preserving Threatened Species.* Madison, University of Wisconsin Press.

Devall, B. and G. Sessions. 1984. The development of natural resources and the integrity of nature. *Environmental Ethics* 6(3):293-315.

Devall, B. and G. Sessions. 1985. *Deep Ecology: Living As If Nature Mattered.* Peregrine Smith Books, Layton, Ut.

Di Castri, F. and M. Hadley. 1985. Enhancing the credibility of ecology: Can research be made more comparable and predictable? *GeoJournal* 11(4):321-338.

Diamond, J.M. 1975. The island dilemma: Lessons of modern biogeographic studies for the design of natural reserves. *Biological Conservation* 7:129-46.

Diamond, J.M. 1984. Biological principles relevant to protected areas design in the New Guinea region. Pp. 330-332 *in:* J.A. McNeely and K.R. Miller (eds.), *National Parks,*

Conservation, and Development: The Role of Protected Areas in Sustaining Society. IUCN/Smithsonian Institution Press, Washington D.C.

Diamond, J.M. 1985. Introductions, extinctions, exterminations, and invasions. Pp. 65-79 *in:* T.J. Case and J.M. Diamond (eds.), *Community Ecology.* Harper & Row, New York.

Diamond, J.M. 1987. Extant unless proven extinct? Or, extinct unless proven extant? *Conservation Biology* 1(1):72-76.

Diamond, J.M. and R.M. May. 1976. Island biogeography and the design of nature reserves. Pp. 163-186 *in:* R.M. May, (ed.), *Theoretical Ecology: Principles and Applications.* Oxford, Blackwell.

Dingwall, Paul (ed). 1987. *Conserving the Natural Heritage of the Antarctic Realm.* IUCN, Gland. 222 pp.

Dixon, John A., R. Carpenter, L. Fallon, P. Sherman, and S. Manopimoke. 1986. *Economic Analysis of the Environmental Impacts of Development Projects.* Asian Development Bank, Manila. 100 pp.

Dixon, John A. and S. Wattanavitukul. 1982. *A Summary of Work on Benefit-Cost Analysis of Natural Systems and Environmental Quality Aspects of Development.* Environment and Policy Institute, East-West Center, Honolulu.

Dony, J.G. and I. Denholm. 1985. Some quantitative methods of assessing the conservation value of ecologically similar sites. *J. Applied Ecology* 22(2):229-238.

Dorfman, R. and Nancy S. Dorfman (eds.). 1977. *Economics of the Environment.* W.W. Norton, New York. 510 pp.

Drewett, J. 1988. Never mind the whale, save the insects. *New Scientist,* 17 December:32-35.

Duane, W.L. 1980. On preserving nature's aesthetic features. *Environmental Ethics* 2(2):293-310.

*Duffey, E. and A.S. Watts (eds.). 1971. *The Scientific Management of Animal and Plant Communities for Conservation.* Blackwell Scientific Publications, Oxford.

Dunlop, Richard C. and Birendra B. Singh. 1978. *A National Parks and Reserves System for Fiji.* The National Trust for Fiji, Suva. 117 pp.

Eachern, J. and E. L. Towle. 1974. *Ecological Guidelines for Island Development.* IUCN, Gland, Switzerland. 65 pp.

Eagles, Paul F.J. 1984. *The Planning and Management of Environmentally Sensitive Areas.* Longman, London and New York. 160 pp.

Ehrenfeld, David. 1972. *Conserving Life on Earth.* Oxford University Press, New York. 360 pp.

Ehrenfeld, David. 1988. Why put a value on biodiversity? Pp. 212-216 *in:* E.O. Wilson and Francis M. Peter (eds.), *Biodiversity.* National Academy Press, Washington D.C.

Ehrlich, Paul R. 1982. Human carrying capacity, extinction, and nature reserves. *BioScience* 32(5):321-326.

Ehrlich, Paul R. 1985. Extinctions and ecosystem functions: Implications for humankind. Pp. 162-176 *in:* R.J. Hoage (ed.), *Animal Extinction: What Everyone Should Know.* Smithsonian Institution Press. Washington, USA.

*Ehrlich, P.R. and A.H. Ehrlich. 1981. *Extinction: The Causes and Consequences of the Disappearance of Species.* Random House, New York. 305 pp.

Ehrlich, Paul R. and H.A. Mooney. 1983. Extinction, substitution, and ecosystem services. *BioScience* 33(4):251-252.

Eisner, Thomas. In press. Chemical exploration of nature: A proposal for action. *In: Ecology, Economics, and Ethics: The Broken Circle.* Yale University Press.

Elkington, J. and T. Burke. 1987. *The Green Capitalists: Industry's Search for Environmental Excellence.* Victor Gollancz, London. 258 pp.

Ellenburg, H. 1986. The effects of environmental factors and use alternatives upon the species diversity and regeneration of tropical rain forest. *Applied Geography and Development* 28:19-36.

Elliott, H.F.L. (ed.). 1964. *The Ecology of Man in the Tropical Environment.* IUCN, Morges. 355 pp.

Elliott, H.F.L. 1972. Island ecosystems and conservation with particular reference to the biological significance of islands in the Indian Ocean and consequential research and conservation needs. *J. Mar. Biol. Ass. India* 14(2):578-608.

*Elliott, H.F.L. (ed.). 1974. *Second World Conference on National Parks.* IUCN, Morges, Switzerland. 504 pp.

Eltringham, S.K. 1984. *Wildlife Resources and Economic Development.* John Wiley, New York. 325 pp.

Endler, J.A. 1982. Pleistocene forest refugia: Fact or fancy? Pp. 179-200 *in:* G. Prance (ed.), *Biological Diversification in the Tropics.* Columbia University Press, New York.

Erwin, Terry L. 1982. Tropical forests: Their richness in Coleoptera and other arthropod species. *Coleopterists Bulletin* 36:74-75.

Erwin, Terry L. 1983. Tropical forest canopies: The last biotic frontier. *Bul. Entomological Soc. America* 29(1):14-19.

Erwin, Terry L. 1988. The tropical forest canopy: The heart of biotic diversity. Pp. 123-129 *in:* E.O. Wilson and Francis M. Peter (eds.), *Biodiversity.* National Academy Press, Washington, D.C.

Faeth, S.H. and E.F. Connor. 1979. Supersaturated and relaxed island faunas: A critique of the species-area relationship. *J. Biogeog.* 6:311-316.

FAO. 1981. *Tropical Forest Resources Assessment Project (GEMS): Tropical Africa, Tropical Asia, Tropical America (4 Vols.)* FAO/UNEP, Rome, Italy.

FAO. 1984. *In Situ Conservation of Wild Plant Genetic Resources: A Status Review and Action Plan.* Forest Resources Division, FAO. Rome.

FAO. 1985. *Yearbook of Forest Products 1972-1983*. FAO, Rome.

FAO/IBRD/WRI/UNDP. 1987. *The Tropical Forestry Action Plan*. FAO, Rome. 32 pp.

FAO/UNEP. 1981. Conservation of the genetic resources of fish. *FAO Fisheries Technical Paper* 217:1-43.

FAO/UNEP. 1982. *Tropical Forest Resources*. FAO, Rome.

Farnsworth, N.R. 1982. The Potential Consequences of Plant Extinction in the United States on the Current and Future Availability of Prescription Drugs. Paper presented at Symposium on Estimating the Value of Endangered Species — Responsibilities and Role of the Scientific Community, Annual Meeting of the AAAS, Washington, D.C.

Farnsworth, N.R. 1988. Screening plants for new medicines. Pp. 83-97 *in:* E.O. Wilson and Francis M. Peter (eds.), *Biodiversity*. National Academy Press, Washington, D.C.

Farnsworth, N.R. and D.D. Soejarto. 1985. Potential consequences of plant extinction in the United States on the current and future availability of prescription drugs. *Economic Botany* 39(2):231-240.

Farnworth, E.G., T.H. Tidrick, W.M. Smathers Jr., and C.F. Jordan. 1983. A synthesis of ecological and economic theory toward more capable valuation of tropical moist forests. *International Environmental Studies* 21(1):11-28.

Fearnside, P.M. 1987. Deforestation and international economic development projects in Brazilian Amazonia. *Conservation Biology* 1(3):214-222.

Fenner, F.J. (ed.). 1975. *A National System of Ecological Reserves in Australia*. Australian Academy of Science, Canberra, 114 pp.

Field, C.R. 1979. Game ranching in Africa. *Applied Ecology* 4:63-101.

Fillon, F.L., A. Jacquemot, and R. Reid. 1985. *The Importance of Wildlife to Canadians*. Canadian Wildlife Service, Ottawa. 19 pp.

Fisher, A.C. 1981a. *Economic Analysis and the Extinction of Species*. Report No. ERG-WP-81-4. Energy and Resources Group, Berkeley, CA. 19 pp.

Fisher, A.C. 1981b. *Resources and Environmental Economics*. Cambridge University Press, Cambridge, UK. 284 pp.

Fisher, A.C and M. Hanemann. 1984. Option values and the extinction of species. *Working Paper No 269*. Giannini Foundation of Agricultural Economics, Berkeley, CA. 39 pp.

Fisher, A.C. and W.M. Hanemann. 1985. Endangered species: The economics of irreversible damage. Pp. 129-138 *in:* D.O. Hall, N. Myers and N.S. Margaris (eds.), *Economics of Ecosystem Management*. W. Junk Publishers, Dordrecht, The Netherlands.

Fisher, A.C. and J. Krutilla. 1972. Determination of optimal capacity of resource-based recreation facilities. *Natural Resources J.* 12(3):417-444.

Fisher, A.C., J.V. Krutilla, and C.J. Cicchetti. 1972. The economics of environmental preservation: A theoretical and empirical analysis. *Am. Econ. Rev.* 62:605-619.

Fisher, J., N. Simon, and J. Vincent. 1969. *Wildlife in Danger*. Viking Press, New York.

Fitter, Richard. 1986. *Wildlife for Man: How and Why We Should Conserve Our Species*. Collins, London. 223 pp.

Fitter, Richard and Maisie Fitter. 1987. *The Road to Extinction*. IUCN, Gland. 121 pp.

Flemming, T.H., R. Breitwisch, and G.H. Whitesides. 1987. Patterns of tropical vertebrate frugivore diversity. *Annual Review of Ecology and Systematics* 18:910109.

Fogden, M.P. 1972. The seasonality and population dynamics of tropical forest birds in Sarawak. *Ibis* 114:307-343.

Foose, T.J., R. Lande, N.R. Flesness, G.Rabb, and B. Read. 1986. Propagation plans. *Zoobiol.* 5:139-46.

Foresta, R.A. 1985. Natural regions and national parks: The Canadian experience. *Applied Geography* 5(3):179-194.

Forester, R.R. 1973. *Planning for Man and Nature in National Parks: Reconciling Perpetuation and Use*. IUCN, Morges, Switzerland. 84 pp.

Forman, Richard T.T., Anne E. Galli, and Charles F. Leck. 1975. Forest size and avian diversity in New Jersey woodlots with some land use implication. *Oecologia* 26:1-8.

Fosberg, F. Raymond. 1988. Artificial diversity. *Environmental Conservation* 15(2):74.

Frankel, O.H. 1974. Genetic conservation: Our evolutionary responsibility. *Genetics* 78:53-65.

Frankel, O.H. and E. Bennett (eds.). 1970. *Genetic Resources in Plants: Their Exploration and Conservation*. Blackwell, Oxford, UK. 554 pp.

Frankel, O.H. and J.G. Hawkes (eds.). 1974. *Plant Genetic Resources for Today and Tomorrow*. Cambridge University Press, London, UK.

*Frankel, O.M. and Michael E. Soulé. 1981. *Conservation and Evolution*. Cambridge University Press, New York. 327 pp.

Freeland, W.J. 1979. Primate social groups as biological islands. *Ecology* 60:719-28.

Freeman, J. 1986. The parks as genetic islands. *National Parks* 60(½):12-17.

Frost, Darrel R. (ed.). 1985. *Amphibian Species of the World: A Taxonomic and Geographical Reference*. Allen Press, Lawrence, Kansas. 732 pp.

Gadgil, M. and V.D. Vartak. 1976. The sacred groves of western Ghats in India. *Economic Botany* 30(2):152-160.

Game, M. 1980. Best shape for nature reserves. *Nature* 287:630-2.

Garcia, Jose Rafael. 1984. Waterfalls, hydropower, and water for industry: Contributions form Canaima National Park. Pp. 588-591 *in:* J.A. McNeely and K.R. Miller (eds.), *National Parks, Conservation, and Development: The Role of Protected Areas in Sustaining Society.* Smithsonian Institution Press, Washington, D.C. 825 pp.

Gardner, J.E. 1981. National parks and native peoples in northern Canada, Alaska, and northern Australia. *Environmental Conservation* 8(3):207-215.

Garratt, Keith. 1984. The relationship between adjacent lands and protected areas: Issues of concern for the protected area manager. Pp. 65-71 *in:* J.A. McNeely and K.R. Miller (eds.), *National Parks, Conservation, and Development: The Role of Protected Areas in Sustaining Society.* Smithsonian Institution Press, Washington, D.C.

Geertz, Clifford. 1983. *Local Knowledge.* Basic Books, New York. 244 pp.

Gehlback, F.R. 1975. Investigation, evaluation and priority ranking of natural areas. *Biological Conservation* 8:79-88.

Gentry, A.H. 1979. Extinction and conservation of plant species in tropical America: A phytogeographical perspective. Pp. 110-126 *in:* I. Hedberg (ed.), *Systematic Botany, Plant Utilization, and Biosphere Conservation.* Almquist and Wiksell International, Stockholm.

Gentry, A.H. 1982a. Neotropical floristic diversity: Phytogeographical connections between Central and South America, Pleistocene climatic fluctuations, or an accident of the Andean orogeny? *Annuals of the Missouri Botanical Garden* 69:557-593.

Gentry, A.H. 1982b. Patterns of neotropical plant species diversity. *Evol. Biol.* 15:1-84.

Gentry, A.H. 1984. *An Overview of Neotropical Phytogeographic Patterns with an Emphasis on Amazonia.* Missouri Botanical Garden, St. Louis.

Gentry, A. H. 1986. Endemism in tropical versus temperate plant communities. Pp. 153-181 *in:* Michael E. Soulé (ed.), *Conservation Biology: The Science of Scarcity and Diversity.* Sinauer, Sunderland, MA.

Gentry, A.H. and C.H. Dodson. 1987. Contribution of nontrees to species richness of a tropical rain forest. *Biotropica* 19:149-156.

Gilbert, L.E. 1980. Food web organization and conservation of neotropical diversity. Pp. 11-34. *in:* M.E. Soulé and B.A. Wilcox (eds.), *Conservation Biology.* Sinauer Associates, Sunderland, Mass.

Giles, Robert H. 1971. *Wildlife Management Techniques (3rd edition).* The Wildlife Society, Washington, D.C. 633 pp.

*Giles, Robert H. 1978. *Wildlife Management.* W.H. Freeman and Co., San Francisco. 416 pp.

Gillis, M. 1986. *Non-wood Forest Products in Indonesia.* Department of Forestry, University of North Carolina, Chapel Hill, North Carolina.

Gilpin, M.E. and J.M. Diamond. 1980. Subdivision of nature reserves and the maintenance of species diversity. *Nature* 285:567-8.

Gilpin, M.E. and J.M. Diamond. 1981. Immigration and extinction probabilities for individual species: Relation to incidence functions and species colonization curves. *Proc. Nat. Acad. Sci.* 78:392-396.

Ginzburg, L.R., L.B. Slobodkin, K. Hohnson, and A.G. Bindman. 1982. Quasiextinction probabilities as a measure of impact on population growth. *Risk Analysis* 2:171-81.

Glick, D. and J. Betancourt. 1983. The Rio Platano biosphere reserve: Unique resource, unique alternative. *Ambio* 12(3/4):168-73.

Godfrey-Smith, W. 1979. The value of wilderness. *Environmental Ethics* 1(4):309-319.

Goeden, G.B. 1979. Biogeographic theory as a management tool. *Environ. Conserv.* 6:27-32.

Goldsmith, F.B. 1973. The ecologist's role in the development for tourism: A case study in the Caribbean. *Biological J. Linnean Society* 5(3):265-287.

Goldsmith, F.B. 1975. The evaluation of ecological resources in the countryside for conservation purposes. *Biological Conservation* 8:89-96.

Gómez-Pompa, A. 1988. Tropical deforestation and Maya silviculture: An ecological paradox. *Tulane Studies in Zoology and Botany* 26:19-37.

Gómez-Pompa, A., C. Vasquez-Yanes, and S. Guevara. 1972. The tropical rain forest: A nonrenewable resource. *Science* 177:762-765.

Gómez-Pompa, A., J.S. Flores, and V. Sosa. 1987. The "Pet Kot": A man-made tropical forest of the Maya. *Interciencia* 12:10-15.

Goodland, R. 1988. A major new opportunity to finance biodiversity preservation. Pp. 437-445 *in:* E.O. Wilson and Francis M. Peter (eds.), *Biodiversity.* National Academy Press, Washington D.C.

Goodman, D. 1975. The theory of diversity-stability relationships in ecology. *Quarterly Review of Biology* 50(2):237-265.

Goodman, D. 1987a. Consideration of stochastic demography in the design and management of biological reserves. *Natural Res. Modeling* 1:205-34.

Goodman, D. 1987b. How do any species persist? Lessons for conservation biology. *Conservation Biology* 1(1):59-63.

Goodson, J. 1988. *Conservation and Management of Tropical Forests and Biodiversity in Zaire.* Unpublished document, USAID, Zaire.

Gordon, H.S. 1954. The economic theory of a common property resource: The fisher. *J. Pol. Econ.* 124-142.

Goulding, M. 1980. *The Fishes and the Forest: Explorations in Amazonian Natural History.* University of California Press, Berkeley. 280 pp.

Government of India. 1983. *Eliciting Public Support for Wildlife Conservation.* Report of Indian Board for Wildlife Task Force, Department of Environment, New Delhi.

Graham, N.E. and W.B. White. 1988. The El Niño cycle: A natural oscillator of the Pacific Ocean-atmosphere system. *Science* 240:1293-1302.

Grassle, J.F. 1985. Hydrothermal vent animals: Distribution and biology. *Science* 229:713-717.

Grassle, J.F. 1989. Species diversity in deep-sea communities. *Trends in Ecology and Evolution* 4(1):12-15.

Greenwood, P.H. 1974. The cichlid fishes of Lake Victoria, East Africa: The biology and evolution of a species-flock. *Bull. British Museum (Natural History) Supplement* 6:1-134.

Groves, R.H. and J.I. Burdon. 1986. *Ecology of Biological Invasions.* Cambridge University Press, New York. 166 pp.

Grubb, P.J. 1977. The maintenance of species richness in plant communities: The importance of the regeneration niche. *Biol Rev.* 52:107-145.

Gulick, P., C. Hershey, and J. Esquinas Alcazar. 1983. *Genetic Resources of Cassava and Wild Relatives.* International Board for Plant Genetic Resources 82/111, Rome.

Gupta, T.A. and A. Guleria. 1982. *Non-Wood Forest Products from India.* IBH Publishing Co., New Delhi.

Haffer, J. 1985. Avian zoogeography of the neotropical lowlands. Pp. 113-146 *in:* P.A. Buckley, M.S. Foster, C.S. Morton, R.S. Ridgely and F.G. Buckley (eds.), *Neotropical Ornithology.* Ornithological Monographs No. 36, American Ornithologists Union, Washington DC.

Hahn, C. 1982. *The Economic Rationale for Protection and Management of Natural Areas in Developing Countries.* Natural Resources Defence Council, Washington, D.C. 47 pp.

Hair, Jay D. 1988. The economics of conserving wetlands: A widening circle. Paper presented at Workshop on Economics, IUCN General Assembly, 4-5 February 1988, Costa Rica.

Hall, Charles A., C.J. Cleveland, and R. Kaufmann. 1986. *Energy and Resource Quality: The Ecology of the Economic Process.* John Wiley, New York. 577 pp.

*Hall, D.O., N. Myers, and N.S. Margaris (eds.). 1985. *Economics of Ecosystem Management.* W. Junk Publishers, Dordrecht, The Netherlands.

Hamilton, A., D. Taylor and J.C. Vogel. 1986. *Early forest clearance and environmental degradation in south-west Uganda.* Nature 320:164-167.

Hamilton, L.S. and Jeff M. Fox. 1987. Protected area systems and local people. Paper presented at Workshop on Fields and Forests, Xishuangbanna, Yunnan, China.

Hamilton, L.S., and S.C. Snedaker (eds.). 1984. *Handbook for Mangrove Area Management.* Honolulu, Hawaii, East-West Centre. 123 pp.

Hammack, J. and G.M. Brown, Jr. 1974. *Waterfowl and Wetlands: Toward Bioeconomic Analysis.* Johns Hopkins University Press, Baltimore, MD.

Hanemann, W. Michael. 1988. Economics and the preservation of biodiversity. Pp. 193-199 *in:* E.O. Wilson and Francis M. Peter (eds.), *Biodiversity.* National Academy Press, Washington, D.C.

Harcourt, A.H., H. Pennington, and A.W. Weber. 1986. Public attitudes to wildlife and conservation in the Third World. *Oryx* 20(3):152.

Harmon, David. 1987. Cultural diversity, human subsistence, and the national park ideal. *Environmental Ethics* 9(2):147-158.

Harrah, D.F. and B.K. Harrah. 1975. *Conservation Ecology: Resources for Conservation Education.* Scarecrow Press, Metuchen, New Jersey. 323 pp.

Harris, Larry D. 1984. *The Fragmented Forest: Island Biogeography Theory and the Preservation of Biotic Diversity.* University of Chicago Press, Chicago. 211 pp.

Harris, R.B., L.A. Maguire, and Mark L. Shaffer. 1987. Sample sizes for minimum viable population estimation. *Conservation Biology* 1:72-77.

Harris, Stuart. 1985. The economics of ecology and the ecology of economics. *Search* 16(9-12):284-290.

Harrison, J., K. Miller and J. McNeely. 1982. The world coverage of protected areas: Development goals and environmental needs. *Ambio* 11(95):238-245.

Hart, W.J. 1966. *A Systems Approach to Park Planning.* IUCN, Morges, Switzerland. 118 pp.

Hartshorn, G.S. 1978. Tree falls and tropical forest dynamics. Pp. 617-638 *in:* P.B. Tomilson and M.H. Zimmer (eds.), *Tropical Trees as Living Systems,* Cambridge University Press, Cambridge.

Hartwick, J.M. and N.D. Olewiler. 1986. *The Economics of Natural Resource Use.* Harper and Row, New York.

Hatley, T. and M. Thompson. 1985. Rare animals, poor people, and big agencies: A perspective on biological conservation and rural development in the Himalaya. *Mountain Research and Development* 5(4):365-377.

Haulot, A. 1985. The environment and the social value of tourism. *International J. Environmental Studies* 25(4):215-218.

Hawkes, J.G. 1983. *The Diversity of Crop Plants.* Harvard University Press, Cambridge. 184 pp.

Hayden, B.P., G.C. Ray, and R. Dolan. 1984. Classification of coastal and marine environments. *Environmental Conservation* 11(3):199-207.

Hazlewood, Peter T. 1989. The Tropical Forestry Action Plan: Opportunities and Challenges. Paper submitted to the 8th Meeting of the TFAP Advisers Group, 9-11 May 1989, Paris. World Resources Institute, Washington, DC.

Hazlewood, Peter T. In Press. *Cutting our Losses: Policy Reform to Sustain Tropical Forest Resources.* World Resources Institute, Washington, DC.

Heaney, L.R. 1986. Biogeography of mammals in Southeast Asia: Estimates of rates of colonization, extinction and speciation. *Biological Journal of the Linnean Society* 28:127-165.

Hedberg, I. (ed.). 1978. *Systematic Botany, Plant Utilization, and Biosphere Conservation.* Slmqvist and Wiksell International, Stockholm.

Helliwell, D.R. 1969. Valuation of wildlife resources. *Regional Studies* 3:41-47.

Helliwell, D.R. 1973. Priorities and values in nature conservation. *J. Environmental Management* 1(1):85-127.

Helliwell, D.R. 1975. Discount rates and environmental conservation. *Environmental Conservation* 2(2):199-201.

Helliwell, D.R. 1976. The extent and location of nature conservation areas. *Environmental Conservation* 3(4):255-258.

Helliwell, D.R. 1982. Assessment of conservation values of large and small organisms. *J. Environmental Management* 15(2):273-277.

Heywood, Vernon. 1989. Conservation biology of threatened species. manuscript.

Higgs, A.J. 1981. Island biogeography theory and nature reserve design. *J. Biogeog.* 8(2):117-124.

Higgs, A.J. and M.B. Usher. 1980. Should nature reserves be large or small? *Nature.* 285:568-9.

Hill, M. 1983. Kakadu National Park and the aboriginals: Partners in protection. *Ambio* 12(3/4):158-67.

Hodgson, G. and J.A. Dixon. 1988. Logging versus fisheries and tourism in Palawan: An environmental and economic analysis. *East-West Environment and Policy Institute Occasional Paper* 7:1-95.

Holdgate, M.W. 1982. The environmental information needs of the decision-maker. *Nature and Resources* 18(1):5-10.

Holdgate, Martin W. 1989. The implications of climatic change and rising sea level. In: W. Verwey, *Proceedings of the International Congress on Nature Management and Sustainable Development.* International Organizing Services, Amsterdam, the Netherlands.

Holling, C.S. 1973. Resilience and stability of ecological systems. *Annual Review of Ecology and Systematics* 4:1-23.

Holt, W.V. and H.D.M. Moore. 1988. Semen banking: Is it now feasible for captive endangered species? *Oryx* 22(3):172-178.

Honacki, James H., K.E. Kinman, and J.W. Koeppl (eds.). 1982. *Mammal Species of the World: A Taxonomic and Geographic Reference.* Allen Press, Lawrence, Kansas. 694 pp.

Hoose, P.M. 1981. *Building an Ark: Tools for the Preservation of Natural Diversity Through Land Protection.* Island Press, Covelo, CA. 212 pp.

Hough, J.L. 1984. An approach to an integrated land use system on Michiru Mountain, Malawi. *Parks* 9(3/4):1-3.

Hough, Walter. 1926. *Fire as an Agent in Human Culture.* U.S. National Museum Bulletin 139, Government Printing Office, Washington D.C. 270 pp.

Houseal, Brian, C. MacFarland, G. Archibold, and A. Chiari. 1985. Indigenous cultures and protected areas in Central America. *Cultural Survival Quarterly* March:10-19.

Houston, D. 1971. Ecosystems of national parks. *Science* 172:648-651.

Hoyt, Erich. 1988. *Conserving the Wild Relatives of Crops.* IBPGR, IUCN, WWF, Rome. 45 pp.

Hubbell, S.P. 1979. Tree dispersion, abundance, and diversity in a tropical dry forest. *Science* 203:1299-1309.

Hubbell, S.P. and R.B. Foster. 1983. Diversity of canopy trees in a neotropical forest and implications for conservation. *in:* S.L. Sutton, T.C. Whitmore and A.C. Chadwick (eds.), *Tropical Rain Forest: Ecology and Management,* Blackwell, Oxford.

Hubbell, S.P. and R.B. Foster. 1986. Commonness and rarity in a neotropical forest: Implications for tropical tree conservation. Pp. 205-231 *in:* M.E. Soulé (ed.), *Conservation Biology: The Science of Scarcity and Diversity.* Sinauer Associates, Sunderland, Massachusetts. 584 pp.

Hueting, R. 1985. Results of an economic scenario that gives top priority to saving the environment instead of encouraging production growth. *Environmentalist* 5(4):253-262.

Hufschmidt, Maynard M. *et al.* 1983. *Environment, Natural Systems, and Development: An Economic Valuation Guide.* Johns Hopkins University Press, Baltimore, MD. 338 pp.

Hufschmidt, Maynard M. and Ruangdej Srivardhana. 1986. The Nam Pong water resources project in Thailand. Pp. 141-162 *in:* John A. Dixon and M. Hufschmidt (eds.), *Economic Valuation Techniques for the Environment — A Case Study Workbook.* Johns Hopkins University Press, Baltimore, MD.

Humphreys, W.F. and D.J. Kitchener. 1982. The effect of habitat utilization on species-area curves: Implications for optimal reserve area. *J. Biogeog.* 9(5):391-396.

Huntley, B.J. 1978. Ecosystem conservation in southern Africa. *Biogeography and Ecology of Southern Africa.* 41:1333-1384.

Huston, M. 1979. A general hypothesis of species diversity. *American Naturalist.* 113:81-101.

Hutto, R.L, S. Reel and P.B. Landres. 1987. A critical evaluation of the species approach to biological conservation. *Endangered Species Update* 4(12):1-4.

Isakov, Y.A. and V.V. Krinitsky. 1986. The system of protected natural areas in the U.S.S.R. and prospects for its development. *Soviet Geography* 27(2):102-114.

Island Resources Foundation. 1981. *Economic Impact Analysis for the Virgin Islands National Park.* U.S. Dept. of the Interior, National Park Service, Washington, D.C.

IUCN. 1979. *The Biosphere Reserve and its Relationship to other Protected Areas.* IUCN, Gland, Switzerland. 26 pp.

*IUCN. 1980. *World Conservation Strategy: Living Resource Conservation for Sustainable Development.* IUCN-UNEP-WWF, Gland. 44 pp.

IUCN. 1985. *1985 United Nations List of National Parks and Protected Areas.* IUCN, Gland, Switzerland. 174 pp.

IUCN. 1986. *Directory of Wetlands of Intrnational Importance.* IUCN, Gland, Switzerland.

IUCN. 1987a. *IUCN Directory of Afrotropical Protected Areas.* IUCN, Gland, Switzerland. 1034 pp.

IUCN. 1987b. *Captive Breeding: IUCN Policy Statement.* IUCN, Gland. 3 pp.

IUCN. 1987c. *The IUCN Position Statement on Translocation of Living Organisms.* IUCN, Gland. 20 pp.

IUCN. 1987d. *Centres of Plant Diversity: A Guide and Strategy for their Conservation.* IUCN Threatened Plants Unit, Kew, Richmond, U.K.

IUCN. 1987e. *Population and Sustainable Development: Task Force Report.* IUCN, Gland. 63 pp.

IUCN. 1988a. *1988 IUCN Red List of Threatened Animals.* IUCN, Gland. 154 pp.

IUCN. 1988b. *Significant Trade in Wildlife: A Review of Selected Species in CITES Appendix II,* Volume 1: Mammals (183 pp); Volume 2: Reptiles and Invertebrates (306 pp); Volume 3: Birds (351 pp). IUCN, Gland, and CITES, Lausanne.

IUCN. 1989a. *Conservation et Utilization Rationelle des Ecosystemes Forestiers en Afrique Centrale.* Report to EEC. 240 pp.

IUCN. 1989b. The impact of climatic change and sea level rise on ecosystems. Paper submitted to Commonwealth Secretariat.

IUCN/ELC. 1984. *Species Mentioned in Legislation.* IUCN, Bonn, FRG. 20 pp.

IUCN/Unesco. 1987. *Directory of Biosphere Reserves.* Unesco, Paris.

IUCN/UNEP. 1986a. *Review of the Protected Areas System in Oceania.* IUCN, Gland, Switzerland. 239 pp.

IUCN/UNEP. 1986b. *Review of the Protected Areas System in the Afrotropical Realm.* IUCN, Gland, Switzerland. 259 pp.

IUCN/UNEP. 1986c. *Review of the Protected Areas System in the Indo-Malayan Realm.* IUCN, Gland, Switzerland. 284 pp.

IUCN/UNEP. 1988. *Coral Reefs of the World.* Volume 1: Atlantic and Eastern Pacific (373 pp); Volume 2: Indian Ocean, Red Sea and Gulf (389 pp); Volume 3: Central and Western Pacific (329 pp). IUCN, Gland, Switzerland and Cambridge, U.K./UNEP, Nairobi.

IUCN/WWF Plants Conservation Programme. 1988. *Centres of Plant Diversity: A Guide and Strategy for their Conservation.* IUCN, Gland. 40 pp.

Izac, A.-M.N. 1986. Resources policies, property rights and conflicts of interest. *Australian J. Agricultural Economics* 30(1):23-27.

Jacobs, P. and D. Munro (eds.). 1987. *Conservation With Equity: Strategies for Sustainable Development.* IUCN, Gland, Switzerland. 466 pp.

Jain, S.K. and K.L. Mehra (eds.). 1983. *Conservation of Tropical Plant Resources.* Botanical Survey of India, Howrah, India.

Jain, S.K. and A.R.K. Sastry. 1982. Threatened plants and habitats: A review of work in India. *Plants Conservation Bulletin* 2:1-9.

Janzen, D.H. 1983. No park is an island: Increase in interference from outside as park size increases. *Oikos* 41:402-410.

Janzen, Dan. 1988. The use of economic incentives in Costa Rica's Guanacaste National Park. Paper presented at Workshop on Economics, IUCN General Assembly, 4-5 February 1988, Costa Rica.

Jarvinen, O. 1982. Conservation of endangered plant populations: Single large or several small reserves? *Oikos* 38(3):301-307.

Jenkins, R.E., Jr. 1977. Classification and inventory for the perpetuation of ecological diversity. Pp. 41-52 *in:* A. Marmelstein (ed.), *Classification, Inventory, and Analysis of Fish and Wildlife Habitat.* U.S. Fish and Wildlife Service, Washington, D.C.

Jenkins, Robert E., Jr. 1985. The identification, acquisition, and preservation of land as a species conservation strategy. Pp. 131-140 *in:* R.J. Hoage (ed.), *Animal Extinctions: What*

Everyone Should Know. Smithsonian Institution Press, Washington D.C.

Jenkins, R.E., Jr. and W.B. Bedford. 1973. The use of natural areas to establish environmental baselines. *Biological Conservation* 5(3):168-174.

Johannes, R.E. 1978. Traditional marine conservation methods in Oceania and their demise. *Annual Review of Ecology and Systematics* 9:49-64.

Johannes, R.E. 1982. Traditional conservation methods and protected marine areas in Oceania. *Ambio* 11(3):258-261.

Johansson, P-O. 1987. *The Economic Theory and Measurement of Environmental Benefits.* Cambridge University Press, London. 238 pp.

Johns, A.D. 1985. Selective logging and wildlife conservation in tropical rainforest: Problems and recommendations. *Biological Conservation.* 31:355-75.

Johnson, M.P., L.G. Mason, and P.H. Raven. 1968. Ecological parameters and plant species diversity. *American Naturalist.* 102:297-306.

Jones, D.M. 1982. Conservation in relation to animal disease in Africa and Asia. Pp. 271-285 *in:* M.A. Edwards and U. McDonnell (eds.), *Animal Disease in Relation to Animal Conservation.* Symposia of the Zoological Society of London.

Jones, J. Greg, W.G. Beardsley, D.W. Countryman, and Dennis L. Schweitzer. 1978. Estimating economic costs of allocating land to wilderness. *Forest Sci.* 24(3):410-422.

Jones-Lee, M. W. 1976. *The Value of Life: An Economic Analysis.* University of Chicago Press, Chicago.

Jordan, C.F. 1985. *Nutrient Cycling in Tropical Forest Ecosystems.* Wiley, Chichester. 190 pp.

Jordan, W.R. and M.E. Gilpin (eds.). 1987. *Restoration Ecology.* Cambridge University Press, Cambridge, UK.

Juday, G.P. 1983. The problem of large mammals in natural areas selection: Examples from the Alaska ecological reserves system. *Natural Areas Journal* 3(3):24-30.

Juvik, J.O. and S.P. Juvik. 1984. Mauna Kea and the myth of multiple use: Endangered species and mountain management in Hawaii. *Mountain Research and Development* 4(3):191-202.

Kangas, P. 1987. The use of species-area curves to predict extinctions. *Bulletin of the Ecological Society of America* 68:158-162.

Karr, J.R. 1976. Within- and between-habitat avian diversity in Africa and neotropical lowlands habitats. *Ecological Monographs* 46:457-481.

Kasran, Baharuddin. 1988. Effect of logging on sediment yield in a hill dipterocarp forest in peninsular Malaysia. *J. Tropical Forest Science* 1(1):56-66.

Kassam, A.H. and G.M. Higgins. 1980. *Land Resources for Populations of the Future.* FAO/UNFA, Rome 369 pp.

Keiter, R.B. 1985. On protecting the national parks from the external threat dilemma. *Land and Water Law Review* 20(2):355-420.

Kellert, S.R. 1984. Assessing wildlife and environmental values in cost-benefit analysis. *J. Environmental Management* 18(4):355-363.

Kelman, Steven. 1981. *What Price Incentives: Economics and the Environment.* Auburn House, Boston, MA.

Kempton, R.A. 1979. The structure of species abundance and measurement of diversity. *Biometrics* 35(2):307-321.

*Kenchington, R.A. and B.E.T. Hudson (eds.). 1984. *Coral Reef Management Handbook.* Unesco Regional Office for Science and Technology in South-East Asia. Jakarta, Indonesia.

Kennedy, Duncan. 1980. Cost-benefit analysis of entitlement problems: A critique. *Standard Law Review* 33(2):419-431.

Kennedy, J.L. 1985. Conceiving forest management as providing for current and future social value. *Forest Ecology and Management* 13(1/2):121-132.

Kepler, C.B. and J.M. Scott. 1985. Conservation of island ecosystems. Pp. 255-271 *in:* P.O. Moors (ed.), *Conservation of Island Birds.* International Council of Bird Preservation, Cambridge, UK.

Ketchum, B.H. (ed.). 1972. *The Water's Edge: Critical Problems of the Coastal Zone.* MIT Press, Cambridge, MA. 303 pp.

King, A.W. and S.W. Pimm. 1983. Complexity, diversity and stability: A reconciliation of theoretical and empirical results. *American Naturalist.* 122:229-239.

Kirkpatrick, J.B. 1983. An iterative method for establishing priorities for the selection of nature reserves: An example from Tasmania. *Biological Conservation.* 25:127-34.

Klee, G.A. (ed.). 1980. *World Systems of Traditional Resource Management.* John Wiley and Sons, New York. 290 pp.

Kleiman, Devra G. 1989. Reintroduction of captive mammals for conservation. *BioScience* 39(3): 152-161.

Klopfer, P.H. 1959. Environmental determinants of faunal diversity. *American Naturalist* 93:337-342.

Kloppenburg, J., Jr. and D.L. Kleinman. 1987. The plant germplasm controversy: Analyzing the distribution of the world's plant genetic resources. *BioScience* 37(3):190-198.

Konstant, W.R. and R.A. Mittermeier. 1982. Introduction, reintroduction and translocation of Neotropical primates: Past experiences and future possibilities. *International Zoo Yearbook* 22:69-77.

Krebs, C.J. 1984. *Ecology: The Experimental Analysis of Distribution and Abundance.* 3rd Ed. Harper & Row, New York.

Kristensen, R.M. 1983. Loricifera, a new phylum with *Aschelminthes* characters from the meiobenthos. *Z. Zool. Syst.* 21(3):163-180.

Krutilla, J.V. and A.C. Fisher. 1975. *The Economics of Natural Environments: Studies in the Valuation of Commodity and Amenities Resources.* Resources for the Future/Johns Hopkins University Press, Baltimore, MD. 292 pp.

Kux, Molly. 1986. Land use options to conserve living resources and biological diversity in developing countries. (MS), University of Florida.

Kwapena, N. 1984. Wildlife management by the people. Pp. 315-321 *in:* J.A. McNeely and K.R. Miller (eds.), *National Parks, Conservation, and Development: The Role of Protected Areas in Sustaining Society.* Smithsonian Institution Press, Washington, D.C.

Lacy, R.C. 1987. Loss of genetic diversity from managed populations: Interacting effects of drift, mutation, immigration, selection, and population subdivision. *Conservation Biology* 1:143-159.

Lamprey, H. 1974. The distribution of protected areas in relation to the needs of biotic community conservation in eastern Africa. Proceedings of a Regional Meeting on the Creation of a Coordinated System of National Parks and Reserves in Eastern Africa. Supplementary Paper No 45. IUCN, Gland, Switzerland.

Lande, R. 1988. Genetics and demography in biological conservation. *Science* 241:1455-1460.

Landres, P.B., J. Verner, and J.W. Thomas. 1988. Ecological uses of vertebrate indicator species: A critique. *Conservation Biology* 2:316-328.

Langford, William A., and Donald J. Cocheba. 1978. The wildlife valuation problem: A critical review of economic approaches. *Can. Wildl. Serv. Occ. Paper* 37:1-35.

Lanly, J.P. 1982. *Tropical Forest Resources.* Food and Agriculture Organization, Rome. 106 pp.

Lausche, B.J. 1980. Guidelines for protected area legislation. *IUCN Environmental Policy and Law Paper* 16:1-108.

Laut, P. and T.A. Paine. 1982. A step towards an objective procedure for land classification and mapping. *Applied Geography* 2(2):109-126.

Leader-Williams, N. and S.D. Albon. 1988. Allocation of resources for conservation. *Nature* 336:533-535.

Lecomber, J.R.C. 1979. *The Economics of Natural Resources.* Macmillan, London.

Ledec, G. 1987. Effects of Kenya's Bura irrigation settlement project on biological diversity and other conservation concerns. *Conservation Biology* 1:247-259.

Ledec, G. and R. Goodland. 1986. Epilogue. *in:* D.A. Schumann and W. L. Partridge (eds.), *The Human Ecology of Tropical Land Settlement in Latin America.* Westview Press, Boulder, CO.

Ledec, G. and R. Goodland. 1988. *Wildlands: Their Protection and Management in Economic Development.* The World Bank, Washington, D.C. 278 pp.

Ledig, F.T. 1986. Conservation strategies for forest gene resources. *Forest Ecology and Management* 14(2):77-90.

Ledig, F.T. 1988. The conservation of diversity in forest trees. *BioScience* 38(7):471-479.

Lehmann, S. 1981. Do wildernesses have rights? *Environmental Ethics* 3(1):129-146.

Lehmkuhl, J.F. 1984. Determining size and dispersion of minimum viable populations for land management planning and species conservation. *Environmental Management* 8(2):167-176.

Leigh, E.G. 1981. The average lifetime of a population in a varying environment. *J. Theor. Biol.* 90:213-39.

Leigh, E.G., Jr., A.S. Rand, and D.M. Windsor (eds.). 1982. *The Ecology of a Tropical Forest: Seasonal Rhythms and Long-Term Changes.* Smithsonian Institution Press, Washington. D.C.

Leigh, J.H., J.D. Briggs, and W. Hartley. 1982. The conservation status of Australian plants. Pp. 13-25 *in:* R.H. Groves and W.D.L. Ride (eds.), *Species at Risk: Research in Australia.* Springer-Verlag, New York.

Lemons, J. 1986. Research in the National Parks. *Environmental Professional* 8(2):127-137.

Lesley, J.T. 1978. *Genetics of Livestock Improvement.* Prentice-Hall, New Jersey.

Lesslie, R.G., B.G. Mackey, and K.M. Preece. 1988. A computer-based method of wilderness evaluation. *Environmental Conservation* 15(3):225-232.

Lewin, R. 1984. Parks: How big is big enough? *Science* 225:611-612.

Lewin, R. 1986. In ecology, change brings stability. *Science* 234:1071-1073.

Lewin, R. 1986. Damage to tropical forests, or why were there so many kind of animals? *Science* 234:149-150.

Lewis, Dale M., G.B. Kaweche, and Ackim Mwenya. 1987. *Wildlife conservation outside protected areas: Lessons from an experiment in Zambia.* Lupande Research Project Publication 4:1-14.

Lewis, Harrison F. 1951. Wildlife in today's economy: Aesthetic and recreational values of wildlife. *Trans. N. Am. Wildl. Conf.* 16:13-16.

Lewontin, R.C. 1974. *The Genetic Basis of Evolutionary Change.* Columbia University Press, New York.

Lisboa, P.L.B., U.N. Maciel, and G.T. Prance. 1987. Some effects of colonization on the tropical flora of Amazonia: A case study from Rondonia. *Ciencia Hoje* 6:48-56.

Livingstone, Ian. 1986. The common property problem and pastoralist economic behavior. *J. Development Studies* 23:5-19.

Loomis, J.B. 1986. Assessing wildlife and environmental values in cost-benefit analysis: State of the art. *J. Environmental Management* 22:125-131.

Lothian, Andrew. 1985. A cost-benefit study of national parks on Kangaroo Island, South Australia. *In: Proceedings of the Conference on Conservation and the Economy 1984.* Australian Government Publishing Service, Canberra.

Lovejoy, T.E. 1976. We must decide which species will go forever. *Smithsonian* 7(4):52-59.

Lovejoy, T.E. 1980. Discontinuous wilderness: Minimum areas for conservation. *Parks* 5(2) 13-15.

Lovejoy, T.E. 1984. Application of ecological theory to conservation planning. Pp. 402-413 *in:* F. di Castri, F.W.G. Baker and M. Hadley (eds.), *Ecology in Practice: Part 1 Ecosystem Management.* Tycooly, Dublin.

Lovejoy, T.E. 1985. Strategies for preserving species in the wild. Pp. 97-113 *in:* R.J. Hoage (ed.), *Animal Extinctions: What Everyone Should Know.* Smithsonian Institution Press, Washington, D.C.

Lovejoy, T.E. 1986. Species leave the ark one by one. Pp. 16-17 *in:* Bryan G. Norton (ed.), *The Preservation of Species: The Value of Biological Diversity.* Princeton University Press, Princeton, NJ.

Lovejoy, T.E., J.M. Rankin, R.O. Bierregaard, K.S. Brown, L.H. Emmons, and M.E. Van der Voort. 1984. Ecosystem decay in Amazon forest fragments. Pp. 295-325 *in:* M.H. Nitecki (ed.), *Extinctions.* University of Chicago Press, Chicago.

Lugo, Ariel. 1988. Diversity of tropical species: Questions that elude answers. *Biology International* 19:1-37.

Lusigi, Walter J. 1978. *Planning Human Activities on Protected Natural Ecosystems.* Dissertationes Botanicae 48. J. Cramer, Vaduz, Germany. 233 pp.

Lusigi, Walter. 1984. Mt. Kulal biosphere reserve: Reconciling conservation with local human population needs. Pp. 459-469 *in:* J.A. McNeely and D. Navid (eds.), *Conservation, Science, and Society.* Unesco-UNEP, Paris.

*Lyster, S. 1985. International Wildlife Law. *IUCN Environmental Policy and Law Paper* 22:1-470.

MacArthur, R.H. 1972. *Geographical Ecology: Patterns in the Distribution of Species.* Harper and Row, New York. 269 pp.

MacArthur, R.H. and E.O. Wilson. 1967. *The Theory of Island Biogeography.* Princeton University Press. Princeton, NJ. 203 pp.

Machlis, Gary E. and David L. Tichnell. 1985. *The State of the World's Parks: An International Assessment of Resource Management, Policy and Research.* Westview, Boulder, Colorado.

Machlis, Gary E. and David L. Tichnell. 1987. Economic development and threats to national parks: A preliminary analysis. *Environmental Conservation* 14(2):154.

MacKinnon, J.R. 1983. Irrigation and watershed protection in Indonesia. Report to the World Bank.

*MacKinnon, J.R., K. MacKinnon, G. Child, and J. Thorsell. 1986. *Managing Protected Areas in the Tropics.* IUCN, Gland. 295 pp.

MacKinnon, J.R. and S. Stuart. 1989. *The Kouprey: An Action Plan for its Conservation.* IUCN, Gland. 19 pp.

Maguire, L.A. 1986. Using decision analysis to manage endangered species. *J. Environmental Management* 22(4):345-360.

Mahar, Dennis J. 1988. *Government Policies and Deforestation in Brazil's Amazon Region.* World Bank, Washington D.C. 56 pp.

Maheshwari, J.K. 1980. Plant resources of the Himalaya and their conservation. *J. Himalayan Studies and Regional Development* 4:3-9.

Malingreau, J.-P. and C.J. Tucker. 1988. Large-scale deforestation in the southern Amazon Basin of Brazil. *Ambio* 17:49-55.

Malla, K.B. 1986. *Report on Land Use Changes with Emphasis on Forest Cover in Nepal.* National Remote Sensing Center, Kathmandu, Nepal.

Malone, S., R.A. Mittermeier, K. Mohadin, M. Werkhoven, M.J. Plotkin, J. MacKnight, and T.B. Werner. 1990. *A Preliminary Action Plan for Conservation in Suriname.* WWF, Washington, D.C.

Maltby, Edward. 1986. *Waterlogged Wealth.* Earthscan, London. 200 pp.

Margules, C., A.J. Higgs and R.W. Rafe. 1982. Modern biogeographic theory: Are there any lessons for nature reserve design? *Biological Conservation.* 24:115-128.

Margules, C., A.O. Nicholls and R.L. Pressey. 1988. Selecting networks of reserves to maximize biological diversity. *Biological Conservation.* 43:63-76.

Margules, C. and M.B. Usher. 1981. Criteria used in assessing wildlife conservation potential: A review. *Biological Conservation.* 21:79-109.

Margulis, L. and K. V. Schwartz. 1982. *Five Kingdoms: An Illustrated Guide to the Phyla of Life On Earth.* W.H. Freeman and Co., New York. 376 pp.

Marks, Stuart A. 1984. *The Imperial Lion: Human Dimensions of Wildlife Management in Central Africa.* Westview Press, Boulder, CO. 210 pp.

Marietta, D.E., Jr. 1979. The interrelationship of ecological science and environmental ethics. *Environmental Ethics* 1(2):195-207.

Martens, J. 1982. Forests and their destruction in the Himalayas of Nepal. *Plant Research and Development* 15:66-96.

Martin, P.S. 1984. Prehistoric overkill: The global model. Pp. 354-403 *in:* P.S. Martin and R.G Klein (eds.), *Quaternary Extinctions: A Prehistoric Revolution.* University of Arizona Press. 892 pp.

*Martin, P.S. and R.G Klein (eds.). 1984. *Quaternary Extinctions: A Prehistoric Revolution.* University of Arizona Press. 892 pp.

Martin, R.B. 1986. *Communal Areas Management Programme for Indigenous Resources (CAMPFIRE).* Branch of Terrestrial Ecology, Working Document No. 1/86, Department of National Parks and Wildlife Management. 110 pp.

Martin, R.B. and V. Clarke. 1988. *Predicted Returns for Wildlife Management in the Omay Communal Land.* Annex to the Land Use Study in Omay Communal Land, Zimbabwe, Agricultural and Rural Development Authority, Harare.

Mascarenhas, A. 1983. Ngorongoro: A challenge to conservation and development. *Ambio* 12(3/4):146-152.

May, R.M. 1974. *Stability and Complexity in Model Ecosystems.* 2nd ed. Princeton University Press, Princeton, NJ.

May, R.M. 1975. Island biogeography and the design of wildlife preserves. *Nature* 254:177-178.

May, R.M. 1988. How many species are there on earth? *Science* 241:1441-1449.

Mayr, E. and R.J. O'Hara. 1986. The biogeographic evidence supporting the Pleistocene forest refuge hypothesis. *Evolution* 40(1):55-66.

McConnell, K.E. and J.G. Sutinen. 1979. Bioeconomic models of marine recreational fishing. *J. Env. Econ. and Mgt.* 6:127-139.

McCoy, E.D. 1983. The application of island-biogeographic theory to patches of habitat: How much land is enough? *Biological Conservation.* 25(1):53-61.

McEachern, J and E. L. Towle. 1974. *Ecological Guidelines for Island Development.* IUCN, Morges, Switzerland.

McElroy, M.B. and R.J. Salawitch. 1989. Changing composition of the global stratosphere. *Science* 243:763-770.

McGonigle, R.M. 1981. The economizing of ecology: Why big, rare whales still die. *Ecology Law Quarterly* 9(1):119-237.

McKinney, M. L. 1987. Taxonomic selectivity and continuous variation in mass and background extinction of marine taxa. *Nature* 325:143-145.

McLellan, C.H., A.P. Dobson, D.S. Wilcove, and J.M. Lynch. 1986. Effects of forest fragmentation of New and Old World bird communities: Empirical observations and theoretical implications. *In:* J. Verner, M. Morrison, and C.J. Ralph (eds.), *Modeling Habitat Relationships of Terrestrial Vertebrates.* University of Wisconsin Press, Madison.

McMahan, L. 1980. Legal protection for rare plants. *American University Law Review* 29(3):515-569.

McMichael, D.F. (ed.). 1971. *Society's Demand for Open Air Recreation, Wilderness, and Scientific Reference Areas.* Institute of Australian Foresters, Canberra.

McNaughton, S.J. 1977. Diversity and stability of ecological communities. *American Naturalist* 111(3):515-525.

McNeely, J.A. (ed.). 1981a. *Conserving Africa's Natural Heritage.* IUCN, Gland. 271 pp.

McNeely, J.A. (ed.). 1981b. *Conserving the Natural Heritage of Latin America and the Caribbean.* IUCN, Gland, Switzerland. 329 pp.

McNeely, J.A. 1982. *The World's Greatest Natural Areas.* IUCN, Gland. 49 pp.

McNeely, J.A. 1987. How dams and wildlife can co-exist: natural habitats, agriculture, and major water resource development projects in tropical Asia. *Conservation Biology* 1: 228-238.

*McNeely, J. A. 1988. *Economics and Biological Diversity: Developing and Using Economic Incentives to Conserve Biological Diversity.* IUCN, Gland, Switzerland. 200 pp.

McNeely, J. A. 1989a. Agriculture and biological diversity: International policy issues. Paper resulting from workshop on Agriculture and Conservation Biology, Asilomar, CA. 20-23 Nov. 1988.

McNeely, J.A. 1989b. Minimum quality criteria for ecologically sensitive areas. Report prepared for the Asian Development Bank. IUCN, Gland, Switzerland. 43 pp.

McNeely, J.A. and J. Harrison. in prep. Protected areas and climate change: Assessing the adaptability of the existing network.

*McNeely, J. A. and K.R. Miller (eds.). 1984. *National Parks, Conservation, and Development: The Role of Protected Areas in Sustaining Society.* Smithsonian Institution Press, Washington, D.C. 838 pp.

McNeely, J. A., K. R. Miller, and James W. Thorsell. 1987. Objectives, selection, and management of protected areas in tropical forest habitats. Pp. 181-204 *in:* C. Marsh and R.A. Mittermeier (eds.), *Primate Conservation in the Tropical Rain Forest.* Alan R. Liss, Inc., New York.

McNeely, J.A. and D. Navid (eds.). 1984. *Conservation,*

Science and Society: The Proceedings of the First International Congress on Biosphere Reserves. Minsk, Byelorussia, U.S.S.R. 600 pp.

McNeely, J.A. and David Pitt (eds.). 1984. *Culture and Conservation: The Human Dimension in Environmental Planning.* Croom Helm, London. 308 pp.

McNeely, J.A., E. Sumardja, and D. Rabor (eds.). 1978. *Wildlife Management in Southeast Asia.* Biotrop, Bogor, Indonesia. 236 pp.

McNeely, J.A. and J.W. Thorsell (eds.). 1985. *People and Protected Areas in the Hindukush-Himalaya.* ICIMOD, Kathmandu. 250 pp.

McNeely, J.A. and J.W. Thorsell. 1987. *Guidelines for Development of Terrestrial and Marine National Parks for Tourism and Travel.* World Tourism Organization, Madrid. 29 pp.

Menges, E.S. 1986. Predicting the future of rare plant populations: Demographic monitoring and modeling. *Natural Areas Journal* 6(3):13-25.

Mercer, D.E. and L.S. Hamilton. 1984. Mangrove ecosystems: some economic and natural benefits. *Nature and Resources* 20(2):14-19.

Merriam, Larry C. 1964. The Bob Marshall wilderness areas of Montana: Some socioeconomic considerations. *J. Forestry* 62(11):789-795.

Messerschmidt, Don. 1985. People's participation in park resource planning and management. Pp. 133-140 *in:* J.A. McNeely and J.W. Thorsell (eds.), *People and Protected Areas in the Hindukush-Himalaya.* ICIMOD, Kathmandu. 250 pp.

Middleton, J. and G. Merriam. 1985. The rationale for conservation: Problems from a virgin forest. *Biological Conservation* 33(2):133-145.

Miller, Daniel J. 1989. Introductions and extinction of fish in the African great lakes. *Trends in Ecology and Evolution* 4(2):56-59.

Miller, David L. 1986. Technology, territoriality and ecology: The evolution of Mexico's Caribbean spiny lobster fishery. Paper presented at Workshop on Ecological Management of Common Property Resources, IV International Congress of Ecology, Syracuse, New York.

Miller, J.R. 1981. Irreversible land use and the preservation of endangered species. *J. Environmental Economics and Management* 8:19-26.

Miller, J.R. and F.C. Menz. 1979. Some economic considerations for wildlife preservation. *Southern Economic J.* 45(3):718-729.

Miller, K.R. 1973. Development and training of personnel: The foundation of national park programs in the future. Pp. 326-347 *in:* H. Elliott (ed.), *Second World Conference on National Parks.* IUCN Gland, Switzerland.

Miller, K.R. 1975. Guidelines for the management and development of national parks and reserves in the American Humid Tropics. Pp. 94-105 *in: Proceedings IUCN Meeting on the Use of Ecological Guidelines for Development in the American Humid Tropics,* IUCN, Morges.

*Miller, K.R. 1980. *Planning National Parks for Ecodevelopment.* University of Michigan, Ann Arbor.

Miller, R.I. 1979. Conserving the genetic integrity of faunal populations and communities. *Environmental Conservation* 6:297-304.

Miller, R.I. and L.D. Harris. 1977. Isolation and extirpations in wildlife reserves. *Biological Conservation* 12(4):311-315.

Miller, R.S. and D.R. Botkin. 1974. Endangered species: Models and predictions. *American Scientist* 62(1):172-181.

Mishra, H.R. 1984. A delicate balance: tigers, rhinoceros, tourists and park management vs. the needs of the local people in Royal Chitwan National Park, Nepal. Pp. 197-205 *in:* J.A. McNeely and K.R. Miller (eds.), *National Parks, Conservation and Development: The Role of Protected Areas in Sustaining Society.* Smithsonian Institution Press, Washington, D.C.

Mittermeier, R.A. 1986. *An Action Plan for Conservation of Biological Diversity in Madagascar.* World Wildlife Fund-US, Washington, D.C.

Mittermeier, R.A. 1988. Primate diversity and the tropical forest: Case studies from Brazil and Madagascar and the importance of the megadiversity countries. Pp. 145-154 *in:* E.O. Wilson and Francis M. Peter (eds.), *Biodiversity.* National Academy Press, Washington, D.C. 521 pp.

Mittermeier, R.A. and T.B. Werner. 1990. Wealth of Plants and Animals Unites "Megadiversity" countries. *Tropicus:* 4(1):1,4-5.

Moen, A.N. 1973. *Wildlife Ecology: An Analytical Approach.* W.H. Freeman and Co., San Francisco. 458 pp.

Mooney, H. (ed.) 1985. *Ecological Consequences of Biological Invasions.* Springer Verlag, New York.

Mooney, H. 1988. Lessons from Mediterranean-climate regions. Pp. 157-165 *in:* E.O. Wilson and Francis M. Peter (eds.), *Biodiversity.* National Academy Press, Washington, D.C.

Mooney, H. and J.A. Drake. 1987. The ecology of biological invasions. *Environmentalist* 19(5):10-37.

Morauta, L., J. Pernetta, and W. Heaney (eds.). 1982. *Traditional Conservation in Papua New Guinea: Implications for Today.* Institute of Applied Social and Economic Research, Boroko, Papua New Guinea.

Morony, J.L., W. Bock, and J. Farrand. 1975. *Reference List of Birds of the World.* American Museum of Natural History, New York.

Movcan, J. 1982. Development and economics in Plitvice National Park. *Ambio* 11(3):282-285.

Myers, N. 1972. National parks in savannah Africa. *Science* 178:1255-1263.

Myers, N. 1976. An expanded approach to the problem of disappearing species. *Science* 193:198-202.

Myers, N. 1979. *The Sinking Ark: A New Look at the Problem of Disappearing Species.* Pergamon Press, Oxford, UK.

Myers, N. 1980. *Conversion of Tropical Moist Forests.* National Research Council, Washington D.C.

Myers, N. 1982. Forest refuges and conservation in Africa with some appraisal of survival prospects for tropical moist forests throughout the biome. Pp. 658-672 *in:* G.T. Prance, (ed.), *Biological Diversification in the Tropics.* Columbia University Press, New York.

Myers, N. 1983a. *A Wealth of Wild Species.* Westview Press, Boulder, CO. 272 pp.

Myers, N. 1983b. Tropical moist forests: Over-exploited and under-utilized? *Forest Ecology and Management* 6(1):59-79.

Myers, N. 1983c. A priority-ranking strategy for threatened species? *Environmentalist* 3(1):97-120.

Myers, N. 1984. *The Primary Source: Tropical Forests and Our Future.* W.W. Norton & Co., New York. 399 pp.

*Myers, N. (ed.). 1985. *The Gaia Atlas of Planet Management.* Pan Books, London. 272 pp.

Myers, N. 1987a. Tackling mass extinction of species: a great creative challenge. The Horace M. Albright Lecture in Conservation. University of California, Berkeley, California.

Myers, N. 1987b. The extinction spasm impending: Synergisms at work. *Conservation Biology* 1:(1)14.

Myers, N. 1988a. Tropical deforestation and climatic change. Paper for Conference on Climate and the Geo Sciences, May 22-27, Louvain University Belgium.

Myers, N. 1988b. Tropical forests: much more than stocks of wood. *J. Tropical Ecology* 4:209-221.

Myers, N. 1988c. Threatened biotas: "Hotspots" in tropical forests. *Environmentalist* 8(3):1-20.

Myers, N. 1988d. Natural resource systems and human exploitation systems: Physiobiotic and ecological linkages. World Bank Environment Department Working Paper 12:1-61.

Namkoong, G. 1982. The management of genetic resources: A neglected problem in environmental ethics. *Environmental Ethics* 4:377-378.

Namkoong, G. 1986. Genetics and the forest of the future. *Unasylva* 38(152):2-18.

NAS (National Academy of Sciences). 1972. *Genetic Vulnerability of Major Crops.* National Academy Press, Washington, D.C. 307 pp.

NAS. 1980. *Research Priorities in Tropical Biology.* Committee on Research Priorities in Tropical Biology, National Academy of Sciences, Washington, D.C. 116 pp.

NAS. 1986. *Ecological Knowledge and Environmental Problem-Solving.* National Academy Press, Washington, D.C. 388 pp.

Nature Conservancy Council. 1979. *Nature Conservation in the Marine Environment.* Report of the NCC/NERC Joint working Party on marine wildlife conservation. Natural Conservancy Council, London. 64 pp.

Navid, D. 1984. International cooperation for wetland conservation: The Ramsar Convention. *Transactions North American Wildlife and Natural Resources Conference* 49:33-41.

Newman, James R. and R. Kent Schereiber. 1984. Animals as indicators of ecosystem responses to air emissions. *Environmental Management* 8(4):309-324.

Nichol, J.E. 1982. Parameters for conservation evaluation. *J. Environmental Management* 14:181-194

Nicholls, Yvonne I. (ed.). 1973. Source book: Emergence of proposals for recompensing developing countries for maintaining environmental quality. *IUCN Environmental Policy and Law Paper 5.* IUCN, Gland, Switzerland.

Nicoll, M.E. and O. Langrand. 1989. *Madagascar: Revue de la conservation et des aires protégées.* WWF, 1989. 374 pp.

Niklas, K.J., B.H. Tiffney, and A.H. Knoll. 1985. Patterns in vascular land plant diversification: An analysis at the species level. *In:* J.W. Valentine (ed.), *Phanerozoic Diversity Patterns: Profiles in Macroevolution.* Princeton University Press, Princeton, N.J.

Nilson, G. 1983. *The Endangered Species Handbook.* Animal Welfare Institute, Washington, D.C. 245 pp.

Nitecki, M.H. 1984. *Extinctions.* University of Chicago Press, Chicago.

Norgaard, R.B. 1984. Environmental economics: an evolutionary critique and a plea for pluralism. Division of Agricultural Sciences, University of California, Berkeley, CA. Working Paper 299:1-24.

Norgaard, R.B. 1987. The economics of biological diversity: Apologetics or theory? *In:* D.D. Southgate and J.F. Disinger (eds.), 1987. *Sustainable Resource Development in the Third World.* Westview Press, Boulder, CO.

Norgaard, R.B. 1988. The rise of the global exchange economy and the loss of biological diversity. Pp. 206-211 *in:* E.O. Wilson and Francis M. Peter (eds.), *Biodiversity.* National Academy Press, Washington, D.C.

Norton, Bryan. 1983. On the inherent danger of undervaluing species. Manuscript, Center for Philosophy and Public Policy, College Park, MD.

Norton, Bryan. 1986. *The Preservation of Species: The Value of Biological Diversity.* Princeton University Press, Princeton, NJ.

Norton, Bryan. 1987. Biodiversity and the public lands: The spiral of life and how it all works, with examples. *Wilderness* 50(176):17-38.

Norton, Bryan. 1988. Commodity, amenity, and morality: The limits of quantification in valuing biodiversity. Pp. 200-205 *in:* E.O. Wilson and Francis M. Peter (eds.), *Biodiversity.* National Academy Press, Washington, D.C.

Noss, R.F. 1983. A regional landscape approach to maintain diversity. *BioScience* 33:700-706.

Noss, R. and L. Harris. 1986. Nodes, networks, and MUM's: Preserving diversity at all scales. *Environmental Management.* 10:299-309.

NRC (National Research Council). 1975. *Underexploited Tropical Plants with Promising Economic Value.* Board on Science and Technology for International Development Report 16. National Academy of Sciences, Washington, D.C. 187 pp.

NRC (National Research Council). 1982. *Ecological Aspects of Development in the Humid Tropics.* National Academy Press, Washington, D.C. 297 pp.

Oates, J.F. 1985. *Action Plan for African Primate Conservation: 1986-1990.* IUCN, Gland. 41 pp.

O'Brien, Stephen J. and J.F. Evermann. 1988. Interactive influence of infectious disease and genetic diversity in natural populations. *Trends in Ecology and Evolution* 3(10):254-259.

Odum, W.E. 1976. *Ecological Guidelines for Tropical Coastal Development.* IUCN, Gland, Switzerland. 61 pp.

OECD (Organization for Economic Cooperation and Development). 1982. *Economic and Ecological Interdependence.* OECD, Paris.

Oka, H.I. and W.T. Chang. 1961. Hybrid swarms between wild and cultivated rice species *Oryza perennis* and *O. sativa. Evolution* 15:418-430.

*Oldfield, Margery. 1984. *The Value of Conserving Genetic Resources.* U.S. Department of Interior, National Park Service. Washington, D.C. 360 pp.

Oldfield, M.L. and J.B. Alcorn. 1987. Conservation of traditional agroecosystems: Can age-old farming practices effectively conserve crop genetic resources? *BioScience* 37(3):199-208.

Oldfield, Sara. 1988. Buffer zone management in tropical moist forests. *IUCN Tropical Forest Paper* 5:1-49.

Olembo, R. 1984. UNEP and protected areas. Pp. 861-684 *in:* J.A. McNeely and K.R. Miller (eds.), *National Parks, Conservation, and Development: The Role of Protected Areas in Sustaining Society.* Smithsonian Institution Press, Washington, D.C. 825 pp.

Olson, S.L. and H.F. James. 1982. Fossil birds from the Hawaiian islands: Evidence for wholesale extinction by man before western contact. *Science* 217:633-635.

Olwig, K.F. and K. Olwig. 1979. Underdevelopment and the development of "natural" park ideology. *Antipode* 11(2):16-25.

Opdam, P., G. Rijsdijk, and F. Hustings. 1985. Communities in small woods in an agricultural landscape: Effects of area and isolation. *Biological Conservation* 34(4):333-352.

Oren, D.C. 1987. Grande Carajás, international financing agencies, and biological diversity in Southeastern Brazilian Amazonia. *Conservation Biology* 1:222-228.

Organization of American States. 1978. *Final Report on Conservation of Major Terrestrial Ecosystems of the Western Hemisphere.* San José, Costa Rica.

Organization of American States. 1987. *Minimum Conflict: Guidelines for Planning the Use of American Humid Tropic Environments.* OAS, Washington, D.C. 198 pp.

O'Riordan, T. 1981. Problems encountered when linking environmental management to development aid. *Environmentalist* 1(1):15-24.

*OTA (U.S. Congress, Office of Technology Assessment). 1987. *Technologies to Maintain Biological Diversity.* U.S. Government Printing Office, Washington, D.C. 334 pp.

Oxley, D.J., M.B. Fenton, and G.R. Carmody. 1974. The effects of roads on populations of small mammals. *J. Applied Ecology* 11(1):51-59.

Padua, M.T.J. and A.T.B. Quintão. 1982. Parks and biological reserves in the Brazilian Amazon. *Ambio* 11(5):309-314.

Padua, M.T.J. and A.T.B. Quintão. 1984. A system of national parks and biological reserves in the Brazilian Amazon. Pp. 565-571 *in:* J.A. McNeely and K.R. Miller (eds.), *National Parks, Conservation, and Development: The Role of Protected Areas in Sustaining Society.* Smithsonian Institution Press, Washington, D.C. 825 pp.

Palmberg, Christel and J.T. Esquinas-Alcazar. 1988. The role of international organizations in the conservation of plant genetic resources. Paper presented to Symposium on the Conservation of Genetic Diversity, Davis, CA, 25 July.

Park, C.C. 1981. Man, river systems, and environmental impacts. *Progress in Physical Geography* 5(1):1-31.

Parker, H.D. 1988. The unique qualities of a geographic information system: A commentary. *Photogrammatic Engineering and Remote Sensing* 54(11):1547-1549.

Partridge, E. (ed.). 1981. *Responsibilities to Future Generations: Environmental Ethics.* Prometheus Books, Buffalo, NY. 319 pp.

Pearce, D.W. 1975. *The Economics of Natural Resource Depletion.* Macmillan, London.

Pearce, D.W. 1976. *Environmental Economics.* Longmans, London.

Pearce, D.W. 1987a. The sustainable use of natural resources in developing countries. *In:* R.K. Turner (ed), *Sustainable Environmental Management: Principles and Practice.* Frances Pinter, London.

Pearce, D.W. 1987b. Economic values and the natural environment. *University College London Discussion Papers in Economics* 87(8): 1-20.

Peet, R.K. 1974. The measurement of species diversity. *Annual Review of Ecology and Systematics* 5:285-307.

Peeters, J.P. and J.T. Williams. 1984. Towards better use of gene-banks with special reference to information. *Plant Genetic Resource News (FAO)* 60:22-32.

Perrings, C.A. 1987. *Economy and Environment: A Theoretical Essay on the Interdependence of Economic and Environmental Systems.* Cambridge University Press, New York. 192 pp.

Perrings, C.A. 1988. An optimal path to extinction? Poverty and resource degradation in the open agrarian economy. *J. Development Economics.*

Perrings, Charles, H. Opschoor, J. Arntzon, A. Gilbert, and D. Pearce. 1988. *Economics and the Environment: A Contribution to the National Conservation Strategy for Botswana.* IUCN, Gland. 171pp.

Peskin, H.M. 1981. National income accounts and the environment. *Natural Resources J.* 21:511-537.

Peterkin, G.F. 1968. International selection of areas for preserves. *Biological Conservation* 1(1):55-61.

Peters, Charles M., A.H. Gentry, and R. Mendelsohn. 1989. Valuation of a tropical forest in Peruvian Amazonia. *Nature* 339:655-656.

Peters, R.L. and J.D.S. Darling. 1985. The greenhouse effect and nature reserves. *BioScience* 35:707-717.

Peterson, D. 1976. Survey of livestock and wildlife: Seasonal distribution in areas of masailand adjacent to Tarangire Park. Final report to the regional livestock development department and the Masai range development project. Mimeo.

Peterson, George L. and Alan Randall. 1984. *Valuation of Wildlife Resource Benefits.* Westview Press, Boulder, CO. 258 pp.

Pickett, S.T.A. and J.N. Thompson. 1978. Patch dynamics and the design of nature reserves. *Biological Conservation* 13:27-37.

Picton, H.D. 1979. The application of insular biogeographic theory to the conservation of large mammals in the northern Rocky Mountains. *Biological Conservation* 15(1):73-79.

Pigram, J.J. 1980. Environmental implications of tourism development. *Annals of Tourism Research* 7(4):554-583.

Pimental, D. 1987. Technology and natural resources. Pp. 679-695 *in:* D.J. McLaren and B.J. Skinner (eds.), *Resources and World Development.* John Wiley, London.

Pimlott, Douglas H. 1969. The value of diversity. *Trans. N. Am. Wildl. and Nat. Res. Conf.* 34:265-273.

Pimm, S.L. 1984. The complexity and stability of ecosystems. *Nature* 307(5949):321-326.

Pimm, S.L. 1987. Determining the effects of introduced species. *Trends in Ecology and Evolution* 2(4):106-108.

Pimm, S.L., H. Lee Jones, and J. Diamond. 1988. On the risk of extinction. *American Naturalist.* 132:757-785.

Pister, E.P. 1979. Endangered species: Costs and benefits. *Environmental Ethics.* 1:341-352.

Plotkin, Mark J. 1988. The outlook for new agricultural and industrial products from the tropics. Pp. 106-116 *in:* E.O. Wilson and Francis M. Peter (eds.), 1988. *Biodiversity.* National Academy Press, Washington, D.C. 521 pp.

Plourde, C. 1975. Conservation of extinguishable species. *Natural Resources J.* 15:791-798.

*Plucknett, D.L.., N.J.H. Smith, J.T. Williams, and N. M. Anishetty. 1987. *Gene Banks and the World's Food.* Princeton University Press, Princeton, N.J.

Polunin, N.V.C. 1983. Marine "genetic resources" and the potential role of protected areas in conserving them. *Environmental Conservation* 10(1):31-41.

Polunin, N. and H.K. Eidsvik. 1979. Ecological principles for the establishment and management of national parks and reserves. *Environmental Conservation* 6(1):21-26.

Pontificiae Academiae Scientiarvm. 1988. Study week on a modern approach to the protection of the environment. *Pontificiae Academiae Scientiarvm Docvmenta* 23:1-24.

Poore, Duncan, and J. Sayer. 1987. *The Management of Tropical Moist Forest Lands: Ecological Guidelines.* IUCN, Gland. 63 pp.

Poore, M.E.D. 1984. Planning reserves in densely-populated areas: Examples from Europe and from the Mediterranean region. Pp. 511-524 *in:* F. Di Castri, F.W.G. Baker, and M. Hadley (eds.), *Ecology in Practice* (Part I), Tycooly International, Dublin. 524 pp.

Population Reference Bureau (PRB). 1989. World Population Data Sheet. [Computer diskette]. Population Reference Bureau, Washington, D.C.

Porter, R.C. 1982. The new approach to wilderness preservation through benefit-cost analysis. *J. Environmental Economics and Management* 9(1):59-80.

Prance, G. 1982a. *Biological Diversification in the Tropics.* Columbia University Press, New York. 714 pp.

Prance, G. 1982b. Forest refuges: Evidence from woody

angiosperms. Pp. 137-157 *in:* G Prance (ed.), *Biological Diversification in the Tropics.* Columbia University Press, New York.

Prance, G.T. and T.S. Elias. 1977. *Extinction is Forever.* New York Botanical Garden, Bronx, New York. 437 pp.

Prance, G.T., W. Balée, B.M. Boom, and R.L. Carneiro. 1987. Quantitative ethnobotany and the case for conservation in Amazonia. *Conservation Biology* 1(4):296-310.

*Prescott-Allen, C. and R. Prescott-Allen. 1986. *The First Resource: Wild Species in the North American Economy.* Yale University Press, New Haven, CT. 529 pp.

Prescott-Allen, R. 1986. *National Conservation Strategies and Biological Diversity.* Report to IUCN, Gland, Switzerland. 67 pp.

Prescott-Allen, R. and C. Prescott-Allen. 1982a. *What's Wildlife Worth? Economic Contributions of Wild Plants and Animals to Developing Countries.* International Institute for Environment and Development (Earthscan), London. 92 pp.

Prescott-Allen, R. and C. Prescott-Allen. 1982b. The case for in-situ conservation of crop genetic resources. *Nature and Resources* 18(1):15-20.

Prescott-Allen, R. and C. Prescott-Allen, 1983. *Genes from the Wild.* Earthscan, London. 101 pp.

Preston, F.W. 1960. Time and space and the variation of species. *Ecology* 41(2):611-627.

Principe, Peter P. 1988a. Valuing diversity of medicinal plants. Paper presented at IUCN/WHO/WWF International Consultation on the Conservation of Medicinal Plants, Chiangmai, Thailand.

Principe, Peter P. 1988b. *The Economic Value of Biological Diversity Among Medicinal Plants.* OECD, Paris.

Quisumbing, E. 1967. Philippine species of plants facing extinction. *Araneta Journal of Agriculture* 14:135-162.

Rabe, F.W. 1984. Selection of high mountain lakes as natural areas. *Natural Areas Journal* 4(1):24-29.

Rabinowitz, D., S. Cairnes, and T. Dillon. 1986. Seven forms of rarity and their frequency in the flora of the British Isles. Pp. 182-204 *in:* M.E. Soulé (ed.), *Conservation Biology: The Science of Scarcity and Diversity.* Sinauer Associates, Sunderland, Massachusetts. 584 pp.

Ragozin, D.L. and G. Brown, Jr. 1985. Harvest policies and nonmarket valuation in a predator-prey system. *J. Environmental Economics and Management* 12:155-168.

Ralls, K. and J. Ballou. 1983. Extinction: Lessons from zoos. Pp. 164-84 *in:* C.M. Schonewald-Cox, S.M. Chambers, B. MacBryde and L. Thomas (eds.), *Genetics and Conservation: A Reference for Managing Wild Animal and Plant Populations.* Benjamin-Cummings, Menlo Park, CA.

Ralls, K., K. Brugger, and J. Ballou. 1979. Inbreeding and juvenile mortality in small populations of ungulates. *Science* 206:1101-1103.

Rambo, A.T. 1979. Primitive man's impact on genetic resources of the Malaysian tropical rain forest. *Malaysian Appl. Biol.* 8:59-65.

Ramsay, W. 1976. Priorities in species preservation. *Environmental Affairs* 4:595-616.

Randall, Alan. 1979. *Resource Economics: An Economic Approach to Natural Resource and Environmental Policy.* Grid Publishing, Columbus, Ohio. 321 pp.

Randall, A. 1986. Human preferences, economics, and the preservation of species. Pp. 79-109 *in:* B.G. Norton (ed.), *The Preservation of Species.* Princeton University Press, Princeton, N.J.

Randall, Alan. 1988. What mainstream economists have to say about the value of biodiversity. Pp. 217-223 *in:* E.O. Wilson and Francis M. Peter (eds.), *Biodiversity.* National Academy Press, Washington, D.C.

Randall, Alan and Hohn R. Stoll. 1983. Existence value in a total valuation framework. *In:* Robert D. Row and L.G. Chestnut (eds), *Managing Air Quality and Scenic Resources at National Parks and Wilderness Areas.* Westview Press, Boulder, CO. 314 pp.

Rapoport, E.H., G. Borioli, J.A. Monjeau, J.E. Puntieri, and R.D. Oviedo. 1986. The design of nature reserves: A simulation trial for assessing specific conservation value. *Biological Conservation* 37(3):269-290.

Ratcliffe, D.A. 1971. Criteria for the selection of nature reserves. *Advancement of Science* 27(2):294-296.

Ratcliffe, D.A. 1977. *A Nature Conservation Review.* Cambridge University Press, Cambridge.

Raup, D.M. 1981. Extinction: Bad genes or bad luck? *Acta Geologica Hispanica* 16(1-2):25-33.

Raup, D.M. 1986. Biological extinction in earth history. *Science* 231:1528-1533.

Raup, D.M. 1987. Diversity crises in the geologic past. *in:* E.O. Wilson and Francis M. Peter (eds.), *Biodiversity.* National Academy, Washington, D.C.

Raup, D.M. and J.J. Sepkoski, Jr. 1982. Mass extinction in the marine fossil record. *Science* 215:1501-1503.

Raven, P.H. 1981. Research in botanical gardens. *Bot. Jahrb. Syst.* 102:53-72.

Raven, P.H. 1987. The scope of the plant conservation problem world-wide. Pp. 19-29 *in:* D. Bramwell, O. Hamann, V. Heywood, and H. Synge (eds.), *Botanic Gardens and the World Conservation Strategy.* Academic Press, London.

Raven, P.H. 1988. Biological resources and global stability. Pp. 3-27 *in:* S. Kawano, J.H. Connell, and T. Hidaka (eds.), *Evolution and Coadaptation in Biotic Communities.* University of Tokyo Press, Tokyo.

Raven, P.H., R.F. Evert, and S.E. Eichhorn. 1986. *Biology of Plants.* Worth Publishers, New York, USA. 775 pp.

Raven, P.H. and G.B. Johnson. 1989. Biology. Times Mirror/Mosby College Publishing, Boston. 1229 pp.

Ray, G.C. 1988. Ecological diversity in coastal zones and oceans. Pp. 36-50 *in:* E.O. Wilson and Francis M. Peter (eds.), *Biodiversity.* National Academy Press, Washington, D.C. 521 pp.

Ray, G.C., J.A. Dobbin, and R.V. Salm. 1978. Strategies for protecting marine mammal habitats. *Oceanus* 21(3):55-67.

Regier, H.A. and E.B. Crowell. 1972. Application of ecosystem theory, succession, diversity, stability, and stress to conservation. *Biological Conservation* 4(2):83-88.

Reid, Walter V., J.N. Barnes, and B. Blackwelder. 1988. *Bankrolling Successes: A Portfolio of Sustainable Development Projects.* Environmental Policy Institute and National Wildlife Federation, Washington, D.C. 48 pp.

*Reid, Walter V. and Kenton R. Miller. 1989. *Keeping Options Alive: The Scientific Basis for Conserving Biodiversity.* World Resources Institute, Washington, D.C.

Rendel, J. 1975. The utilization and conservation of the world's animal genetic resources. *Agriculture and Environment* 2(2):101-119.

Rennie, J.K. and C. Convis. 1989. *Natural Resources Information in Southern Africa.* Report to IUCN.

Repetto, Robert. 1987. Economic incentives for sustainable production. *Annals of Regional Science* 21(3):44-59.

Repetto, Robert. 1988. *The Forest for the Trees? Government Policies and the Misuse of Forest Resources.* World Resources Institute, Washington, D.C. 105 pp.

Repetto, Robert and Malcolm Gillis (eds.). 1988. *Public Policies and the Misuse of Forest Resources.* Cambridge University Press, Cambridge. 432 pp.

Repetto, Robert, William Magrath, Michael Wells, Christine Beer, and Fabrizio Rossini. 1989. *Wasting Assets: Natural Resources in the National Income Accounts.* World Resources Institute, Washington, D.C.

Ribbink, A.J., B.A. Marsh, A.C. Marsh, A.C. Ribbink, and B.J. Sharp. 1983. A preliminary survey of the cichlid fishes of rocky habitats in Lake Malawi. *South African J. Zoology* 18(3):149-310.

Richter-Dyn, N. and N.S. Goel. 1972. On the extinction of a colonizing species. *Theor. Pop. Biol.* 3:406-33.

Ricklefs, R.E. 1987. Community diversity: Relative roles of local and regional processes. *Science* 235(4785):167-171.

Ricklefs, R.E., Z. Naveh, and R.E. Turner. 1984. *Conservation of Ecological Processes.* IUCN, Gland, Switzerland. 34 pp.

Riney, T. 1982. *The Study and Management of Large Mammals.* J. Wiley and Sons, New York.

Roberts, J.O.M. and B.D.G. Johnson. 1985. ''Adventure'' tourism and sustainable development: Experience of the Tiger Mountain group's operations in Nepal. Pp. 81-84 *in:* J.A. McNeely and J.W. Thorsell (eds.), 1985. *People and Protected Areas in the Hindukush-Himalaya.* ICIMOD, Kathmandu. 250 pp.

Roberts, T.H. and L.J. O'Neil. 1985. Species selection for habitat assessments. *Transaction of the North American Wildlife And Natural Resources Conference* 50:352-362.

Rogers, D.L. and S.E. Randolph. 1988. Tsetse flies in Africa, bane or boon? *Conservation Biology* 2:57-66.

Rolston, H. III. 1985a. Duties to endangered species. *BioScience* 35(11):718-726.

Rolston, H., III. 1985b. Valuing wildlands. *Environmental Ethics* 7:23-48.

Ronsivalli, L.J. 1978. Sharks and their utilization. *Marine Fisheries Review* 40(2):1-13.

Roome, N.J. 1984. Evaluation in nature conservation decision making. *Environmental Conservation* 11(3):247-252.

Roth, R.R. 1976. Spatial heterogeneity and bird species diversity. *Ecology* 57(3):773-782.

Roughgarden, J. 1979. *Theory of Population Genetics and Evolutionary Ecology: An Introduction.* Macmillan, New York.

Rowe, J.S. and J.W. Sheard. 1981. Ecological land classification: A survey approach. *Environmental Management* 5(5):451-464.

Rowell, G. 1989. Annapurna, Sanctuary for the Himalaya. *National Geographic.* 176(3):390-405.

Ruddle, K. 1986. No common property problem: village fisheries in Japanese coastal waters. Paper presented at Workshop on Ecological Management of Common Property Resources, IV International Congress of Ecology, Syracuse, New York.

Ruggieri, C.D. 1976. Drugs from the sea. *Science* 194:491-497.

Saenger, P., E.J. Hegerl, and J.D.S. Davie (eds.). 1983. *Global Status of Mangrove Ecosystems.* IUCN, Gland, Switzerland.

Sagoff, M. 1974. On preserving the natural environment. *Yale Law Journal* 84(2):205-267.

Sahni, K.C. 1979. Endemic, relict, primitive and spectacular taxa in Eastern Himalayan flora. *Indian Journal of Forestry* 2:181-190.

Sale, J.B. 1981. *The Importance and Values of Wild Plants and Animals in Africa.* IUCN, Gland, Switzerland. 44 pp.

Salm, R.V. 1984. Ecological boundaries for coral-reef reserves: Principles and guidelines. *Environmental Conservation* 11(30:199-207.

*Salm, R. and J. Clark. 1984. *Marine and Coastal Protected Areas: A Guide for Planners and Managers.* IUCN, Gland. 302 pp.

Samples, Karl, John Dixon, and Marcia Gowen. 1986. Information disclosure and endangered species evaluation. *Land Economics* 62(2):306-312.

Samson, F.B. 1983. Minimum viable populations: A review. *Natural Areas Journal.* 3(3):15-23.

Sargent, F.O. 1969. A resource economist views a natural area. *Journal of Soil and Water Conservation* 24(1):8-11.

Sattaur, Omar. 1987. Trees for the people. *New Scientist* 10 September.

Savidge, J.A. 1987. Extinction of an island forest avifauna by an introduced snake. *Ecology* 68(3):660-668.

Sax, J.L. 1980. *Mountains Without Handrails: Reflection on the National Parks.* University of Michigan Press, Ann Arbor. 152 pp.

Schall, J.J. and E.R. Pianka. 1978. Geographical trends in numbers of species. *Science* 201:679-686.

Scheuer, P.J. 1973. *Industry of Marine Natural Products.* Academic Press, New York.

Schneider, S.N. 1989. The greenhouse effect: science and policy. *Science* 243:771-781.

Schoener, Amy. 1974. Experimental zoogeography: Colonization of marine mini-islands. *American Naturalist* 108:715-38.

Schoener, T.W. 1983. Rate of species turnover decreases from lower to higher organisms: A review of the data. *Oikos* 41(3):372-377.

*Schonewald-Cox, Christine, S.M. Chambers, B. MacBryde, and L. Thomas. 1983. *Genetics and Conservation: A Reference for Managing Wild Animal and Plant Populations.* Benjamin/Cummings Publishing, Menlo Park, CA. 722 pp.

Schonewald-Cox, C.M. and J.W. Bayless. 1986. The boundary model: A geographical analysis of design and conservation of nature reserves. *Biological Conservation* 34(4):305-322.

Schreiber, R. 1977. Landscape planning and protection of the environment: The contribution of landscape ecology. *Applied Sciences and Development* 9:128-139.

Schulze, W., R. D'Arge, and D. Brookshire. 1981. Valuing environmental commodities: Some recent experiments. *Land Econ.* 151-171.

Schwartzman, S. 1987. Extractive production in the Amazon rubber tappers' movement. Paper presented to "Forests, Habitats, and Resources: A Conference in World Environmental History," 30 April, Duke University, Durham, NC.

Scott, J.M., B. Csuti, J.D. Jacobi and J.E. Estes. 1987. Species richness: a geographic approach to protecting future biological diversity. *BioScience* 37:782-788.

Scott, J.M., C.B. Kepler, P. Stine, H. Little, and K. Taketa. 1987. Protecting endangered forest birds in Hawaii: the development of a conservation strategy. Pp. 683-696 *in:* R.D. McCabe, (ed.), *Transactions of the 52nd North American Wildlife and Natural Resources Conference,* Wildlife Management Institute, Washington, D.C.

Scott, M.E. 1988. The impact of infection and disease on animal populations: Implications for conservation biology. *Conservation Biology* 2(1):40-57.

Scott, Margaret. 1988. Loggers and locals fight for the heart of Borneo. *Far Eastern Economic Review* 28 April:44-48.

Seal, U.S. 1988. Intensive technology in the care of *ex situ* populations of vanishing species. Pp. 289-295 *in:* Wilson, E.O. and Francis M. Peter (eds.), *Biodiversity.* National Academy Press, Washington, D.C. 521 pp.

Seidensticker, J. 1984. *Managing Elephant Depredation in Agriculture and Forestry Development Projects.* World Bank Technical Paper, Washington, D.C. 33 pp.

Sevilla, Roque Larrea. 1988. Debt swap for conservation: the Ecuadorean case. Paper presented at Workshop on Economics, IUCN General Assembly, 4-5 February 1988, Costa Rica.

Shaffer, M.L. 1981. Minimum population sizes for species conservation. *BioScience* 31:131-134.

Shaffer, M.L. 1983. Determining minimum viable population sizes for the grizzly bear. *Int. Cong. Bear Res. Manage.* 5:133-9.

Shaffer, M.L. and F.B. Samson, 1985. Population size and extinction: a note on determining critical population sizes. *American Naturalist.* 125: 144-52.

Shane, D.R. 1986. *Hoofprints on the Forest: Cattle Ranching and the Destruction of Latin America's Tropical Forests.* Philadelphia Institute for the Study of Human Issues, Philadelphia, PA. 159 pp.

Shaw, W.W. and E.H. Zube (eds.). 1980. *Wildlife Values.* University of Arizona, School of Renewable Natural Resources, Tucson, AZ.

Sheail, J. 1984. Wildlife conservation: an historical perspective. *Geography* 69(2):119-127.

Shmida, A. and M.V. Wilson. 1985. Biological determinants of species diversity. *J. Biogeog.* 12(1):1-20.

Siegfried, W.R. and B.R. Davies. 1982. Conservation of ecosystems: Theory and practice. *SANSP* 61:1-97.

Simberloff, D.S. 1982. Big advantages of small refuges. *Natural History* 91(4):6-14.

Simberloff, D.S. 1983. Island biogeography theory and the citing of wildlife refuges. *Sov. J. Ecol.* 13:215-25.

Simberloff, D.S. 1984. Mass extinction and the destruction of moist tropical forests. *J. General Biol. (Moscow)* 45:767-778.

Simberloff, D.S. 1986. Are we on the verge of a mass extinction in tropical rain forests? Pp. 165-180 *in:* David K. Elliott (ed.), *Dynamics of Extinction.* John Wiley, New York.

Simberloff, D.S. and L.G. Abele. 1982. Refuge design and island biogeographic theory: effects of fragmentation. *American Naturalist.* 120:41-50.

Simberloff D.S. and James Cox. 1987. Consequences and costs of conservation corridors. *Conservation Biology* 1(1):63-71.

Simberloff, D.S. and E.O. Wilson. 1969. Experimental zoogeography of islands: the colonization of empty islands. *Ecology* 50:278-96.

Simpson, G.G. 1964. Species diversity of North American recent mammals. *Systematic Zoology* 13(1):57-73.

Sinden, J. 1981. Estimating the value of wildlife for preservation: A comparison of approaches. *J. Environmental Management* 12:11-125.

Sinden, J. and A. Worrell. 1979. *Unpriced Values: Decisions Without Market Prices.* J. Wiley, New York.

Sinden, J.A. and G.K. Windsor. 1981. Estimating the value of wildlife for preservation: A comparison of approaches. *J. Environmental Management* 12(1):111-125.

Slatkin, M. 1987. Gene flow and the geographical structure of natural populations. *Science* 236:787-792.

Slatyer, R.O. 1975. Ecological reserves: size, structure, and management. Pp. 22-38 *in:* F. Fenner (ed.), *A National System of Ecological Reserves in Australia.* Australian Academy of Sciences, Canberra.

Smith, N.J.H. 1987. Genebanks: a global payoff. *Professional Geographer* 39(1):1-12.

Smith, P.G.R. and J.B. Theberge. 1986. A review of criteria for evaluating natural areas. *Environmental Management* 10(6):715-734.

Smith, V.K. and J.V. Krutilla. 1979. Endangered species, irreversibilities and uncertainty: A comment. *American J. Agricultural Economics* 61: 371-375.

Sokolov, V. 1985. The system of biosphere reserves in the USSR. *Parks* 10(3):6-8.

Sondaar, P.Y. 1977. Insularity and its effect on mammal evolution. Pp. 671-707 *in:* M.K. Hecht, R.C. Goody, and B.M. Hecht (eds.), *Major Patterns in Vertebrate Evolution.* Plenum, New York.

Soulé, M.E. 1985. What is conservation biology? *BioScience* 35(11):727-734.

*Soulé, M.E (ed.). 1986. *Conservation Biology: The Science of Scarcity and Diversity.* Sinauer Associates, Sunderland, Massachusetts. 584 pp.

Soulé, M.E. 1987. *Viable Populations for Conservation.* Cambridge University Press, Cambridge. MA.

Soulé, M.E. and K.A. Kohm (eds.). 1989. *Research Priorities for Conservation Biology.* Island Press, Covelo, CA.

Soulé, M.E. and D. Simberloff. 1986. What do genetics and ecology tell us about the design of nature reserves? *Biological Conservation.* 35:19-40.

Soulé, M.E., B.A. Wilcox, and C. Holtby. 1979. Benign neglect: A model of faunal collapse in the game reserves of East Africa. *Biological Conservation* 15:259-272.

*Soulé, M.E. and B.A. Wilcox. 1980. *Conservation Biology: An Evolutionary-Ecological Approach.* Sinauer Associates, Sunderland, Massachusetts. 395 pp.

Sousa, W.P. 1984. The role of disturbance in natural communities. *Annual Review of Ecology and Systematics* 15:353-392.

SPC/IUCN. 1985. *Action Strategy for Protected Areas in the South Pacific Region.* SPC, Noumea. 21 pp.

Spence, A.M. 1984. Blue whales and applied control theory. Pp. 43-71 *in:* Y. Ahmad, P. Dasgupta and K-G Maler (eds.), *Environmental Decision Making.* Hodder and Stoughton, London.

Spencer, J.E. 1966. *Shifting Cultivation in Southeast Asia.* University of California Press, Berkeley, CA. 247 pp.

Stamps, J.A., M. Buechner, and V.V. Krishnan. 1987. The effects of edge permeability and habitat geometry on emigration from patches of habitat. *American Naturalist* 129(4):533-552.

Stanley, Steven M. 1985. Extinction as part of the natural evolutionary process: a paleobiological perspective. Pp. 31-46 *in:* R.J. Hoage (ed.), *Animal Extinction: What Everyone Should Know.* Smithsonian Institution Press, Washington, D.C.

Stanley, S.M. 1987. Periodic mass extinction of the Earth's species. *Bull. American Academy of Arts and Sciences* 60:29-48.

Stanton, Nancy L. and J.D. Lattin. 1989. In defense of species. *BioScience* 39(2):67.

Stenseth, N.C. 1979. Where have all the species gone? On the nature of extinction and the red queen hypothesis. *Oikos* 33(1):196-227.

Stevens, Joe B. 1969. Measurement of economic values in sport fishing: An economist's views on validity, usefulness, and propriety. *Trans. Am. Fish. Soc.* 98(2):352-357.

Stocking, M. and N. Abel. 1981. Ecological and environmental indicators for the rapid appraisal of natural resources. *Agricultural Administration* 8(6):473-484.

Stoll, J.R. and L.A. Johnson. 1984. Concepts of value, nonmarket values, and the case of the Whooping Crane. *Trans. North Am. Wildl. Nat. Resour. Conf.* 49:382-393.

Stone, C. 1972. Should trees have standing? Towards legal rights for natural objects. *Southern California Law Review* 45:450-501.

Stott, P. 1981. *Historical Plant Geography: An Introduction.* Allen and Unwin, London.

Strain, Boyd R. 1987. Direct effects of increasing atmospheric carbon dioxide on plants and ecosystems. *Trends in Ecology and Evolution* 2(1):18-21.

Stuart, Simon. 1987. Why we need action plans. *Species* 8:11-12.

Sullivan, A.L. and M.L. Shaffer. 1975. Biogeography of the megazoo. *Science* 189:13-17.

Sumner, F.B. 1921. The responsibility of the biologist in the matter of preserving natural conditions. *Science* 54:39-43.

Sutton, S.L., T.C. Whitmore and A.C. Chadwick (eds.). 1984. *Tropical Rain Forests: Ecology and Management.* Blackwell, Oxford.

Tarrant, James, E. Barbier, R. Greenberg, M. Higgins, S. Lintner, C. Mackie, L. Murphy, and H. van Veldhuisen 1987. *Natural Resources and Environmental Management in Indonesia: An Overview.* USAID, Jakarta. 58 pp.

Taylor, P.W. 1986. *Respect for Nature: A Theory of Environmental Ethics.* Princeton University Press, Princeton, NJ. 329 pp.

Temple, S.A. 1981. Applied island biogeography and the conservation of endangered island birds in the Indian Ocean. *Biological Conservation* 20:147-161.

Tepedino, V.J. and N.L. Stanton. 1976. Cushion plants as islands. *Oecologia* 25:243-56.

Terborgh, J. 1974a. Faunal equilibria and the design of wildlife preserves. Pp. 369-380 *in:* F.B. Golley and E. Medina (eds.), *Tropical Ecological Systems.* Springer-Verlag, New York.

Terborgh, J. 1974b. Preservation of natural diversity: The problem of extinction prone species. *BioScience* 24:715-722.

Terborgh, J. 1983. *Five New World Primates: A Study in Comparative Ecology.* Princeton University Press, Princeton, N.J.

Terborgh, J. 1986. Keystone plant resources in the tropical forest. Pp. 330-344 *in:* M.E. Soulé (ed.), *Conservation Biology: The Science of Scarcity and Diversity.* Sinauer Associates, Sunderland, MA.

Terborgh, J. and B. Winter. 1980. Some causes of extinction. Pp. 119-133 *in:* M.E. Soulé and B.A. Wilcox (eds.), *Conservation Biology: An Evolutionary-Ecological Perspective.* Sinauer Associates, Sunderland, MA.

Terborgh, J. and B. Winter. 1983. A method for citing parks and reserves with special reference to Colombia and Ecuador. *Biological Conservation* 27:45-58.

Thibodeau, F.R. 1983. Endangered species: Deciding which species to save. *Environmental Management* 7(2):101-107.

Thomas, W.L. (ed.). 1956. *Man's Role in Changing the Face of the Earth.* University of Chicago Press, Chicago.

Thorpe, H. 1978. The man-land relationship through time. *In:* J.G. Hawkes (ed.), *Agriculture and Conservation.* Duckworth, London.

Thorsell, J.W. 1984. *Managing Protected Areas in Eastern Africa: A Training Manual.* College of African Wildlife Management, Mweka, Tanzania.

Thorsell, J.W. 1986. Parks on the borderline. *IUCN Bulletin* 16:128-130.

Thorsell, J.W. 1989. *Directing research programmes in protected areas: Some suggested guidelines.* IUCN, Gland. 11 pp.

Thresher, P. 1981. The present value of an Amboseli lion. *World Animal Review* 40:30-33.

The Times. 1988 (seventh edition). *The Times Atlas of the World.* Times Books, New York.

Tisdell, C.A. 1972. Provision of parks and the preservation of nature: Some economic factors. *Australian Economic Papers* 11:154-162.

Tisdell, C.A. 1982. *Wild Pigs: Economic Resource or Environmental Pest?* Pergamon Press, Sydney, Australia.

Tisdell, C.A. 1983. Conserving living resources in Third World countries: Economic and social issues. *J. Environmental Studies* 22(1):11-24.

TPU (Threatened Plants Unit). 1988. *IUCN Plant Information Plan.* IUCN, Gland, Switzerland.

Tsun-Shen, Y. and Z. Zhi-Song. 1984. Endemism in the flora of China: studies on the endemic genera. *Acta Phytotaxonomica Sinica* 22:259-268.

Tubbs, C.R. and J.W. Blackwood. 1971. Ecological evaluation of land for planning purposes. *Biological Conservation* 3(3):169-172.

Udvardy, Miklos. 1969. *Dynamic Zoogeography.* Van Nostrand Reinhold, New York. 445 pp.

Ulph, A.M. and I.K. Reynolds. 1981. *An Economic Evaluation of National Parks.* Centre for Resource and Environmental Studies, Australian National University, Canberra. 221 pp.

UNDP/ESCAP and SEAMP-BIOTROP (South East Asian Regional Centre for Tropical Biology). 1985. *Remote Sensing in Vegetation Studies: Report of the ESCAP-BIOTROP Training Course on Remote Sensing Techniques Applied to Vegetation Studies.* Bangkok, Thailand. 339 pp.

UNEP/FAO. 1982. *The Global Assessment of Tropical Forest Resources.* GEMS PAC Information series No.3, Nairobi.

Unesco. 1972. *Convention Concerning the Protection of the World Cultural and Natural Heritage.* Paris. 10 pp.

Unesco. 1985. Action plan for biosphere reserves. *Environmental Conservation* 12(1):17-27.

Unkel, W.C. 1985. Natural diversity and national forest planning. *Natural Areas Journal* 5(4):8-13.

USAID. 1987. *AID Manual for Project Economic Analysis.* USAID Bureau for Program and Policy Coordination, Washington, D.C. 207 pp.

Usher, M.B. 1973. *Biological Management and Conservation: Ecological Theory, Application and Planning.* Chapman and Hall, London. 394 pp.

Usher, M.B. 1985. Implications of species-area relationships for wildlife conservation. *J. Environmental Management* 21(2):181-191.

*Usher, M.B. 1986. *Wildlife Conservation Evaluation.* Chapman and Hall, London. 394 pp.

van Lavieren, L.P. 1983. *Wildlife Management in the Tropics with Special Emphasis on South-East Asia: A Guidebook for the Warden.* Handbook prepared for School of Environmental Conservation Management. Bogor, Indonesia.

van Steenis, C.G.G.J. 1958. *Vegetation Map of Malaysia.* Unesco Humid Tropics Research Project, Paris.

Vedder, Amy. 1989. In the hall of the mountain king. *Animal Kingdom* 92(3):31-43.

Vermeij, G.J. 1978. *Biogeography and Adaptation: Patterns of Marine Life.* Harvard University Press, Cambridge, MA.

Vida, G. 1978. Genetic diversity and environmental future. *Environmental Conservation* 5(2):127-132.

Vietmeyer, N.D. 1986. Lesser-known plants of potential use in agriculture and forestry. *Science* 232:1379-1384.

Vitousek, P.M., P.R. Ehrlich, A. H. Ehrlich, and P.A. Matson. 1986. Human appropriation of the products of photosynthesis. *BioScience* 36:368-373.

Volk, Tyler. 1989. Sensitivity of climate and atmospheric carbon dioxide to deep-ocean and shallow-ocean carbonate burial. *Nature* 337:637-640.

von Droste, B. and W.P. Gregg, Jr. 1985. Biosphere reserves: demonstrating the value of conservation in sustaining society. *Parks* 10(3):2-5.

Walsh, Richard G., J.B. Loomis, and R.A. Gillman. 1984. Valuing option, existence, and bequest demands for wilderness. *Land Economics* 60(1):14-19.

Warford, J. 1987a. *Environment, Growth and Development.* Economic Development Committee, World Bank, Washington, D.C.

Warford, J. 1987b. Nature resource management and economic development. Pp. 71-85 *in:* P. Jacobs and D. Munro (eds.), *Conservation With Equity: Strategies for Sustainable Development.* IUCN, Gland, Switzerland. 466 pp.

Warford, J. 1987c. Natural resources and economic policy in developing countries. *Annals of Regional Science* 21(3):3-17.

Warner, R.E. 1968. The role of introduced diseases in the extinction of the endemic Hawaiian avifauna. *Confor* 70:101-120.

Warren, D.M., L.J. Slikkerveer, and S.O. Titilola. 1989. *Indigenous Knowledge Systems: Implications for Agriculture and International Development.* Studies in Technology and Social Change 11, Iowa State University, Ames, IO. 186 pp.

Warrick, R.A., P.D. Jones, and J.E. Russell. 1988. The greenhouse effect, climatic change and sea level: An overview. Paper prepared for Commonwealth Expert Group on Climatic Change and Sea Level Rise, London, May 1988.

West, P.C. and S. Brechin (eds.). In press. *Resident Populations and National Parks in Developing Nations: Interdisciplinary Perspectives and Policy Implications.* University of Arizona Press, Tucson, AZ.

Western, D. 1984. Amboseli National Park: Human values and the conservation of a savanna ecosystem. Pp. 93-100 *in:* J.A. McNeely and K.R. Miller (eds.), *National Parks, Conservation, and Development: The Role of Protected Areas in Sustaining Society.* Smithsonian Institution Press, Washington, D.C.

Western, D. and W. Henry. 1979. Economics and conservation in Third World national parks. *BioScience* 29(7):414-418.

Western, D. and M. Pearl (eds.). 1989. *Conservation Biology for the Next Century.* Oxford University Press, New York.

Westman, Walter E. 1977. How much are nature's services worth? *Science* 197: 960-964.

Wetterberg, G.B. 1976. *An Analysis of Nature Conservation Priorities in the Amazon.* Brazilian Institute for Forestry Development, Brasilia, Brazil.

Wetterberg, G.B., G.T. Prance, and T. Lovejoy. 1981. Conservation progress in Amazonia: A structural review. *Parks* 6(2):5-10

Wharton, Charles H. 1968. Man, fire, and wild cattle in southeast Asia. *Proc. Ann. Tall Timbers Fire Ecol. Conf.* 8:107-167.

Whitcomb, R.F., J.F. Lynch, P.A. Opler, and C.S. Robbins. 1976. Island biogeography and conservation: strategies and limitations. *Science* 193:1030-1032.

White, F. 1983. *The Vegetation of Africa: A Descriptive Memoir to Accompany the Unesco/AETFAT/UNSO Vegetation Map of Africa.* Unesco, Paris.

White, P.S. and S.P. Bratton. 1980. After preservation: philosophical and practical problems of change. *Biological Conservation* 18(3):241-255.

Whitmore, T.C. 1984. *Tropical Rainforests of the Far East.* Oxford University Press, London.

Whitmore, T.C., R. Peralta, and K. Brown. 1985. Total species count in a Costa Rican tropical rain forest. *J. Tropical Ecology* 1:375-378.

Whittaker, R.H. 1972. Evolution and the measurement of species diversity. *Taxon* 21(2):213-51.

Whittaker, R.H. and G.E. Likens. 1975. The biosphere and man. *in:* H. Lieth and R.H. Whittaker (eds.), *Primary Productivity of the Biosphere.* Springer-Verlag, New York.

Whitten, A.J. 1987. Indonesia's transmigration program and its role in the loss of tropical rain forests. *Conservation Biology* 1(3):239.

Whitten, A.J., K.D. Bishop, S.V. Nash, and L. Clayton. 1987. One or more extinctions from Sulawesi, Indonesia? *Conservation Biology* 1(1):42-48.

Whitten, A.J., M. Mustafa and G.S. Henderson. 1987. *The Ecology of Sulawesi.* Gadjah University Press, Indonesia. 777 pp.

Wiess, E.B. 1984. The planetary trust: Conservation and intergenerational equity. *Ecology Law Quarterly* 11(4):495-582.

Wilcove, D.S. and R.M. May. 1986. National park boundaries and ecological realities. *Nature* 324:206-207.

Wilcox, B.A. 1980. Insular ecology and conservation. Pp. 95-117 *in:* M.E. Soulé and B.A. Wilcox (eds.), *Conservation Biology: An Evolutionary-Ecological Perspective.* Sinauer Associates, Sunderland, Mass.

Wilcox, B.A. and D.D. Murphy, 1985. Conservation strategy: The effects of fragmentation on extinction. *American Naturalist.* 125:879-887.

Willard, L.D. 1980. On preserving nature's aesthetic features. *Environmental Ethics* 2(4):293-310.

Williams, S.B. 1984. Protection of plant varieties and parts as intellectual property. *Science* 225:18-23.

Wilson, C.C. and W.L. Wilson. 1975. The influence of selective logging on primates and some other animals in East Kalimantan. *Folia Primatol.* 23:245-274.

Wilson, E.O. 1985. The biological diversity crisis. *Bioscience* 35:700-706.

*Wilson, E.O. and Francis M. Peter (eds.). 1988. *Biodiversity.* National Academy Press, Washington, D.C. 521 pp.

Wilson, E.O. 1988a. The current state of biological diversity. Pp. 3-18 *in:* E.O. Wilson and Francis M. Peter (eds.), *Biodiversity.* National Academy Press, Washington, D.C. 521 pp.

Wilson, E.O. 1988b. The diversity of life. Pp. 68-81 *in:* H.J. de Blij (ed.), *Earth '88: Changing Geographic Perspectives.* National Geographic Society, Washington, D.C. 392 pp.

Wilson, E.O. and E.O. Willis. 1975. Applied biogeography. Pp. 522-536 *in:* M.L. Cody and J.M. Diamond (eds.), Ecology and Evolution of Communities. Cambridge Mass. Harvard University Press.

Wirth, C.L. 1980. *Parks, Politics, and the People.* University of Oklahoma Press, Norman Oklahoma. 397 pp.

Wolf, Edward C. 1987. *On the Brink of Extinction: Conserving the Diversity of Life.* Worldwatch Institute, Washington, D.C.

Wood, D. 1988. Introduced crops in developing countries: a sustainable agriculture? *Food Policy* May:167-177.

Woodwell, G.M., H.E. Hobby, R.A. Houghton, J.M. Melillo, B. Moore, B.J. Peterson, and G.R. Shaver. 1983. Global deforestation: contribution to atmospheric carbon dioxide. *Science* 222:1081-1086.

World Bank. 1981. *Economic Development and Tribal Peoples: Human Ecological Considerations.* International Bank for Reconstruction and Development (World Bank), Washington, D.C.

*World Bank. 1988. *Wildlands: Their Protection and Management in Economic Development.* World Bank, Washington, D.C. 278 pp.

World Commission on Environment and Development (WCED). 1987. *Our Common Future.* Oxford University Press, Oxford, UK.

Worster, Donald. 1985. *Nature's Economy.* Cambridge University Press, Cambridge, UK.

WRI (World Resources Institute). 1989a. *NGO Consultation on the Implementation of the Tropical Forestry Action Plan.* Summary Report. Washington, D.C.

WRI (World Resources Institute). 1989b. *The International Conservation Financing Project.* Working Paper. Final Draft, June 30, 1989. World Resources Institute, Washington, D.C.

WRI/IIED. 1986. *World Resources 1986.* Basic Books, N.Y.

WRI/IIED. 1987. *World Resources 1987.* Basic Books, N.Y.

*WRI/IIED. 1988. *World Resources 1988-1989.* Basic Books, N.Y.

Wright, D.F. 1977. A site evaluation scheme for use in the assessment of potential nature reserves. *Biological Conservation* 11(3):293-305.

Wright, S.J. and S.P. Hubbell. 1983. Stochastic extinction and reserve size: a focal species approach. *Oikos* 41:466-76.

WWF. 1986. *Conservation Yearbook.* WWF, Gland. 587 pp.

WWF-US. 1988. Debt-for-nature swaps: A new conservation tool. *World Wildlife Fund Letter* 1:1-9.

*Yeatman, C.W., D. Krafton, and G. Wilkes (eds.). 1984. *Plant Genetic Resources: A Conservation Imperative.* Westview Press, Boulder, Colorado.

Zimmerman, B.L. and R.O. Bierregaard, 1986. Relevance of the equilibrium theory of island biogeography and species-area relations to conservation with a case from Amazonia. *J. Biogeog.* 13:133-43.

Zube, E.H. 1986. Local and extra-local perceptions of national parks and protected areas. *Landscape and Urban Planning* 13(1):11-18.

INDEX

Action plans
biodiversity-promoting strategies and, 14, 109-15
cross-sectoral, 112-115
see also Strategies
Africa
biodiversity maintenance habitats in, 99-100
biodiversity rich areas in, 90
consumptive use values in, 28
elephant protection in, 80
forest products' economic value to, 30
forest status in, 79
germplasm exchange by, 57
habitat changes impact on sub-Saharan, 44-46
land area protected in (by percentage), 104
national data information center in, 76
trophy hunting in, 122
see also individual countries in
Afrotropical Action Plan (IUCN/UNEP), 112
Afrotropical biodiversity strategy, 110, 111, 112-13
Agency for International Development, U.S. (USAID)
biodiversity programs backed by, 95
investment in conservation through greater use of local currencies, 125
utilitarian value assessments by, 27
Agriculture
biodiversity rich sites of plants whose germplasm may be necessary to, 100-3
food security and genetic variety, 29-30
IUCN's category VI and, 60
protecting traditional forms, 60
seed banks maintained to protect, 65-66
traditional strategies for maximum conservation of biodiversity, 50
Amazon. *See* countries within
American Association of Zoological Parks and Aquaria (AAZPA), 63, 110
Antarctica, ozone depletion over, 66
Army Corps of Engineers, U.S.
estimate of non-consumptive benefits from conserving wetlands, 32
Asia
biodiversity of countries in, 88, 89-90
forest products economic value to, 30
germplasm exchange by, 57
habitat changes' impact on tropical, 44, 45, 47
land area protected in (by percentage), 104
see also individual countries in
Asian Development Bank
global conservation strategy involvement by, 21-22
Australia
biodiversity in, 46, 91
major human caused extinctions by, 20
Bali Action Plan
funding and implementation of, 115
objectives, activities, and priority projects of, 59, 140-46
Bali Declaration, 58
Bangladesh, national data information center in, 76
Biological resources
contribution to "non-conservation" sectors by, 129-31
dimensions of lack of data concerning, 39-47, 103-4
direct values of, 28-31
economic factors causing overexploitation of, 38-39, 47-49
economic solution to conserving global, 38, 39
ex situ mechanisms to conserve, 62-66
factors threatening, 12, 37-52
formulating responses to problems facing the conservation of, 115
global strategy for conservation of, 113-15
indirect values of, 31-34
international conventions on further conservation of, 68-69
linkages among sectors affecting, 113-15
local knowledge on conservation of, 73-74
maintenance of biodiversity in marine, 88
major obstacles to greater progress in conserving, 51-52
management to ensure sustainable development, 18-19, 20, 28-29, 31, 32, 132
non-sustainable management and use of, 19, 31, 32, 33, 38-39, 47-49
policy shifts to encourage conservation of, 55-56
profit return from exploitation of, 121-22
property rights and, 118
scientific knowledge on conservation of, 71-73, 114
social factors threatening, 49-50
Biosphere Reserve Action Plan (Unesco), funding and implementation of, 115
Biosphere Reserve network, 72
Biosphere Reserves Program, 58
Birds, habitat sites of threatened African, 99-100
Botanic Gardens
conservation role of, 64-65, 109-10
Botanic Gardens Conservation Secretariat (BGCS) — IUCN, 62, 65, 109
Botswana
consumptive use values in, 28
Research needs in, 75
Brazil
biodiversity richness in, 92
Germplasm exchange by, 57
processes leading to Brazil nut production in, 32-33
removing subsidies for forest clearing in, 55
Brazilian Space Research Institute
deforestation estimates by, 44
British Petroleum (BP), 104

California, biodiversity in, 46
Canada
consumptive use values in, 28
National park features in, 59
Cape Floristic Kingdom, species diversity within, 87
Captive Breeding Specialist Group (IUCN), 63-64
Center for Marine Conservation, 110
Central America
germplasm exchange by, 57
long-term sustainable land use developed by native people in, 50
national CDC in, 76
see also Latin America; South America; individual countries in
Chicago Zoological Society, 110
China
national conservation database in, 76, 77
Climate change
biodiversity maintenance and, 38
management action in response to, 66-68
Colombia, biodiversity richness in, 93
Commission on National Parks and Protected Areas (CNPPA), 112
Committee on Research Priorities in Tropical Biology, 86
need for trained help by, 18
research priorities available from, 71
systematists estimate by, 72
Conservation
biodiversity-promoting approaches to, 12-13, 55-69
data required for biodiversity, 13-14, 71-81
economics of biodiversity, 14-15, 117-26
importance of biodiversity, 11, 17-22
major obstacles to biodiversity, 51-52
modern approaches to biodiversity, 20-21
national approaches to, 88-98
''non-conservation'' partners for biodiversity, 15, 129-32
priority establishment for biodiversity, 14, 83-106
priority guidelines for action on, 104-6
regional approach to, 88
strategies and action plans promoting biodiversity, 14, 109-15
Conservation concessions, 125
Conservation data centers (CDCs)
establishment of national, 76, 77
Conservation International (CI)
Brazilian data publications work by, 104
CDC development aid by, 76
debt-for-nature swap involvement by, 124, 125
global conservation strategy involvement by, 21-22
linkage approach to Ecosystem Conservation by, 124
Species Action Plans backing by, 110
Consultative Group on International Agriculture Research (CGIAR)
ex situ conservation contributions by, 62
germplasm conservation by, 118
Convention for the Protection of the Ozone Layer, 66
Convention on Conservation of Migratory Species of Wild Animals, 56
Convention Concerning the Protection of the World Cultural and Natural Heritage, 57-58, 137-39
Convention on International Trade in Endangered Species of Wild Fauna and Flora, 56
Convention on the International Trade in Endangered Species (CITES), 80
Convention on the Preservation of Marine Pollution by Dumping of Wastes and Other Matter, 66
Convention on Wetlands of International Importance, 57
Colombia, botanic gardens in, 65
Cooperation, international
for elimination of biodiversity loss, 56-62, 68-69
of *ex situ* germplasm by, 62
managing biological resources, 132
Costa Rica
Conservation Data Center in, 76
species diversity in, 44
taxes on biological resources in, 120
Costs
benefit of natural areas consideration in developments', 48
of forest assessments and deforestation rates in the tropics, 44
see also Economics; Funding; Market price
Côte d' Ivoire, deforestation estimates for, 44
Cross-sectoral strategies, action plans and, 112-15

Data
exchange among national CDCs, 76
lack of biodiversity conservation, 39-47, 103-4
needs at international level for sustainable policy development, 77-80, 115
needs at national and local levels for sustainable policy development, 74-77
required to conserve biodiversity, 13-14, 71-81
Debt, external
negotiating swaps of protected areas in return for reduction of, 124-25
option values and, 35
Debt-for-nature swaps, arranging, 124-25
Developing countries
exchanges of genetic material among, 30
funding conservation in, 119-20, 121-23
''hotspots'', 87
see also individual countries
Development
biodiversity and 19-20
effects of non-sustainable, 19, 31, 32, 33, 34-35, 47-49
projects funding biodiversity conservation, 120-21
sustainable, on a global scale, 113-14

Ecologically sensitive areas (ESAs), 83
criteria for designating and managing, 85-86
identifying, 84-85
Economics
of biodiversity maintenance, 14-15, 25-27, 30-31, 32, 34, 47-49, 117-26
of bioreserve management, 104
conserving global biodiversity through appropriate use of, 38, 39
of *ex situ* conservation, 62
methods for assigning values to natural biological resources using, 27-35
of a regional action plan, 113
of remote sensing technology, 73
of various uses/managements of forests, 30-31
see also Costs; Funding; Market price
Ecosystems
biologically diverse types of, 46-47
maintaining marine, 88, 110
Ecuador, debt-for-nature swap involving, 124-25
Education
ex situ conservation programs used for public, 62, 64, 65
of parataxonomists for collecting and documenting new specimens, 72
Environmental Impact Assessments, 56
Environmental Law Centre (ELC) — IUCN, 80
Ethics, biodiversity and, 25-26, 27
Ethiopia, germplasm exchange by, 57
Europe
forest die off in, 38
see also individual countries in
European Economic Community (EEC), Action Plan funding by, 113
Extinction
causes and rates of species, 39-47
climate change and, 67
Ex situ, contributions to biodiversity, 62-66

Fish and Wildlife Service, U.S., 110
Food and Agriculture Organization (FAO) — U.N.
biodiversity plant data collected by, 104
deforestation estimates by, 44
ex situ conservation contributions by, 62
global conservation strategy involvement by, 21-22, 46
world reliance on tropical forest data coverage by, 72-73
Forests
biodiverse sites of tropical, 92, 103-4
conservation of temperate region, 87
critical areas in, 86-87
containing largest varieties of threatened birds, 99-100
developing countries dependence on economic value of, 30-31
estimates of remaining tropical moist, 44
FAO estimate of the extent of tropical, 72-73
government overexploitation of tropical, 49
Funding
global approach to biodiversity conservation, 68-69, 123
implementing coordinated conservation activities and, 115
international biodiversity conservation, 117, 123-26
methods for biodiversity conservation at local and national levels, 118-23
protected areas through NGO foundations, 122-23
zoological community, 64
see also Costs; Economics; Market price

Geographic Information Systems (GIS), 73
Ghana, consumptive use values in, 28
Global Environment Monitoring System (GEMS) — UNEP, 81, 105
Global Resource Information Database (GRID), 81
Global Strategy for the Conservation of Biodiversity
major issues of emphasis in development of, 113-15
planning to ensure wise use of investments in, 115
Greenhouse effect. *See* Climate change
Greenland National Park, 58
"Green parties", increased public support for, 11, 21
Gross Domestic Product (GDP), consumptive use value inclusion into, 28, 30-32
Guidelines, conservation action priority, 104-6

Habitats
action plans for conserving, 111-12
African sites of threatened bird, 99-100
biodiversity maintenance and alteration of, 38
causes and extent of lost, 43-47
integrated approach to protecting, 56-62
pre-industrial alteration of "natural", 51
WCMC data on protected, 78-79
Heritage Species Centers, 124
Honduras, non-consumptive benefits from conservation in, 32
Hotspots, tropical forest area, 87

Incentives
to encourage global biodiversity conservation, 113-14, 115
to ensure sustainable use of resources, 35
to local people to respect protected areas, 117-18, 121-22
see also Subsidies
India, economics of forest use in, 30-31
Indomalayan Action Plan (IUCN/UNEP), 112
Indonesia
biodiversity richness in, 94
non-consumptive benefits from conservation in, 32
Indonesian Wildlife Fund, 122
Industrial countries
existence value of species and habitats to populations in, 34
hotspot areas in, 87
see also individual countries
Instituto Brasileiro de Recursos Naturais Renovaveis e Meio Ambiente (IBAMA), 104

International Board on Plant Genetic Resources (IBPGR) — (CGIAR)
ex situ plant conservation by, 62
location and conservation of wild relatives encouraged by, 29
International Conservation Financing Program (WRI), 117
International Convention for the Regulation of Whaling, 56
International Council for Bird Preservation (ICBP)
Afrotropical biodiversity strategy work by, 112
priority site location hunting by, 100
International Fur Trade Federation, 110
International Species Inventory System (ISIS), 63
International Tropical Timber Organization (ITTO), 80
International Trust for Nature Conservation (Nepal), 122
International Union for Conservation of Nature and Natural Resources (IUCN), 77, 112
ethical foundation for conservation produced by, 25-26, 27
global conservation strategy involvement by, 21-22, 68-69, 113, 117
location and conservation of wild crop relatives encouraged by, 29
mini-reserves project supported by, 61
plant diversity sites in book prepared by, 102-3
protected areas review by, 103
protected categories of, 91, 97, 105
Species Action Plans of, 110-11
International Union of Directors of Zoological Parks, 63

Java, botanic gardens in, 65

Kenya, profits from resource exploitations of protected areas returned to local people in, 121
Kesterson National Wildlife Refuge, 66
Kouprey Action Plan, 63

Land tenure, sustainable investment encouragement through giving rural people, 55, 131
Land use, policy shifts promoting biodiversity of, 55-56, 60
Latin America
land area protected in (by percentage), 104
see also South America; individual countries in
Legislation
biodiversity protection, 56, 66, 80, 137-39
pollution control, 66, 130-31
species and habitat protecting, 56-57
see also individual statutes
Linkages
between development and conservation funding, 120-21
between international systems on status of biological resources, 75
between various policy sectors' effect on biodiversity, 55-56, 112-15
biodiversity-threatening economic, 37-39, 47-49
see also Cooperation, International

Madagascar
biodiversity richness in, 95
germplasm exchange by, 57
major human-caused extinctions on, 20
National Action Plan of, 110, 112
Malawi, consumptive use values in, 28
Malawi Lake, biodiversity in, 46
Malaysia
consumptive use values in, 28
non-consumptive benefits from conservation in, 32
Management
of conservation data, 75-81
of Ecological Sensitive Areas, 85-86
economics of priority area, 104
to ensure survival of existing species, 18-19, 20
local information, 75-76
national conservation, 76
necessity in "natural" habitats of, 51
new approaches for sustainable biological resource, 132
objectives and categories of protected areas, 59-62
of protected areas that exclude people, 51
responses to pollution and climate change, 66-68
requirements for a global conservation strategy, 114-15
traditional strategies of *in situ* crop, 50
Marine ecosystems, 88, 110
Market price, as an accurate representation of the value of natural resources, 31, 48
Mauritius, 61
Media, nature's abstract importance communicated by, 25-26
Megadiversity, country concept, 88-99
Megadiversity countries
concept and workings of, 88-90
profiles of important, 91-97
Mexico
biodiversity richness in, 96
germplasm exchange by, 57
Military, promoting conservation interests in, 131
Models
lack of attention given to non-monetary values in standard biological resource conservation, 25, 47-49
for sustainable land use in tropical forests, 50
Montreal Protocol on Substances That Deplete the Ozone Layer, 66
Myers, Norman, 87, 91

National Conservation Strategy (NCS), 56
National Nature Reserves network (U.K.), 103
National Marine Fisheries Service, U.S., 31
National Parks. *See* Protected Areas
Nepal
consumptive use values in, 28
non-consumptive benefits from conservation in, 32
Netherlands Antilles, national CDC center in, 76
New Zealand, major human-caused extinctions in, 20

Nigeria, consumptive use values in, 28
Non-governmental organizations (NGOs)
  Conservation Strategy input from, 56, 80-81
  global strategy input of, 113-15
  incentives for conservation provided by, 122
North America
  major human-caused extinctions in, 20
  see also individual countries in

Oceania
  Action Plan, 112
  major human-caused extinctions in, 38, 40-41, 51
Oceans. *See* Marine ecosystems
Overexploitation
  economic factors causing biological resource, 38-39, 47-49
  of wildlife stimulated by consumptive use value, 28-29

Peru, economics of forest uses in, 31
Plants
  biodiverse sites of, 100-3
  cost of ensuring evolution of wild species of, 118
  see also Agriculture
Policy
  biodiversity maintenance through changes in non-biological, 19-20, 113-15
  biological resources investment, 125-26
  effect on Action Plan success, 113
  guidelines for conservation, 104-6
  maintaining *ex situ* continuity needed in, 62
  new sustainable production, 132
  other resource management issues' impact on biodiversity conservation, 38, 39, 47
  shifts encouraging biodiversity, 19-20, 55-56
Pollution, chemical
  biodiversity maintenance and, 38
  management action in response to, 66-68
Population
  increase and biodiversity maintenance, 38-39
  managing captive breeding animal, 63
Priorities
  biodiversity conservation, 14, 83-106
  establishing national, 83-86
  international approaches to determining a system of, 86-104
Property rights, for biological resources, 118
Private sector. *See* Non-government organizations (NGOs)
Protected areas
  abuse of areas surrounding, 49
  Action Plans for, 112
  altitudinal range of global, 67-68
  categories and management objectives of, 59-62
  designed to conserve traditional land use, 50
  entry and other user fees to finance, 119
  funding IUCN strategies for, 115
  returning profits to local people from exploitation of, 121-22
  WCMC data on, 77-80
  where they are and where they are needed, 58
Puerto Rico, National CDC in, 76

Rain Forest Imperative Campaign (CI)
  megadiversity concept's use in, 90
Red Data Books (IUCN), 41-42, 43, 77-78, 97, 99-100
Regional Seas Programs, 58
Royal Botanic Gardens, 104
Rwanda, "gorilla tourism" in, 119

Scientific knowledge, required to conserve biodiversity, 71-73, 114
Senegal, consumptive use values in, 28
Seed banks, wild relatives of major crops held in, 65-66
South Africa, biodiversity in, 46
South America
  forest products economic value to, 30
  major human-caused extinctions in, 20
  national data information center in, 76
  see also individual countries in
Species
  action plans of IUCN, 110-11
  biodiversity maintenance and introduced, 38
  dimension of biodiversity loss among, 39-47
  estimates on number of global, 18
  integrated approach to protecting, 56-62
  management to ensure biodiversity, 18-19
  strategies and action plans for conserving, 109-11
  WCMC data on, 77-78
Species Survival Commission (SSC) — IUCN, action plans of, 63, 110-11
Sri Lanka
  botanic gardens in, 65
Strategies
  biodiversity-promoting action plans and, 14, 109-15
  cross-sectoral promotion, biodiversity enhancing, 55-56, 113-15
  developing global biodiversity conservation, 21-22, 112-15
  see also Action Plans; individual strategies
Subsidies
  encouraging deforestation, 55
  negative impacts on biodiversity from government, 117
  see also Incentives

Tanganyika, Lake, biodiversity in, 46
Tanzania
  consumptive use values in, 28
  non-sustainable hunting results in, 33
Technologies
  contributing to increased biodiversity, 56
  data management, 72, 73
  remote sensing, 73

Thailand
forest products trade by, 44
national data information center in, 76
non-consumptive benefits from conservation in, 32
The Nature Conservancy (TNC)
debt-for-nature swaps involvement by, 124, 125
national data center establishment work by, 76
Tiger Mountain Group, 122
Tropical Forestry Action Plan, funding and implementation of, 115
Trade, wildlife, 79-80
TRAFFIC Network, 80
Training
need for increased professional systematists, 18
see also Education
Tropical Forest Resource Atlases, preparation of, 104
Tropical Forestry Action Plan (TFAP) — FAO, establishment and workings of, 46, 111-12
Tropical Wilderness Areas, national approach to managing major remaining, 98
Tropics
government-promoted overexploitation of forests in, 49
need for international help to conserve species in, 21
priority for basic inventory work in, 80-81
research centers to train personnel in, 72
undescribed species in, 18, 39

Unesco, global conservation strategy involvement by, 21-22, 58
United Kingdom
Environmentally Sensitive Areas in, 83
national park features in, 59
United Nations Development Programme (UNDP), global conservation strategy involvement by, 46
United Nations Environment Programme (UNEP), 58, 81, 110
biodiversity convention promoted by, 68-69
conservation database aid by, 76
deforestation estimates by, 44
funding and implementation of Regional Seas action plans of, 115
Global Strategy for the Conservation of Biodiversity work by, 21-22, 113, 117
United States (U.S.)
forest products economic value to, 30
germplasm exchange by, 57
non-consumptive benefits from conservation in, 32

Values
consumptive use, 28-29
of biological diversity, 11-12, 25-35
existence, 34
non-consumptive use, 31-33
option, 33-34
productive use, 29-31

Venezuela, National Action Plan of, 110, 112
Victoria, Lake, biodiversity in, 46

Wetlands, species diversity within, 46
Wildlife, trade data involving, 79-80
Wildlife Conservation International, 110
Wilson, Edward O., 22
World Bank
biodiversity considerations in projects of, 95, 112, 120
global conservation strategy involvement by, 21-22, 46
loan justified by non-consumptive benefits, 32
policy on wildlands, 58, 120-21, 147-52
World Charter for Nature (U.N.), 25, 68,
ethical commitment for biodiversity conservation in, 27
principles, functions, and implementation of, 134-36
societal changes called for by, 51
Woodlands. *see* Forests
World Commission on Environment and Development (WCED)
cross-sectoral policy shifts encouraging sustainable use of biological resources indicated by, 55-56
forest cover loss estimates by, 44
linkages and, 112-13
societal changes proposed by, 51
sustainable development recognition by, 19
World Conservation Monitoring Centre (WCMC)
priority input by, 105
protected habitat data of, 78-79
species data of, 77-78
wildlife trade data by, 79-80
World Conservation Strategy (WCS) — IUCN, 19, 35, 113
preparation and implementation of, 21-22
societal changes called for by, 51
World Resources Institute (WRI)
global conservation strategy involvement by, 21-22, 46, 113, 117
World Wildlife Fund (WWF), 110
biodiversity attention given to projects of, 90
conservation database aid by, 76, 77
debt-for-nature swaps involvement by, 124-25
global conservation strategy involvement by, 21-22
location and conservation of wild relatives encouraged by, 29
mini-reserves project supported by, 61

Yellowstone National Park, 51, 58
profits from resource exploitation of protected areas returned to local people in, 121

Zaire
biodiversity richness in, 97
consumptive use values in, 28
Zambia
Wildlife Fund economics, used in 121, 122
research needs in, 75

Zimbabwe
profits from resource exploitation of protected areas returned to local people in, 121
research needs in, 75
Wildlife Fund economics used in, 121
Zoological gardens
contributions to biodiversity conservation by, 62-64, 124

LIST OF ACRONYMS AND ABBREVIATIONS

AAAS	American Association for the Advancement of Sciences
AAZPA	American Association of Zoological Parks and Aquaria
ADB	Asian Development Bank
AfDB	African Development Bank
BOSTID	Board on Science and Technology for International Development of National Research Council
BGCS	Botanic Gardens Conservation Secretariat of IUCN
CATIE	Centro Agronómico Tropical de Investigación y Enseñanza (Tropical Agricultural Research and Training Center, Costa Rica)
CDC	Conservation Data Center
CGIAR	Consultative Group on International Agricultural Research
CI	Conservation International
CIDA	Canadian International Development Agency
CITES	Convention on International Trade in Endangered Species of Wild Fauna and Flora
cm	centimeter
CMEA	Council for Mutual Economic Assistance
CNPPA	Commission on National Parks and Protected Areas of IUCN
DANIDA	Danish International Development Agency
EEC	European Economic Community
EIA	Environmental Impact Assessment
ELC	Environmental Law Centre of IUCN
ESA	Ecologically Sensitive Area
est.	estimated
FAO	Food and Agriculture Organization of the United Nations
FINNIDA	Finnish International Development Agency
FRG	Federal Republic of Germany
GATT	General Agreement on Tariffs and Trade
GDP	Gross Domestic Product
GEMS	Global Environment Monitoring System of UNEP
GIS	Geographic Information System
GNP	Gross National Product
GRID	Global Resource Information Database
GTZ	Deutsche Gesellschaft für Technische Zusammenarbeit (German Agency for Technical Cooperation)
ha	hectare
IBAMA	Instituto Brasileiro de Recrusos Naturais Renovaveis e Meio Ambiente (Brazilian Institute of Renewable Natural Resources and Environment)
IBPGR	International Board for Plant Genetic Resources
IBRD	International Bank for Reconstruction and Development
ICBP	International Council for Bird Preservation
ICOMOS	International Council on Monuments and Sites
ICSU	International Council of Scientific Unions
IDA	International Development Association
IDB	InterAmerican Development Bank
IGBP	International Geosphere Biosphere Project of ICSU
IIED	International Institute for Environment and Development
IMF	International Monetary Fund
IPAL	Integrated Project on Arid Lands (Kenya)
IRRI	International Rice Research Institute (Philippines)
ISIS	International Species Inventory System
ITTO	International Tropical Timber Organization
IUDZG	International Union of Directors of Zoological Parks
IUCN	International Union for Conservation of Nature and Natural Resources
kg	kilogram
km	kilometer
m	meter
MS	Manuscript
mt	metric ton
NAS	National Academy of Sciences
NCS	National Conservation Strategy
NESDIS	National Environmental Satellite, Data, and Information Service
NG	New Guinea
NGO	Non-Governmental Organization
NOAA	National Oceanic and Atmospheric Administration
NORAD	Norwegian Agency for International Development
ODA	Overseas Development Agency of the United Kingdom
OECD	Organisation for Economic Co-operation and Development
OTA	Office of Technology Assessment of the US Congress
PADU	Protected Areas Data Unit of WCMC
PCBs	Polychlorinated biphenyls
PVO	Private Voluntary Organization
RDB	Red Data Book of IUCN

SADCC	Southern African Development Coordination Conference
SFr	Swiss franc
SIDA	Swedish International Development Authority
sq km	square kilometer
SSC	Species Survival Commission of IUCN
TFAP	Tropical Forestry Action Plan
TNC	The Nature Conservancy
TPU	Threatened Plants Unit
UK	United Kingdom of Great Britain and Northern Ireland
UN	United Nations
UNDP	United Nations Development Programme
UNEP	United Nations Environment Programme
Unesco	United Nations Educational, Scientific, and Cultural Organization
USAID	United States Agency for International Development
USFWS	United States Fish and Wildlife Service
USNPS	United States National Park Service
WCED	World Commission on Environment and Development
WCMC	World Conservation Monitoring Centre
WCS	World Conservation Strategy
WHO	World Health Organization
WMA	Wildland Management Area
WRI	World Resources Institute
WWF	World Wide Fund for Nature (previously World Wildlife Fund, and still World Wildlife Fund in the United States)
yr	year